Handbook of Electrical Engineering

Handbook of Electrical Engineering

Editor: Norman Schultz

NY RESEARCH PRESS

New York

Published by NY Research Press
118-35 Queens Blvd., Suite 400,
Forest Hills, NY 11375, USA
www.nyresearchpress.com

Handbook of Electrical Engineering
Edited by Norman Schultz

International Standard Book Number: 978-1-63238-652-6 (Hardback)

Cataloging-in-Publication Data

Handbook of electrical engineering / edited by Norman Schultz.
    p. cm.
Includes bibliographical references and index.
ISBN 978-1-63238-652-6
1. Electrical engineering. 2. Electrical engineering--Materials. I. Schultz, Norman.
TK145 .H36 2019
621.3--dc23

Contents

Preface

The main aim of this book is to educate learners and enhance their research focus by presenting diverse topics covering this vast field. This is an advanced book which compiles significant studies by distinguished experts in the area of analysis. This book addresses successive solutions to the challenges arising in the area of application, along with it; the book provides scope for future developments.

Electrical engineering is a branch of engineering concerned with the study of electromagnetism, electronics and electricity, and their applications. It applies the theories and principles of physics and mathematics to understand electrical devices. Various sub-fields of this discipline are power engineering, telecommunications, signal processing, instrumentation, etc. A comprehensive knowledge of electrical engineering is of significance in the design, development and operation of electrical systems, like telecommunication systems, electric power stations, electrical control of industrial machinery, etc. This book contains some path-breaking studies in the area of electrical engineering. It traces the progress of this field and highlights some of its key concepts and applications. For all readers who are interested in electrical engineering, the case studies included in this book will serve as an excellent guide to develop a comprehensive understanding.

It was a great honour to edit this book, though there were challenges, as it involved a lot of communication and networking between me and the editorial team. However, the end result was this all-inclusive book covering diverse themes in the field.

Finally, it is important to acknowledge the efforts of the contributors for their excellent chapters, through which a wide variety of issues have been addressed. I would also like to thank my colleagues for their valuable feedback during the making of this book.

Editor

The System of Fast Charging Station for Electric Vehicles with Minimal Impact on the Electrical Grid

Petr CHLEBIS, Martin TVRDON, Katerina BARESOVA, Ales HAVEL

Department of Electronics, Faculty of Electrical Engineering and Computer Science,
VSB–Technical University of Ostrava, 17. listopadu 15, 708 33 Ostrava–Poruba, Czech Republic

petr.chlebis@vsb.cz, martin.tvrdon@vsb.cz, katerina.baresova@vsb.cz, ales.havel@vsb.cz

Abstract. *The searching and utilization of new energy sources and technologies is a current trend. The effort to increase the share of electricity production from renewable energy sources is characteristic for economically developed countries. The use of accumulation of electrical energy with a large number of decentralized storage units is most preferred, as well as the focus on the production of energy at the point of its consumption. Modern cogeneration units are a good example. This paper describes the accumulation of electrical energy for equalizing the power balance of electric charging stations with high instantaneous power. The possibility of re-utilization of electrical energy from the charged vehicle in the case of lack of electricity in the power grid is solved at the same time. This paper also deals with the selection of appropriate concept of accumulation system and its cooperation with both renewable and distribution networks. Details of the main power components including the results obtained from the system implementation are also described in this paper.*

Keywords

Accumulation, electric vehicle, electrical grid, fast charging station, renewable energy sources.

1. Introduction

The growing share of electric vehicles in passenger traffic brings demands for expansion of infrastructure of fast charging stations. The average annual electricity consumption of the personal electric vehicle is reported to be around 3300 kWh/year. This value is very close to the average annual household consumption, which is reported to be in the range of 3700–5400 kWh/year for flats and smaller family houses. The disadvantage in covering the consumption of electric vehicles unlike households is that vehicles can take their charging energy at different places and at different times depending on their actual position. Electrical distribution network for households and other immobile points of consumption is sufficiently dimensioned and there is always a firmly defined maximum of consumed power given in the contractual relationship with the customer. The consumption facilities for charging electric vehicles have their position defined, but their operation is dependent on the number of charged vehicles. This creates an inappropriate load for the grid, especially during a short consumption period associated with fast-charging [1], [2].

1.1. State of the Art

Contemporary fast charging stations require high power connection with electrical energy, and their usage causes large consumption peaks in the grid. With a growing number of electric vehicles and proportionally expanding infrastructure of charging stations there are two possible ways to solve this problem. The first solution is very economically challenging and requires strengthening of the distribution system, the other one is to build charging stations with minimal impact on the electrical system. A brief overview of similar existing solutions in the area of fast charging stations for electric vehicles is presented below.

One of the known solutions is called the high-speed charging station for EV's battery charging which has a high power output converter, whose input and output terminals are connected to respective electrical energy storage device and electrical load. The disadvantage of this solution is that it does not allow bidirectional power flow between the electric vehicle and the distribution network [3].

Another similar patented solution is called the fast charging system of electric vehicle. The accumulation system is composed of at least two accumulators connected in parallel or in series which can be repeatedly charged or discharged. Again, this solution does not support the bidirectional flow of electric power [4].

Last patented solution, which will be mentioned in this paper, is called the electric automobile fast charging station. It contains the charging system with multisource power supply. Again, this solution does not allow to supply the energy back to the grid from connected electrical vehicles or from its own storage system [5].

One of the basic general requirements on the charging station is to work with minimal impacts on the power grid. It is possible to meet this requirement through sufficient amount of energy stored in the storage station. Charging stations with accumulation will then work within the framework of the daily consumption diagram, in the balanced energy budget if possible [6], [7].

Another option is to connect different types of renewable energy sources, especially solar and wind power plants. Active behavior of the system with respect to the distribution network allows supplying the electricity from solar and wind power plants during peak demands of the daily chart according to the needs of the power grid. When the charging station is not used, it is possible to supply power to the grid from connected renewable energy sources (if they are active) and to consume power from the storage station at the same time. Under critical operating conditions of the power grid, the system is also able to consume power (while meeting defined requirements) from the battery of electric vehicle which is connected to the charging station. However, this mode is considerably limited by the properties of the charging station and even by the structure and by defined modes in the connection point. Figure 1 shows the structure which represents the requirements for a conceptual solution of the system for electric vehicle charging.

1.2. The Main Elements of Active Charging Station

The basic elements of the power system of an active charging station are DC/DC converters used for fast charging and the storage system of the station. Linking the battery with the AC power grid is provided by a three-phase pulse rectifier which enables bidirectional

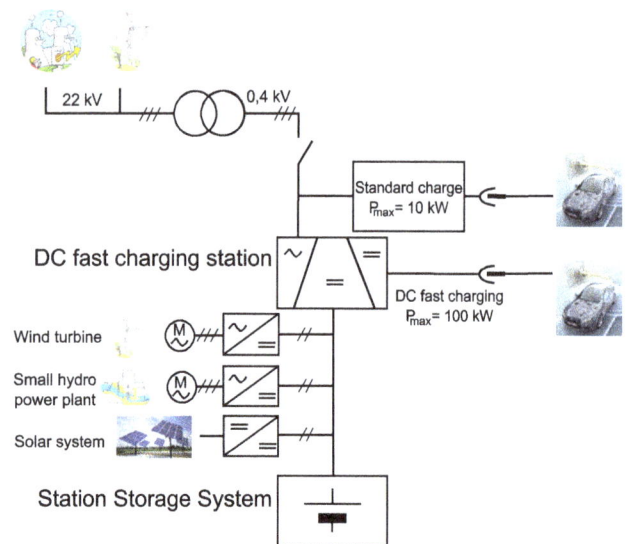

Fig. 1: Active charging station with renewable sources.

flow of electric power. The fast charging process is provided by a two-quadrant DC/DC converter with current reversion. This converter allows bidirectional flow of energy throughout the whole power range. Its disadvantage is that it does not provide galvanic isolation of the charging station from the vehicle's battery, but this can be solved in other ways. The storage system of the station is composed of $LiFePO_4$ batteries containing about 80 kWh of electrical energy. The energy stored in the station's storage system may be sufficient to charge two to four common personal electric vehicles without the use of other connected sources. If the energy is supplied to the system from renewable energy sources, or if it is consumed from the power grid, then it is possible to charge the electric vehicles continuously. The designed conceptual solution fully satisfies the above defined requirements and allows many variations of energy flow according to actual conditions and requirements [8], [9].

2. Description of the System

This paper follows up on implementation of storage systems in the framework of the Technology Center of Ostrava (TCO) built with the support of the project ENET - Energy Units for Utilization of non Traditional Energy Sources. The pilot production facilities were built in the TCO for research and development of energy accumulation systems. One of those systems utilizes lead-acid and $LiFePO_4$ batteries in a large accumulation station, the other one is a mobile accumulation unit [10].

The system of mobile accumulation is designed as a battery container unit equipped with LiFePO$_4$ batteries, which allows the connection of renewable energy sources, energy networks and fast charging stations for road and also rail electric vehicles. The system is designed with respect to development possibilities of various energetic solutions like Smart grids with ability to test these solutions and to put them into practice.

Figure 2 shows a block diagram of the mobile accumulation station. The block called ACB is the basic accumulation element of the station which is in this particular case composed of LiFePO$_4$ batteries. The block called RE represents the renewable energy source, such as a solar power plant or wind power plant, which can be connected to the accumulation system usually via a suitable coupling converter. Another unit, directly connected to the DC link of the accumulator system, is the device for fast charging of electric vehicles with the power up to 100 kW, which allows the vehicle to charge in 20 to 30 minutes, depending on the type and quality of its battery system. That block is marked by EVC – DC. The next part of the mobile accumulation system is a standard network connection using its own charger for vehicles with power up to 10 kW. With the support of energy accumulation it is possible to charge several vehicles at the same time from this terminal for a time period of 2-8 hours (according to the number of installed devices) without undesirable effects on the power grid. That block is marked by EVC – AC. The accumulation unit is connected to the network via a reversible voltage inverter for the possibilities of transmission of electricity produced from renewable energy sources and also for accumulation of energy surplus in the grid.

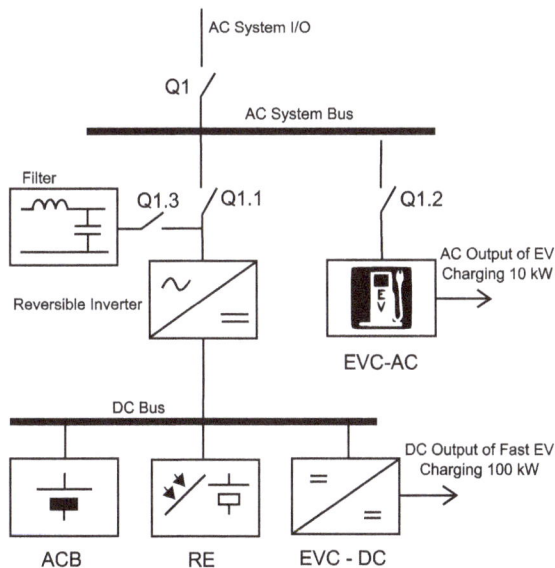

Fig. 2: Topology of mobile accumulation.

3. Modelling of the Charging System and Analysis of Simulation Results

Modelling of the charging system is implemented in two separate parts. The first part consists of a model of the pulse rectifier which creates a link between the grid and the battery station. The second part consists of a pulse converter which creates a link between the battery station and the charged electric vehicle. The function of the entire system in required operation modes was verified by the simulation results which are presented below.

The simulation of the pulse rectifier model verifies its function in the rectifier mode, in the inverter mode and the transition between these two modes. The results of the simulation are shown in Fig. 3 and Fig. 4. The waveforms of voltage and current are simulated on the input and output of the pulse rectifier. The first period of both pictures shows the start of the rectifier with charging current 1 A. After stabilization of circuit parameters the value of the charging current is set to 5 A (second period). The last period presents the transition to the inverter mode with a current value 3 A supplied from the battery station to the grid.

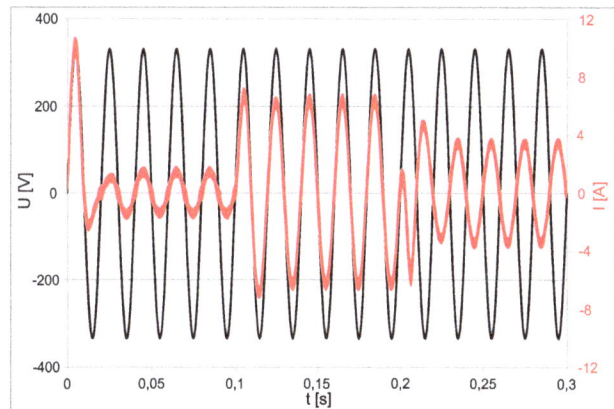

Fig. 3: The AC voltage and current on the input of a pulse rectifier.

The simulations of the standard and fast charging processes were performed in the simulation software Matlab Simulink. In this software, the entire structure of the charging system was designed to resemble the real model as much as possible. The results depicted on the following pictures are demonstrating the functionality of the system in both required modes [8], [9].

Figure 5 shows the waveforms of basic variables during a step-change of charging current supplied to the battery. The black waveform represents the change of the battery's state of charge. This value increases linearly while charging with constant current. Due to the

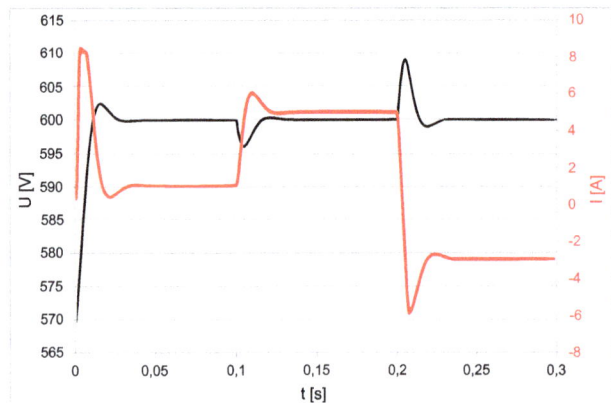

Fig. 4: The voltage and current at the battery station.

transparency, the red curve in the chart was inverted and represents the current supplied to the battery. In the case of negative current value, the battery is charging, and the product of negative current and positive voltage on the battery means that the negative power is "consumed".

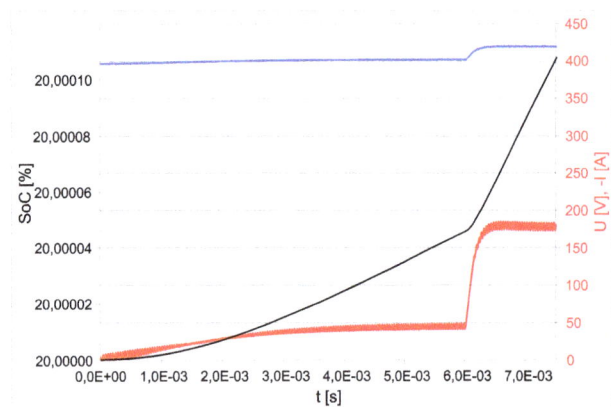

Fig. 5: Waveforms of variables in the battery of charged EV with charging current 45 A in time interval (0–6) ms and 180 A after 6 ms.

Tab. 1: Detailed description of waveforms in Fig. 5.

Color	Variable	Description of the waveform
Black	SoC	The state of charge of the EV battery unit
Red	I_{ACU}	The current supplied to the battery
Blue	U_{ACU}	The battery voltage during charging

The following Fig. 6 shows the waveforms of basic variables informing about the status of electrical energy consumed from the battery of electric vehicle.

In the first time period, the current is set to 45 A and after 6 ms of simulation, the value is increased to 180 A. The battery voltage and current are both positive and their product represents the consumed power – so the battery is discharged. The state of charge curve

(SoC) decreases during discharging of the battery. If the consumed current value is constant, the curve is linear.

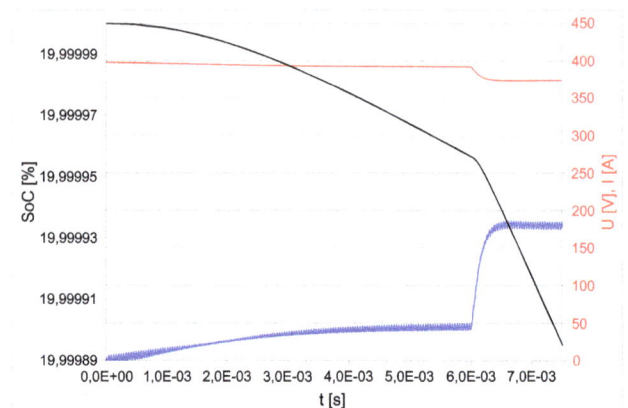

Fig. 6: Waveforms of variables in the battery of discharged EV with discharging current 45 A in time interval (0–6) ms and 180 A after 6 ms.

Tab. 2: Detailed description of waveforms in Fig. 6.

Color	Variable	Description of the waveform
Black	SoC	The state of charge of the EV battery unit
Red	I_{ACU}	The current consumed from the battery
Blue	U_{ACU}	The battery voltage during discharging

4. Experimental Results in Different Modes of Operation

The main purpose of this paper was to verify the functionality of the proposed system in the view of energy flow between various sources or energy storing elements such as electrical grid, battery stations and batteries of electric vehicles. The electrical energy in a form of power flow which is in time supplied or consumed by various sources is shown in Fig. 7. A negative power value of the particular component indicates that this component constitutes a power source and supplies its power to the system. On the other hand, a positive power value indicates the opposite action when this component acts as an electrical appliance and the energy of the system is consumed or supplied to the grid.

Three different sections A, B and C are highlighted in the Fig. 7, where the section A shows the mode of energy consumption from the distribution network, section B represents the off-grid mode when the system operates independently on the distribution network, and the section C displays the mode of energy supply to the distribution network.

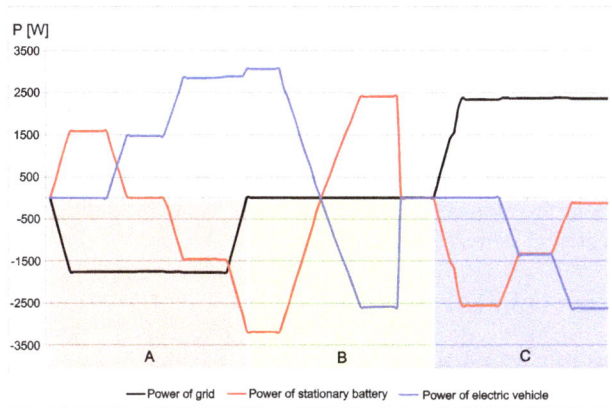

Fig. 7: Waveforms of the power flow in particular system sources in different operation modes.

Tab. 3: Detailed description of waveforms in Fig. 7.

Color	Variable	Description of the waveform
Black	P_{GRID}	The power of the grid
Red	P_{ACU}	The power of stationery battery
Blue	P_{EV}	The power of electric vehicle

5. Conclusion

This paper deals with the energetic structure for accumulation of electrical energy which is utilized as a storage unit in intelligent energy networks, commonly referred to as Smart grids. A section of the fast charging station for electric vehicles is contained within this structure. The entire structure creates a link between the grid and fast charging, which requires large amount of electric power. The energetic structure enables the charging station to consume high power and does not create a demand for high power consumption from the grid at the same time, because the peak power is provided by the storage unit.

The experimental part of the paper demonstrates the functionality of the system in all modes of operation. The efficiency of the entire system in a limited operating range reaches 89 %. It is a product of partial efficiencies of two separated power converters connected in series. The efficiency of individual converters reaches acceptable values: 92.5 % for the pulse rectifier and 96 % for the DC/DC converter.

Acknowledgment

This paper was funded by a grant of SGS No. SP2015/174 from the VSB–Technical University of Ostrava, Czech Republic.

References

[1] SLADECEK, V., P. PALACKY, D. SLIVKA and M. SOBEK. Influence of Power Semiconductor Converters Setup on the Quality of Electrical Energy from Renewable Sources. In: *Proceedings of the 11th International Scientific Conference on Electric Power Engineering.* Brno: Brno University of Technology, 2010, pp. 687–691. ISBN 978-80-214-4094-4.

[2] CHLEBIS, P., A. HAVEL, P. VACULIK and Z. PFOF. Modern Instruments for increasing the Efficiency of the Energy Transfer in Electric Vehicles. In: *14th International Power Electronics and Motion Control Conference (EPE-PEMC).* Ohrid: IEEE, 2010, pp. 89–93. ISBN 978-1-4244-7856-9. DOI: 10.1109/EPEPEMC.2010.5606586.

[3] BOEHM, M. and J. REINSCHKE. High-speed charging station for charging battery of electric vehicle. European patent, DE102010062362 (A1), 6.6.2012.

[4] XINJIA, Z. Fast charging system of electric vehicle. Chinese patent, CN201750168 (U), 16.2.2011.

[5] JIFENG, R. and R. RUI. Electric automobile fast charging station. Chinese patent, CN201191766 (Y), 4.2.2009.

[6] BRIANE, B. and S. LOUDOT. Rapid reversible charging device for an electric vehicle. United States patent, US2011254494 (A1), 20.10.2011.

[7] CHEN, M. and G. A. RINCON-MORA. Accurate electrical battery model capable of predicting runtime and I-V performance. *IEEE Transactions on Energy Conversion.* 2006, vol. 21, iss. 2, pp. 504–511. ISSN 0885-8969. DOI: 10.1109/TEC.2006.874229.

[8] CHLEBIS, P., M. TVRDON, A. HAVEL and K. BARESOVA. Comparison of Standard and Fast Charging Methods for Electric Vehicles. *Advances in Electrical and Electronic Engineering.* 2014, vol. 12, no. 2, pp. 111–116. ISSN 1804-3119. DOI: 10.15598/aeee.v12i2.975.

[9] TVRDON, M., P. CHLEBIS and M. HROMJAK. Design of Power Converters for Renewable Energy Sources and Electric Vehicles Charging. *Advances in Electrical and Electronic Engineering.* 2013, vol. 11, no. 3, pp. 204–209. ISSN 1804-3119. DOI: 10.15598/aeee.v11i3.795.

[10] BRANDSTETTER, P. and M. SKOTNICA. ANN Speed Controller for Induction Motor Drive with Vector Control. *International Review of Electrical Engineering.* 2011, vol. 6, iss. 7, pp. 2947–2954. ISSN 1827-6660.

About Authors

Petr CHLEBIS was born in 1956. He was appointed as professor at Faculty of Electrical Engineering and Computer Science of VSB–Technical University of Ostrava in 2005. His research includes power electrical converters, their control systems and utilization of nontraditional energy sources.

Martin TVRDON was born in 1986. He obtained his Master's degree in electronics in 2011. He is currently pursuing Ph.D. study at the Department of Electronics at VSB–Technical University of Ostrava at Faculty of Electrical Engineering and Computer Science. His research includes power systems, design and modeling of fast charging stations for electric vehicles.

Katerina BARESOVA was born in 1988. She obtained her Master's degree in the field of power electronics systems in 2013. She is currently pursuing Ph.D. study at the Department of Electronic at VSB–Technical University of Ostrava at Faculty of Electrical Engineering and Computer Science. Her research includes power systems and modeling of electrical converters.

Ales HAVEL was born in 1984. He obtained his Master's degree in the field of power electronics systems and electric machines design in 2009, and Ph.D. degree in 2013 at the Department of Electronics at VSB–Technical University of Ostrava at Faculty of Electrical Engineering and Computer Science. His research includes power systems, design and modeling of electric machines.

2

Thermal Experimental Analysis for Dielectric Characterization of High Density Polyethylene Nanocomposites

Ahmed THABET[1], *Youssef MOBARAK*[1, 2]

[1]Nanotechnology Research Center, Faculty of Energy Engineering, Aswan University, 81528 Aswan, Egypt
[2]Department of Electrical Engineering, Faculty of Engineering, Rabigh, King Abdulaziz University, 21589 Jeddah, Kingdom of Saudi Arabia

athm@aswu.edu.eg, ysoliman@kau.edu.sa

Abstract. *The importance of nanoparticles in controlling physical properties of polymeric nanocomposite materials leads us to study effects of these nanoparticles on electric and dielectric properties of polymers in industry In this research, the dielectric behaviour of High-Density Polyethylene (HDPE) nanocomposites materials that filled with nanoparticles of clay or fumed silica has been investigated at various frequencies (10 Hz–1 kHz) and temperatures (20–60 °C). Dielectric spectroscopy has been used to characterize ionic conduction, then, the effects of nanoparticles concentration on the dielectric losses and capacitive charge of the new nanocomposites can be stated. Capacitive charge and loss tangent in high density polyethylene nanocomposites are measured by dielectric spectroscopy. Different dielectric behaviour has been observed depending on type and concentration of nanoparticles under variant thermal conditions.*

Keywords

Dielectric properties, high density polyethylene, insulation, nano-composite, nanoparticles, polymers.

1. Introduction

Among the various types of polymeric dielectrics, High-Density PolyEthylene (HDPE) has been standing out as a raw material for the production of insulators, spacers, and also as a coating for cable conductors used in electrical power distribution networks. For this type of application, the dielectric strength is one of the properties that must be taken into account in order to check the ability to withstand high electric fields. Dielectric strength is defined as a relationship between the breakdown voltage and the dielectric thickness, representing the maximum field which the material can support indefinitely for a specific experimental setup. The use of high purity polymers in engineering applications is technologically not viable. This problem leads to the development of formulations with additives in order to protect the polymers against losses in their properties (for example, mechanical and thermo-mechanical) during the processing stages or in service. These additives used in polymers for electrical insulation may or may not harm the electric properties [1], [2], [3], [4] and [5]. Polymer composite, compared with conventional sing-phase insulation, can improve dielectric properties. Non-linear conductivity in the insulating polymer has been achieved by the introduction of inorganic semi-conductor in particulate.

The polymer composite with field dependent conductivity can be used to improve the distribution of electric field. For different application, the different composites are used. For example, the silicon rubber and EPDM filled with non-linear fillers are used as electric stress grating materials in cable joints and terminations, and the epoxy composites filled with carborundum (SiC) are used for grating the electric field distribution at the end of windings in electric machines. The polymer composite filled with non-linear inorganic fillers, as one kind of non-linear dielectrics, can be called "smart insulating materials", due to the function of grading electric field and restraining the formation of space charge. The properties of the non-linear materials are dependent on the basic materials, and also dependent on the fillers. In any engineering application, it is vital that the selected materials exhibit an appropriate combination of properties throughout the design lifetime of the plant. For polymers, macroscopic properties are determined by two factors; microstructure and composition [6], [7], [8], [9] and [10]. For example, the growth of spherulites may lead to increase or decrease in electrical breakdown strength, depending upon the precise structure of these objects.

The composition of a polymeric material can be varied in many ways, through the addition of antioxidant, plasticizers, crosslinkers, fillers, etc. The addition of inorganic filler to an elastomer will increase its stiffness, albeit at the expense of reduced elongation at break [10], [11], [12], [13] and [14]. As of now, work is underway to examine the physical properties of nanocomposite materials composed of nanoparticles and their compounds stabilized within a polymeric dielectric matrix. In recent years polymer nanocomposites have attracted wide interest with regard to enhancing polymer properties and extending their utility. It has been found that the dielectric properties have a close relationship with the interfacial behaviour between the fillers and the polymer matrix in such composites. The electric and optic properties of these materials have been demonstrated to be highly dependent on the size, structure, and concentration of the nanoparticles, as well as on the type of polymeric matrix [15], [16], [17], [18] and [19].

Great expectations have been focused on effects and importance of costless nanoparticles [20], [21], [22], [23], [24], [25], [26], [27] and [28]. However, it has been concerned in this paper about the effect of types of costless nanoparticles on the electrical properties of a polymeric nanocomposite. With a continual progress in polymer nanocomposites, this research depicts the effects of types and concentration of costless nanoparticles in electrical properties of industrial polymer material. All the experimental results of dielectric spectroscopy have been investigated and discussed to detect all nanoparticles effects on electrical properties of nanocomposite industrial material which fabricated; like High Density PolyEthylene (HDPE) with various nanoparticles of clay and fumed SiO_2.

2.   Experimental Setup

HIOKI 3522-50 LCR Hi-tester device measured electrical parameters of nanocomposite solid dielectric insulation specimens at various frequencies: $|Z|$, $|Y|$, Θ, Rp (DCR), Rs (ESR, DCR), G, X, B, Cp, Cs, Lp, Ls, D (tan δ), and Q. Specification of LCR is Power supply: 100, 120, 220 or 240 V ($\pm$10 %) AC (selectable), 50/60 Hz, Frequency: DC, 1 mHz to 100 kHz, Display Screen: LCD with backlight/99999 (full 5 digits), Basic Accuracy: Z : $\pm$0.08 % rdg. Θ: $\pm$0.05 $^\circ$, and External DC bias $\pm$40 V max.(option)

(3522-50 used alone $\pm$10 V max./using 9268 $\pm$40 V max.).

Finally, all dielectric properties for pure and nanocomposite industrial materials can be measured using HIOKI 3522-50 LCR Hi-tester device. Figure 1 shows HIOKI 3522-50 LCR Hi-tester device for measuring characterization of nanocomposite insulation industrial materials.

Fig. 1: HIOKI 3522-50 LCR Hi-tester device.

3.   Preparation of Nanocomposites and Characterization

The industrial materials studied here are high density polyethylene which has been formulated utilizing variant concentrations of nanoparticles of clay and fumed silica. High density polyethylene nanocomposites have been prepared and fabricated by using recent nanotechnology procedures and devices for melting pure high density polyethylene grains, mixing and penetrating nanoparticles inside the base matrix HDPE by modern ultrasonic devices. Most of all nanocomposite materials are commercial and available already in the manufacturing of High-Voltage (HV) industrial products and their properties detailed in Tab. 1.

Tab. 1: Electric and dielectric properties of pure and nanocomposite materials.

Materials	Dielectric constant at 1 kHz	Resistivity ($\mu\cdot$m)
Pure HDPE	2.3	10^{15}
HDPE + 1 wt% clay	2.23	10^{16}
HDPE + 5 wt% clay	1.99	$10^{16} - 10^{19}$
HDPE + 10 wt% clay	1.76	$10^{19} - 10^{21}$
HDPE + 1 wt% SiO_2	2.32	10^{14}
HDPE + 5 wt% SiO_2	2.39	$10^{14} - 10^{12}$
HDPE + 10 wt% SiO_2	2.49	$10^{12} - 10^{10}$

SEM images illustrate penetration of nanoparticles in polymeric nanocomposites; thus, Fig. 2 shows SEM

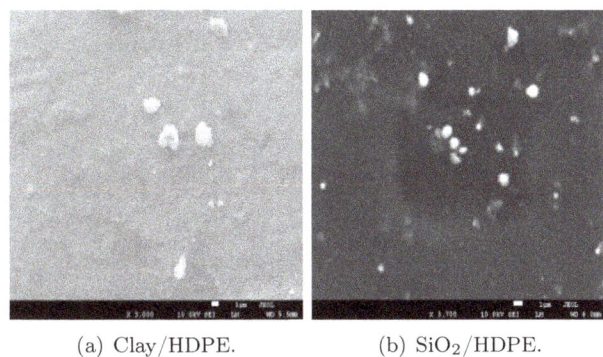

(a) Clay/HDPE.   (b) SiO_2/HDPE.

Fig. 2: SEM images for high density polyethylene nanocomposites.

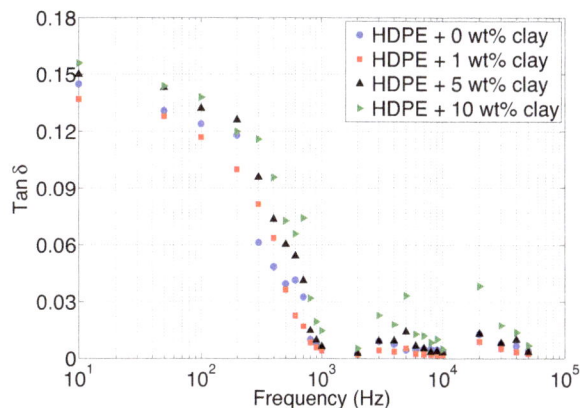

Fig. 3: Measured loss tangent for clay/HDPE nanocomposites at $T = 20$ °C.

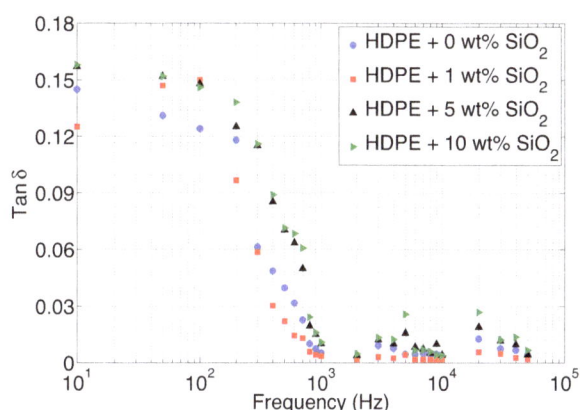

Fig. 4: Measured loss tangent for SiO_2/HDPE nanocomposites at $T = 20$ °C.

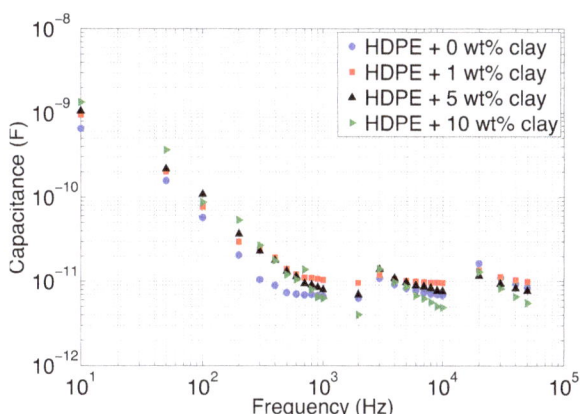

Fig. 5: Measured capacitance clay/HDPE nanocomposites at $T = 20$ °C.

images that illustrate the penetration of cost-fewer nanoparticles in high density polyethylene nanocomposites. It has flakes like morphology with high surface area. Also, it illustrates that the nanoparticles are uniformly dispersed in the polymer matrix.

4.   Results and Discussion

Dielectric Spectroscopy is a powerful experimental method to investigate the dynamical behaviour of a sample through the analysis of its frequency-dependent dielectric response. This technique is based on the measurement of the capacitance as a function of frequency of a sample sandwiched between two electrodes. The tan δ, and capacitance (C) was measured as a function of frequency in the range 10 Hz to 50 kHz at variant temperatures for all the test samples. The measurements were made using high-resolution dielectric spectroscopy.

4.1.   Effect of Nanoparticles at 20 °C

Figure 3 shows loss tangent as a function of frequency for clay/HDPE nanocomposites at room temperature (20 °C). This figure illustrates the loss tangent of clay/HDPE nanocomposites increases with increasing clay nanoparticles concentration up to 1 wt%, especially at low frequencies, but it decreases with increasing clay nanoparticles concentration up to 10 wt%. In addition, Fig. 4 shows loss tangent as a function of frequency for (20 °C). The loss tangent of SiO_2/HDPE nanocomposites decreases with increasing fumed silica concentration nanoparticles up to 1 wt%, especially at high frequencies but it increases with increasing fumed silica concentration nanoparticles (1–10 wt%).

Figure 5 shows capacitance as a function of frequency for clay/HDPE nanocomposites at room temperature (20 °C). It is clear that the measured capacitance of clay/HDPE nanocomposites increases with in-

creasing clay concentration nanoparticles up 10 wt%. On the other hand, Fig. 6 shows capacitance as a function of frequency for SiO_2/HDPE nanocomposites at room temperature (20 °C). Furthermore, the measured capacitance of SiO_2/HDPE nanocomposites increases with increasing fumed silica concentration nanoparticles up to 5 wt% but it decreases with increasing fumed silica concentration nanoparticles up to 10 wt%.

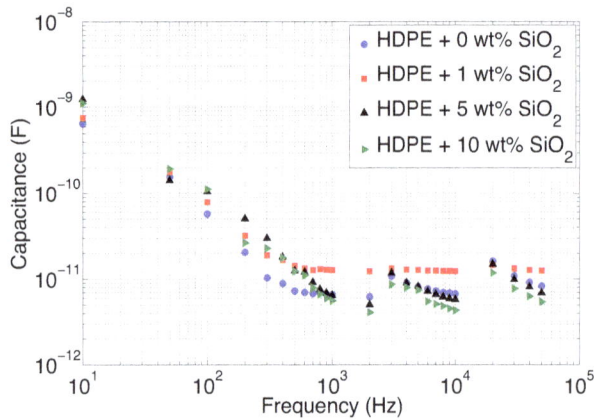

Fig. 6: Measured capacitance SiO_2/HDPE nanocomposites at $T = 20\ °C$.

4.2.   Effect of Nanoparticles on HDPE Characterization at 40 °C

Figure 7 shows loss tangent as a function of frequency for clay/HDPE nanocomposites at a temperature (40 °C). The loss tangent of clay/HDPE nanocomposites increases with increasing clay nanoparticles concentration up to 10 wt%, especially at low frequencies. Figure 8 shows loss tangent as a function of frequency for SiO_2/HDPE nanocomposites at a temperature (40 °C), the measured loss tangent of SiO_2/HDPE nanocomposites increases with high fumed silica concentration nanoparticles up to 1 wt%, especially at low frequencies. Noting that the loss tangent of SiO_2/HDPE nanocomposites decreases with increasing fumed silica concentration nanoparticles (1–10 wt%).

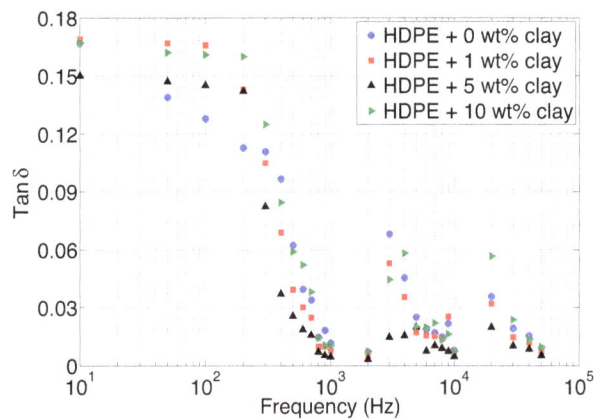

Fig. 7: Measured loss tangent for clay/ HDPE nanocomposites at $T = 40\ °C$.

Figure 9 shows capacitance as a function of frequency for clay/HDPE nanocomposites at a temperature (40 °C). The measured capacitance of clay/HDPE nanocomposites decreases with increasing clay concentration nanoparticles.   Similarly, Fig. 10 shows the

measured capacitance of SiO_2/HDPE nanocomposites decreases with increasing fumed silica concentration nanoparticles capacitance as a function of frequency at a temperature (40 °C).

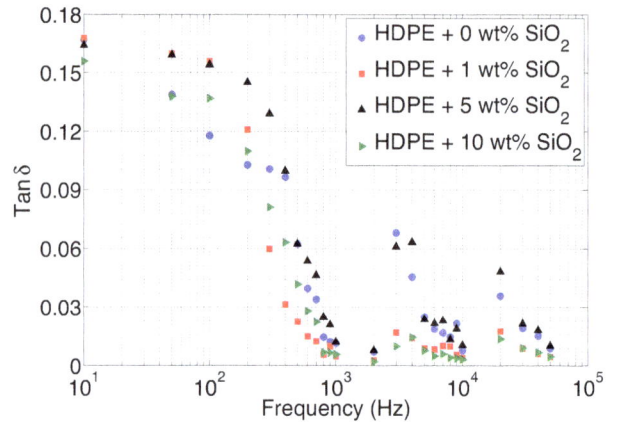

Fig. 8: Measured loss tangent for SiO_2/HDPE nanocomposites at $T = 40\ °C$.

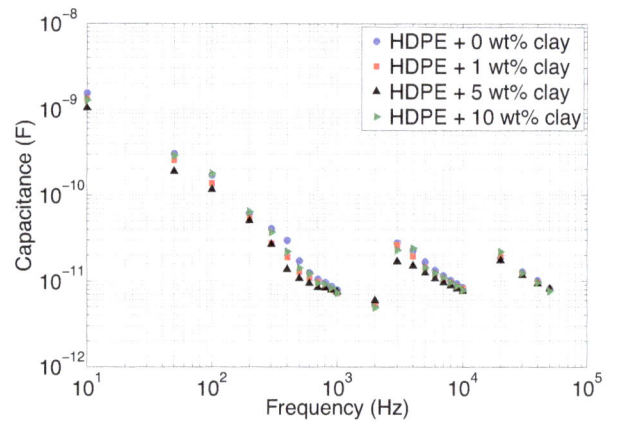

Fig. 9: Measured capacitance for clay/HDPE nanocomposites at $T = 40\ °C$.

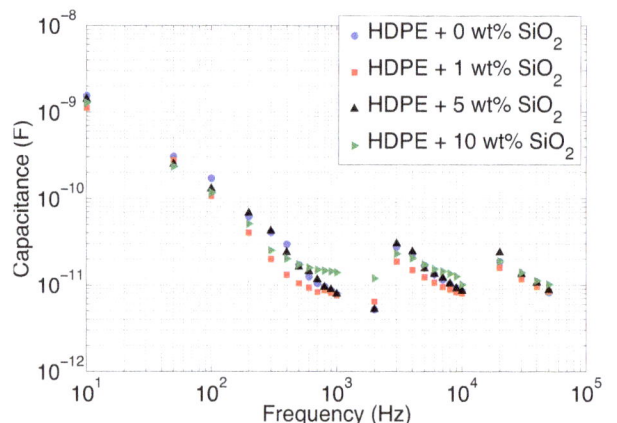

Fig. 10: Measured capacitance for SiO_2/HDPE nanocomposites at $T = 40\ °C$.

4.3. Effect of Nanoparticles at 60 °C

Figure 11 shows loss tangent as a function of frequency for clay/HDPE nanocomposites at a temperature (60 °C). The loss tangent of clay/HDPE nanocomposites decreases with high clay nanoparticles concentration up to 10 wt%, especially at low frequencies. On the other hand, Fig. 12 shows loss tangent as a function of frequency for SiO_2/HDPE nanocomposites at a temperature (60 °C), moreover, it is noticed that the loss tangent of SiO_2/HDPE nanocomposites decreases with high fumed silica concentration nanoparticles up to 10 wt%, specially, at low frequencies.

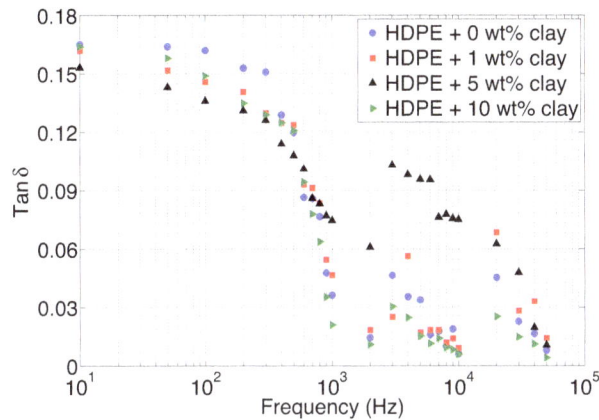

Fig. 11: Measured loss tangent for clay/HDPE nanocomposites at $T = 60$ °C.

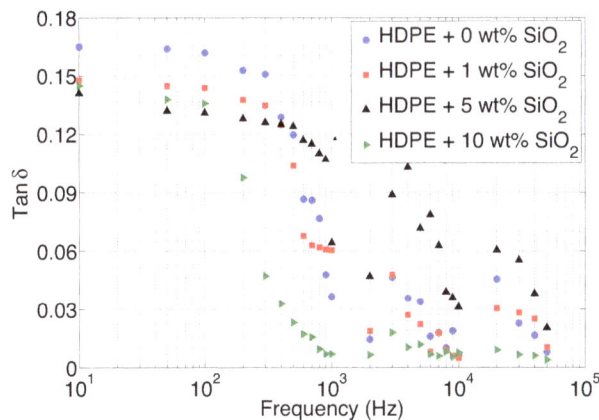

Fig. 12: Measured loss tangent for SiO_2/HDPE nanocomposites at $T = 60$ °C.

Figure 13 shows capacitance as a function of frequency for clay/HDPE nanocomposites at a temperature (60 °C). It illustrates the measured capacitance of clay/HDPE nanocomposites increases with increasing clay concentration nanoparticles up to 10 wt%. On the other hand, Fig. 14 shows capacitance as a function of frequency for SiO_2/HDPE nanocomposites at a temperature (60 °C). Moreover, it is illustrated that the capacitance of SiO_2/HDPE nanocomposites increases

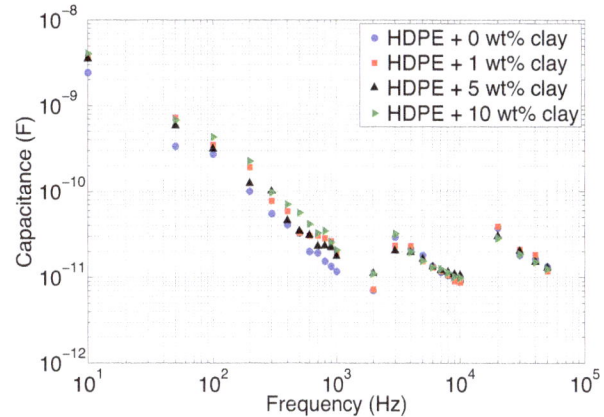

Fig. 13: Measured capacitance for clay/HDPE nanocomposites at $T = 60$ °C.

with high fumed silica concentration nanoparticles up to 1 wt%, in addition, it decreases with the growth of fumed silica concentration nanoparticles (5–10 wt%).

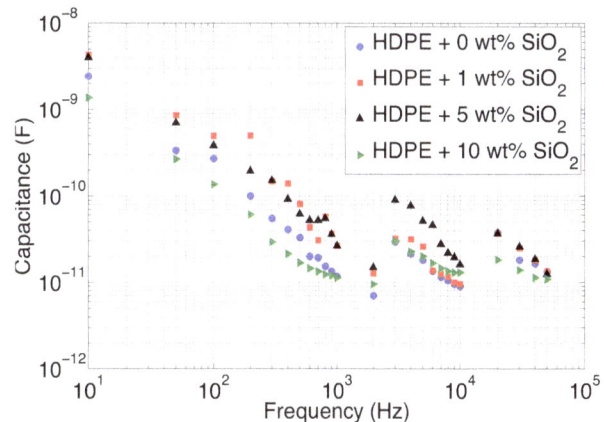

Fig. 14: Measured capacitance for SiO_2/HDPE nanocomposites at $T = 60$ °C.

4.4. Trends of Nanoparticles under Thermal Conditions

All depicted results have cleared that adding fumed silica increases permittivity of High density polyethylene insulation materials; furthermore, adding clay has decreases permittivity of High density polyethylene insulation materials as shown in Tab. 1. Physical interface between high density polyethylene and nanoparticles has been affected on capacitance and dielectric loss angle curves under normal thermal conditions (20 °C) that are pointed out in (Fig. 3, Fig. 4, Fig. 5, Fig. 6). Therefore, the loss tangent of clay/HDPE nanocomposites increases with increasing clay nanoparticles concentration up to 1 wt%, specially, at low frequencies but it decreases with the raise of clay nanoparticles concentration up to 10 wt%. Moreover, the loss tangent of SiO_2/HDPE nanocomposites decreases with fumed silica concentration nanoparticles raised up to 1 wt%, especially, in case of high frequencies but it increases with increasing fumed silica concentration nanoparticles (1–10 wt%). The measured capacitance of clay/HDPE nanocomposites increases with increasing clay concentration nanoparticles up to 10 wt%.Furthermore, the measured capacitance of SiO_2/HDPE nanocomposites increases with increasing fumed silica concentration nanoparticles up to 5 wt% but it decreases with increasing fumed silica concentration nanoparticles up to 10 wt%.

Changing thermal conditions can be affected by physical interface between high density polyethylene and nanoparticles and so will be affected by capacitance and dielectric loss angle curves as pointed out in (Fig. 7, Fig. 8, Fig. 9 and Fig. 10) for high thermal condition (40 °C)Thus, the loss tangent of clay/HDPE nanocomposites increases with increasing clay nanoparticles concentration up to 10 wt%, specially, at low frequencies. Moreover, the measured loss tangent of SiO_2/HDPE nanocomposites increases with increasing fumed silica concentration nanoparticles up to 1 wt%. Noting that the loss tangent of SiO_2/HDPE nanocomposites decreases with the higher concentration of fumed silica nanoparticles(1–10 wt%). Also, the measured capacitance of clay/HDPE nanocomposites decreases with the addition of clay concentration nanoparticles up to 10 wt%. It is clear that the measured capacitance of SiO_2/HDPE nanocomposites raises with the concentration of the fumed silica nanoparticles up to 1 wt% and also increases with the addition of fumed silica concentration nanoparticles up to 10 wt%.

Finally, the effect of raising thermal conditions up to (60 °C) on physical interface between high density polyethylene and nanoparticles is pointed out in (Fig. 11, Fig. 12, Fig. 13 and Fig. 14) where loss tangent and capacitance of new nanocomposite materials are reported for different concentration weights of modified nanoparticles concentration at (60 °C) temperature. Thus, the loss tangent of clay/HDPE nanocomposites decreases with clay nanoparticles concentration raised up to 10 wt%, especially in case of low frequencies. It is noticed that the loss tangent of SiO_2/HDPE nanocomposites decreases with fumed silica concentration nanoparticles up to 10 wt%, especially in case of low frequencies. The measured capacitance of clay/HDPE nanocomposites increases with the addition of the clay nanoparticles concentration up to 10 wt%. The capacitance of SiO_2/HDPE nanocomposites increases with fumed silica concentration nanoparticles increased up to 1 wt% but it decreases with the addition of fumed silica nanoparticles (5–10 wt%).

5.   Conclusion

Adding fumed silica has increased permittivity of the new high density polyethylene nanocomposite materials, but adding clay has decreased permittivity of the new high density polyethylene nanocomposite materials.

Nanoparticles can be controlled in the loss tangent, and capacitance of new high density polyethylene nanocomposites depend on type and concentration of nanoparticles in nanocomposites according to physical interface that created between high density polyethylene and nanoparticles.

Thermal environment is an effective parameter for increasing and decreasing the loss tangent and capacitance of new high density polyethylene nanocomposites with respect to the type and concentration of nanoparticles. Thermal environment conditions are affected by physical interface created between high density polyethylene and nanoparticles because clay nanoparticles are more efficient than fumed silica nanoparticles for decreasing charging capacitance and loss tangent performance under room thermal conditions. However, under the high thermal conditions, the performance of charging capacitance and loss tangent are changed gradually.

Acknowledgment

The present work was supported by Nanotechnology Research Center at Aswan University that is established by aided the Science and Technology Development Fund (STDF), Egypt, Grant No: Project ID 505, 2009-2011.

References

[1] TANIMOTO, G., M. OKASHITA, F. AIDA and Y. FUJIWARA. Temperature dependence of tan δ in polyethylene. In: *Proceedings of the 3rd International Conference on Properties and Applications of Dielectric Materials*. Tokyo: IEEE, 1991, pp. 1068–1071. ISBN 0-87942-568-7. DOI: 10.1109/ICPADM.1991.172259.

[2] UEKI, M. M. and M. ZANIN. Influence of additives on the dielectric strength of high-density polyethylene. *IEEE Transactions on Dielectrics and Electrical Insulation*. 1999, vol. 6, iss. 6, pp. 876–881. ISSN 1070-9878. DOI: 10.1109/94.822030.

[3] KAO, K. C. Electrical conduction and breakdown in insulating polymers. In: *Proceedings of the*

6th InternationalProceedings of the Conference on Properties and Applications of Dielectric Materials. Xi'an: IEEE, 2000, pp. 1–17. ISBN 0-7803-5459-1. DOI: 10.1109/ICPADM.2000.875620.

[4] AIRONG, Z. and C. XIAOLONG. The electron-beam irradiation cross-linking and modifying of high density polyethylene insulation. In: *Proceedings of the 6th International Conference on Properties and Applications of Dielectric Materials.* Xi'an: IEEE, 2000, pp. 213–216. ISBN 0-7803-5459-1. DOI: 10.1109/ICPADM.2000.875668.

[5] XI, B. and G. CHEN. Influence of polymer matrix on the PTC properties of polyethylene/carbon black composites. In: *7th International Conference on Solid Dielectrics.* Eindhoven: IEEE, 2001, pp. 109–112. ISBN 0-7803-5459-1. DOI: 10.1109/ICSD.2001.955527.

[6] OMBELLO, F., G. ATTOLINI, P. CARACINO, M. NASSI, S. SPREAFICO and G. C. MONTANARI. Insulating materials evaluation for Cold Dielectric superconducting cables. *IEEE Transactions on Dielectrics and Electrical Insulation.* 2002, vol. 9, iss. 6, pp. 958–963. ISSN 1070-9878. DOI: 10.1109/TDEI.2002.1115490.

[7] CRUZ, S. A. and M. ZANIN. Evaluation of the incorporation of recycled material in the dielectric properties of high density polyethylene. In: *Proceedings of the 7th International Conference on Properties and Applications of Dielectric Materials.* Nagoya: IEEE, 2003, pp. 503–505. ISBN 0-7803-7725-7. DOI: 10.1109/ICPADM.2003.1218463.

[8] FILIPPINI, J. C., R. TOBAZEON, C. MARTEAU, R. COELHO, J. MATALLANA and H. JANAH. The alternate polarization current method for conduction analysis in polymers. In: *IEEE International Conference on Solid Dielectrics.* Toulouse: IEEE, 2004, pp. 115–118. ISBN 0-7803-8348-6. DOI: 10.1109/ICSD.2004.1350303.

[9] CRUZ, S. A. and M. ZANIN. Electrical properties and morphology of polyethylene produced with a novel catalyst. *IEEE Transactions on Dielectrics and Electrical Insulation.* 2004, vol. 11, iss. 5, pp. 855–860. ISSN 1070-9878. DOI: 10.1109/TDEI.2004.1349791.

[10] YAMAMOTO, Y., M. IKEDA and Y. TANAKA. Assessment of dielectric behavior of recycled/virgin high density polyethylene blends. *IEEE Transactions on Dielectrics and Electrical Insulation.* 2004, vol. 11, iss. 5, pp. 881–890. ISSN 1070-9878. DOI: 10.1109/TDEI.2004.1349794.

[11] VILCKAS, J. H., L. G. ALBIERO, S. A. CRUZ, M. M. UEKI and M. ZANIN. Study of electrical and mechanical properties of recycled polymer blends. In: *International Symposium on Electrical Insulating Materials.* Kitakyushu: IEEE, 2005, pp. 683–686. ISBN 4-88686-063-X. DOI: 10.1109/ISEIM.2005.193462.

[12] GUO, W. M., B. Z. HAN, H. ZHENG and Z. H. LI. The Effect of Basic Resins on Conductivity Properties of the Polyethylene and Carborundum Composite. In: *8th International Conference on Properties & applications of Dielectric Materials.* Bali: IEEE, 2006, pp. 747–750. ISBN 1-4244-0190-9. DOI: 10.1109/ICPADM.2006.284286.

[13] TUNCER, E., I. SAUERS, D. R. JAMES, A. R. ELLIS and M. PACE. Electrical properties of commercial sheet insulation materials for cryogenic applications. In: *Annual Report Conference on Electrical Insulation and Dielectric Phenomena.* Quebec: IEEE, 2008, pp. 301–304. ISBN 978-1-4244-2549-5. DOI: 10.1109/CEIDP.2008.4772931.

[14] GREEN, C. D., A. S. VAUGHAN, G. R. MITCHELL and T. LIU. A Structure property relationships in polyethylene/montmorillonite nanodielectrics. *IEEE Transactions on Dielectrics and Electrical Insulation.* 2008, vol. 15, iss. 1, pp. 134–143. ISSN 1070-9878. DOI: 10.1109/TDEI.2008.4446744.

[15] LYNN, C., A. NEUBER, J. KRILE, J. DICKENS and M. KRISTIANSEN. Electrical conduction in select polymers under shock loading. In: *IEEE Pulsed Power Conference.* Washington: IEEE, 2009, pp. 171–174. ISBN 978-1-4244-4064-1. DOI: 10.1109/PPC.2009.5386199.

[16] SHAH, K. S., R. C. JAIN, V. SHRINET, A. K. SINGH and D. P. BHARAMBE. High Density Polyethylene (HDPE) Clay Nanocomposite for Dielectric Applications. *IEEE Transactions on Dielectrics and Electrical Insulation.* 2009, vol. 16, iss. 3, pp. 853–861. ISSN 1070-9878. DOI: 10.1109/TDEI.2009.5128526.

[17] SAMI, A., E. DAVID and M. FRECHETTE. Dielectric characterization of high density polyethylene/SiO2 nanocomposites. In: *IEEE Conference on Electrical Insulation and Dielectric Phenomena.* Virginia Beach: IEEE, 2009, pp. 689–692. ISBN 978-1-4244-4557-8. DOI: 10.1109/CEIDP.2009.5377742.

[18] BOIS, L., F. CHASSAGNEUX, S. PAROLA, F. BESSUEILLE, Y. BATTIE, N. DESTOUCHES, A. BOUKENTER, N. MONCOFFRE and N. TOULHOAT. Growth of ordered silver nanoparticles in silica film mesostructured

with a triblock copolymer PEO–PPO–PEO. *Journal of Solid State Chemistry*. 2009, vol. 182, iss. 7, pp. 1700–1707. ISSN 0022-4596. DOI: 10.1016/j.jssc.2009.01.044.

[19] DAO, N. L., P. L. LEWIN, I. L. HOSIER and S. G. SWINGLER. A comparison between LDPE and HDPE cable insulation properties following lightning impulse ageing. *IEEE International Conference on Solid Dielectrics*. Potsdam: IEEE, 2010, pp. 1–4. ISBN 978-1-4244-7945-0. DOI: 10.1109/ICSD.2010.5567944.

[20] THABET, A., Y. A. MOBARAK and S. ABOZEID. Exponential Power Law Model for Predicting Dielectric Constant of New Nanocomposite Industrial Materials. In: *International Conference On Materials Imperatives in the New Millennium*. Cairo: MINM, 2010, pp. 1–4. ISBN 978-160876579-9.

[21] THABET, A. Influence of Cost-Less Nanoparticles on Electric and Dielectric Characteristics of Polyethylene Industrial Materials. *International Journal of Electrical Engineering and Technology*. 2013, vol. 4, iss. 1, pp. 58–67. ISSN 0976-6553.

[22] THABET, A. Experimental Investigation on Thermal Electric and Dielectric Characterization for Polypropylene Nanocomposites Using Cost-fewer Nanoparticles. *International Journal of Electrical Engineering and Technology*. 2013, vol. 4, iss. 1, pp. 1–12. ISSN 0976-6545.

[23] GOUDA, O. E.-S., A. THABET, Y. A. MOBARAK and M. SAMIR. Nanotechnology Effects on Space Charge Relaxation Measurements for Polyvinyl Chloride Thin Films. *International Journal on Electrical Engineering and Informatics*. 2014, vol. 6, no. 1, pp. 1–12. ISSN 2085-5830. DOI: 10.15676/ijeei.2014.6.1.1.

[24] THABET, A. Experimental Enhancement for Dielectric Strength of Polyethylene Insulation Materials Using Cost-fewer Nanoparticles. *International Journal of Electrical Power & Energy Systems*. 2015, vol. 64, no. 1, pp. 469–475. ISSN 0142-0615. DOI: 10.1016/j.ijepes.2014.06.075.

[25] THABET, A. and Y. A. MOBARAK. Experimental Dielectric Measurements for Cost-fewer Polyvinyl Chloride Nanocomposites. *International Journal of Electrical and Computer Engineering*. 2015, vol. 5, no. 1, pp. 13–22. ISSN 2088-8708. DOI: 10.11591/IJECE.V5I1.6743.

[26] THABET, A. Experimental study of space charge characteristics in thin films of polyvinyl chloride nanocomposites. *International Journal on Electrical Engineering and Informatics*. 2015, vol. 7, iss. 1, pp. 1–11. ISSN 2085-6830. DOI: 10.15676/ijeei.2015.7.1.1.

[27] THABET, A. Experimental Verification for Improving dielectric strength of polymers by using clay nanoparticles. *Advances in Electrical and Electronic Engineering*. 2015, vol. 13, no. 2, pp. 182–190. ISSN 1336-1376. DOI: 10.15598/aeee.v13i2.1249.

[28] THABET, A. and Y. A. MOBARAK. Predictable Models and Experimental Measurements for Electric Properties of Polypropylene Nanocomposite Films. *International Journal of Electrical and Computer Engineering*. 2016, vol. 6, no. 1, pp. 120–129. ISSN 2088-8708. DOI: 10.11591/ijece.v6i1.9108.

About Authors

Ahmed THABET was born in Aswan, Egypt in 1974. He received the B.Sc. (FEE) Electrical Engineering degree in 1997 and M.Sc. (FEE) Electrical Engineering degree in 2002 both from Faculty of Energy Engineering, Aswan, Egypt. Ph.D. degree had been received in Electrical Engineering in 2006 from El-Minia University, Minia, Egypt. He joined with Electrical Power Engineering Group of Faculty of Energy Engineering in Aswan University as a Demonstrator at July 1999, until; he held Associate Professor Position at October 2011 up to date. His research interests lie in the areas of analysis and developing electrical engineering models and applications, investigating novel nano-technology materials via addition nano-scale particles and additives for usage in industrial branch, electromagnetic materials, electroluminescence and the relationship with electrical and thermal ageing of industrial polymers. On 2009, he had been a Principle Investigator of a funded project from Science and Technology Development Fund "STDF" for developing industrial materials of ac and dc applications by nano-technology techniques. He has been established first Nano-Technology Research Centre in the Upper Egypt. He has many of publications which have been published and under published in national, international journals and conferences and held in Nano-Technology Research Centre website.

Youssef MOBARAK was born in Luxor, Egypt in 1971. He received his B.Sc. and M.Sc. degrees in Electrical Engineering from Faculty of Energy Engineering, Aswan University, Egypt, in 1997 and 2001 respectively and Ph.D. from Faculty of Engineering, Cairo University, Egypt, in 2005. He joined

Electrical Engineering Department, Faculty of Energy Engineering, Aswan University as a Demonstrator, as an Assistant Lecturer, and as an Assistant Professor during the periods of 1998–2001, 2001–2005, and 2005–2009 respectively. He joined Artificial Complex Systems, Hiroshima University, Japan as a Researcher 2007–2008. Also, he joined King Abdulaziz University, Rabigh, Faculty of Engineering 2010 to present. His research interests are power system planning, operation, and optimization techniques applied to power systems. Also, his research interests are Nanotechnology materials via addition nano-scale particles and additives for usage in industrial field.

3

FLICKER CAUSED BY OPERATION OF INDUSTRIAL TECHNOLOGY

Martin KASPIREK[1], *Petr KREJCI*[2], *Pavel SANTARIUS*[3], *Karel PROCHAZKA*[4]

[1]Management of Grid, E.ON, F. A. Gerstnera 2151/6, 37001 Ceske Budejovice, Czech Republic
[2]Department of Electrical Power Engineering, Faculty of Electrical Engineering and Computer Science, VSB–Technical University of Ostrava, 17. listopadu 15/2172, 708 33 Ostrava, Czech Republic
[3]Department of Cybernetics and Biomedical Engineering, Faculty of Electrical Engineering and Computer Science, VSB–Technical University of Ostrava, 17. listopadu 15/2172, 708 33 Ostrava, Czech Republic
[4]EGC EnerGoConsult CB, Cechova 727, 37001 Ceske Budejovice, Czech Republic

martin.kaspirek@eon.cz, petr.krejci@vsb.cz, pavel.santarius@vsb.cz, kprochazka@egc-cb.cz

Abstract. *There are more and more electrical devices operating in industrial plants which impact negatively on the distribution network. The EN 50160 standard defines the voltage characteristics of electricity supplied by the public distribution system and it is a problem for a distribution network operator when the voltage quality in the distribution network does not comply with the requirements of the EN 50160 standard because of complaints about voltage quality. Such a complaint is justified and the distribution network operator has to pay a penalty and has to remedy the situation. This paper describes a problem of flicker in the medium-voltage distribution grid when the flicker is produced by one customer operating a forging press. The remedies from the side of the distribution network operator and the customer with the aim of reducing the flicker level in the grid are described.*

Keywords

EN 50160 standard, flicker, voltage dip, voltage quality.

1. Introduction

The supply territory of the company E.ON Distribution in the Czech Republic (run by the company E.ON Czech Republic) accounts for approximately 1.5 million customers. Most customers are connected to a Low-Voltage (LV) distribution grid and six thousand customers are connected to a Medium-Voltage (MV) distribution grid, with the nominal voltage of the MV grid in the Czech Republic being 22 kV. Every year the Distribution Network Operator (DNO) E.ON monitors Voltage Quality (VQ) at the delivery points of some important customers supplied from the MV distribution grid. A general circuit diagram is shown in Fig. 1. One of the measured customers is a customer called Magna. This customer has its own MV substation (see Fig. 2) with its own 22/0.4 kV transformers T1 and T2. Voltage quality measurement has to be performed at the delivery point of the customer (node U3 in Fig. 1) so that the voltage quality is measured at the output of the instrument transformer 22/0.1 kV. This instrument transformer is a part of the MV switchboard and it is in the possession of the customer.

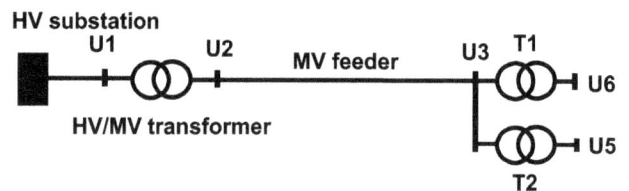

Fig. 1: General circuit diagram of connection of customer to MV grid.

2. Voltage Quality and the EN 50160 Standard

2.1. Supply Voltage Variations

Under normal operating conditions, excluding periods with interruptions, supply voltage variations should not exceed ± 10 % of the declared voltage U_c and the

Fig. 2: MW switchboard of the customer Magna (node U3 according to Fig. 1).

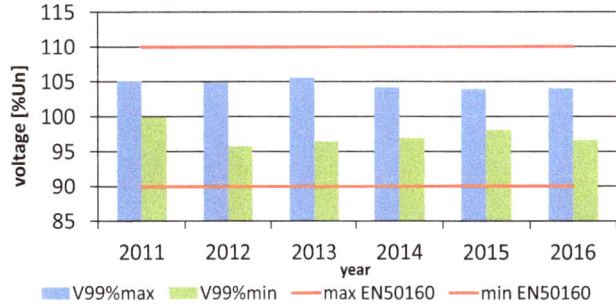

Fig. 3: Evaluation of voltage variations (99 % percentile).

test method according to the standard [1] is the following:

- at least 99 % of the 10-min mean r.m.s. values of the supply voltage must be below the upper limits of +10 %,

- at least 99 % of the 10-min mean r.m.s. values of the supply voltage must be above the lower limits of -10 % given in 5.2.2.1,

- none of the 10-min mean r.m.s. values of the supply voltage must be outside the limits of ± 15 % of U_c.

2.2. Flicker Severity

Under normal operating conditions, during each period of one week the long-term flicker severity P_{lt} caused by voltage fluctuation should be less than or equal to 1 for 95 % of the time [6].

3. Evaluation of VQ Measurements

3.1. Supply Voltage Variations

Figure 3 and Fig. 4 show that voltage variations comply with the EN 50160 standard. Small differences between 100 % and 99 % values were evaluated. The maximal 10-min mean values reach the value of 106.7 % U_n (Fig. 4, year 2013) and the minimal values reach the value of 95.8 % U_n (Fig. 4, year 2012).

3.2. Flicker Severity

Figure 5 shows the flicker increasing from 2012. The limit $P_{lt}=1$ is exceeded and the problem has to be solved.

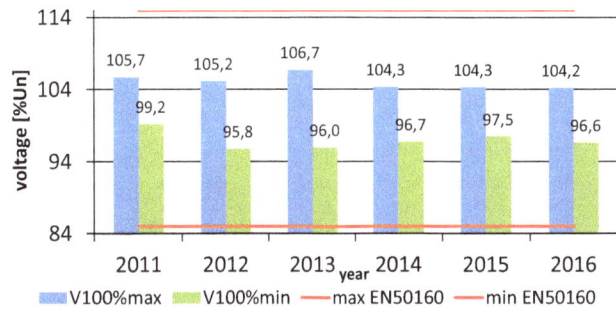

Fig. 4: Evaluation of voltage variations (100 % percentile).

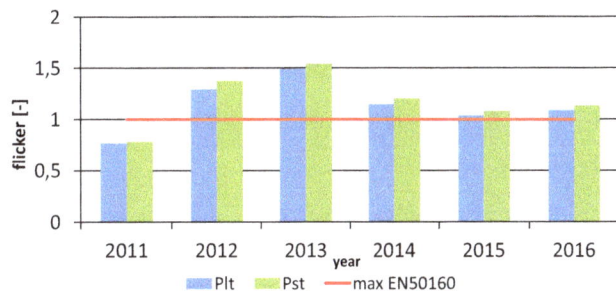

Fig. 5: Evaluation of short-term severity (P_{st}) and long-term severity (P_{lt}).

4. Responsibility for Poor Voltage Quality

4.1. Source of Flicker

Flicker is produced by the operation of new forging presses (see Fig. 7 and Fig. 8) which were put into operation in 2012.

To find the source of flicker it is necessary to measure and evaluate the current and flicker together. When the flicker waveform corresponds to the current waveform (see Fig. 8), the flicker is produced by the customer being measured.

Fig. 6: Forging press (Magna).

Fig. 7: Belt feeder of sheet steel for the forging press.

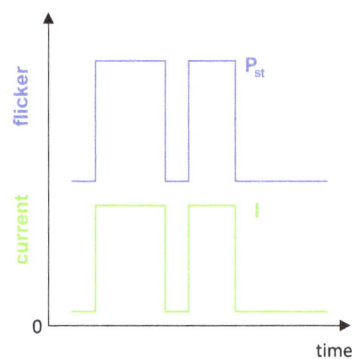

Fig. 8: Dependence of flicker on current.

4.2.   Study if Forging Presses Connectable

A study of the VQ was made before the connecting of the forging presses because it was supposed beforehand that the operation of these forging presses would influence the VQ negatively. The author of this study did not recommend the connection of the forging presses because of flicker when a flicker level above the value 1 was calculated [2].

4.3.   Call from the Side of the DNO

Magna was informed about the problem of flicker because it was proved that flicker was caused by the customer's operation. The flicker intensity depends on the current intensity – see Fig. 8. The customer did not believe the DNO that the flicker was produced by its operations and so it was decided that the university should perform an independent assessment of the cause of the flicker.

4.4.   Assessment of the Cause of the Flicker Performed by the University

Experts from the Technical University of Ostrava made independent VQ measurements and discovered the cause of the flicker. The customer operates two forging presses and the university experts compared the flicker level in two states (see Tab. 1). In the first state both the forging presses were operated and the flicker level reached the value of 1.16 on the MV side. In the second state both the forging presses were out of operation and the flicker level reached the value of 0.63 on the MV side and so the dependence of the flicker on the operation of the forging presses was proved.

Tab. 1: Flicker on the MV level in different states of the customer's operations.

State of the operation	Flicker P_{lt} (-)
Forging presses are in operation	1.16
Forging presses out of operation	0.63

After this independent study the customer accepted that he was responsible for the flicker in the MV grid and he proposed a remedy. This remedy is based on adjusting the forging presses. It is possible to remedy the situation from the side of the DNO too. The DNO suggested building a new HV/MV substation near the customer so that the short circuit power should increase and the flicker level should decrease. This solution would take many years due to the justification of construction of a new high-voltage distribution feeder.

5.   Remedy from the Side of the Customer

The forging presses operate with a definite reserve for performing the pressing. This reserve was 100 % because of defective work. With the reduction of this reserve the current value should decrease. A decrease in the current value should reduce the flicker level. The customer proposed reducing the reserve R from

R=100 % to R=75 % and R=60 %. The defective work has to be evaluated.

Figure 9 shows that the RMS peak of the current is approximately 700 A if R=100 %. Figure 10 shows that the RMS peak of the current is approximately 540 A if R=75 %. Figure 11 shows that the RMS peak of the current is approximately 400 A if R=60 %. A summary evaluation is given in Tab. 2.

Tab. 2: Current peak in dependency of reserve R.

R of the forging press (%)	Current peak (A)
100	700
75	540
60	400

The size of the voltage changes depends on the size of the current peak (see Fig. 12) and the flicker level depends on the size of the voltage changes. Thus the flicker level decreased because of the changes to a lower voltage, but not enough. After the adjustment of both the forging presses to R=60 %, the flicker level reached the value approximately of 1.1 (see Fig. 5, years 2014, 2015 and 2016) and the number of rejects did not increase. The voltage quality in the MV distribution grid does not comply with the requirements of the EN 50160 standard because of flicker but Magna has no other way to solve the problem apart from halting production. Fortunately, no complaints regarding the flicker were received.

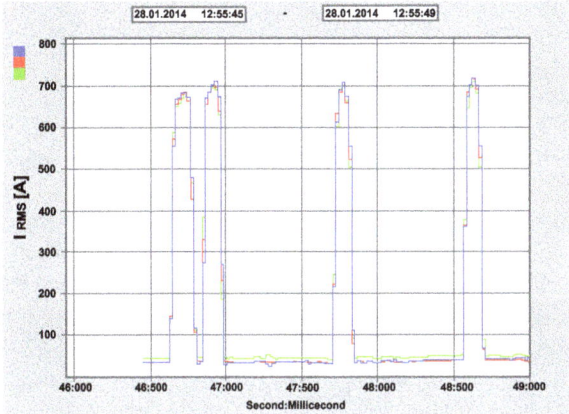

Fig. 9: RMS peak of current, forging press set to R=100 %.

Fig. 10: RMS peak of current, forging press set to R=75 %.

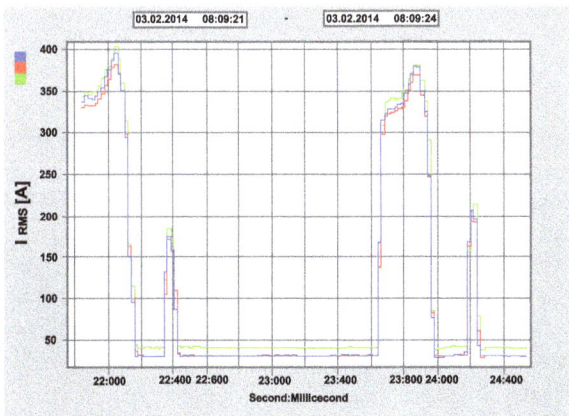

Fig. 11: RMS peak of current, forging press set to R=65 %.

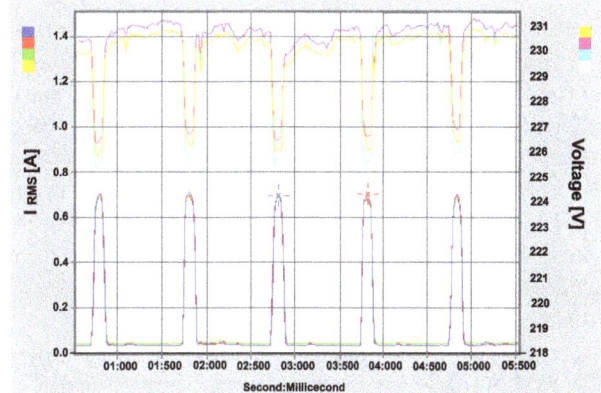

Fig. 12: RMS peak of current (lower curve) and voltage changes (upper curve) on the low-voltage level.

6.   Remedy from the Side of the DNO

It is possible to remedy the situation from the side of the DNO too. The DNO plans to build a new HV/MV substation near the customer so that the short circuit power should grow and the flicker level should drop. The expected value of the short circuit power after the building of the new HV/MV substation is approximately $S_k''=130$ MVA. So the flicker level should drop below the limit according to the EN 50160 standard – see Tab. 3. The investment costs for building the new HV/MV substation, including a supply HV feeder, are approximately one million EUR. This solution will take many years due to the justification of construction of a new high-voltage distribution feeder [5].

Tab. 3: Short circuit power at the delivery point of Magna.

Way of supply	Current MV feeder	New HV/MV substation
S_k'' (MVA)	30	130
Flicker P_{lt} (-)	1.1	0.25

7.   Flicker Propagation

Many other customers are influenced by the operation of the forging presses, or, more precisely, the flicker level $P_{lt}=1$ can be expected to be exceeded at the delivery point of more customers. The highest flicker values should be experienced by customers in the vicinity of Magna. The flicker level should decrease with an increase in the distance from the HV/MV transformer. This theoretical assumption was verified by practical measurement. It was measured at two points of the MV grid. The first measurement was made at the delivery point of Magna (node U4 according to Fig. 13) and the second at the delivery point of another customer (node U4 according to Fig. 13). The distance between the customer C1 and the HV/MV transformer is less than the distance between the customer C2 and the HV/MV transformer. Both the measurements were performed at the same time and the duration of these measurements are one week. The measurements were evaluated according to the EN 50160 standard when the short circuit power was calculated at the delivery points of both the customers.

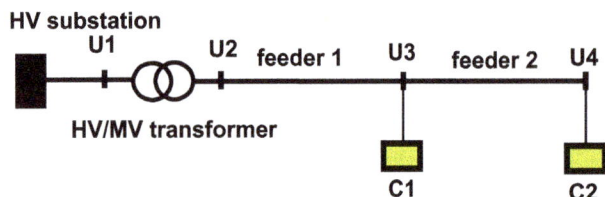

Fig. 13: General circuit diagram of connection of customers to MV grid.

Tab. 4: Short circuit power and measured values of flicker P_{lt} at the delivery point of the customers.

Customer	S_k'' (-)	Flicker P_{lt} (-)
C1	40.4	0.93
C2	29.2	1.32

Table 4 shows the values of satisfactory flicker level at the delivery point of the customer C1 when the flicker value is less than $P_{lt}=1$. If the flicker level at the delivery point of the customer C2 is known, the flicker level at the delivery point of the customer C1 should theoretically reach the value:

$$P_{lt}(\text{C1}) = P_{lt}(\text{C2}) \cdot \frac{S_k''(\text{C2})}{S_k''(\text{C1})} = 1.32 \cdot \frac{29.2}{40.4} = 0.95. \quad (1)$$

The calculated flicker level $P_{lt}=0.95$ at the delivery point of the customer C1 accords with the measured value $P_{lt}=0.93$ and thus the theoretical assumption regarding the flicker propagation was confirmed.

8.   Problems of Voltage Dips

It is interesting that no complaints regarding the flicker were received, although the flicker level exceeds the value $P_{lt}=1$. But complaints about voltage dips were received and one of the complaining customers is Magna. This customer is supplied by a long MV feeder (approximately 30 km) and so many voltage dips occur (see Fig. 14). Voltage dips can cause the same effect as long-term interruption by customers with hard automation. The EN 50160 standard defines no limits for voltage dips but the analysis of historical data can enter the algorithm of the prediction methods used [3]. Voltage dips are typically caused by faults occurring in the public network, so in the case of an outside MV feeder faults and voltage dips mainly occur in summer as a result of the impact of the weather (windstorms). A short voltage interruption on the primary side (on the MV level) results in a short voltage interruption on the secondary side of the MV/LV transformer. Short voltage interruptions of MV feeders with failure are transmitted into other networks as voltage dips [4].

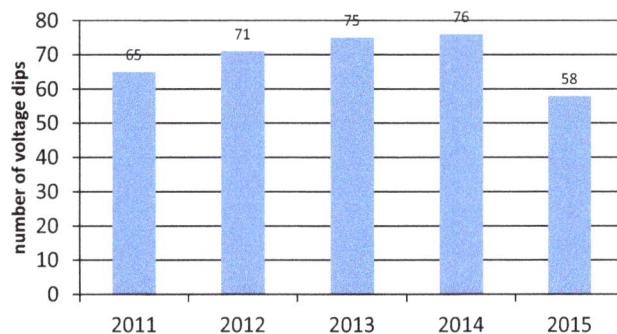

Fig. 14: Number of recorded voltage dips at the delivery point of Magna.

9.   Conclusion

Experts from the Technical University of Ostrava, in cooperation with colleagues from the company E.ON, performed new monitoring measurement in November 2015 (on the MV level at the delivery point of the customer Magna). Figure 5 shows the evaluation according to the EN 50160 standard. The flicker level $P_{lt}=1$ was exceeded but so far no complaints have been received. The next VQ measurement will be performed in March 2017.

It is a problem when the VQ in the MV distribution grid does not comply with the requirements of the EN

50160 standard because it influences many other customers. The question is who is responsible for poor VQ. A simple method based on the correlation of the flicker and current was described in this paper. The remedy does not always have to be expensive from the point of view of the customer but sometimes it might not be sufficiently efficient. In this study the DNO is going to build a new HV/MV substation but it takes about three years. No other affordable solution is possible apart from halting the production but the DNO has limited tools to realise it. Magna has to produce and the flicker causes the firm no problems but Magna is sensitive to voltage dips. DNO has no positive experience of the operation of dynamic compensation because the compensation is too slow when a response time of no more than 5 ms is needed and the use of compensation balance is too expensive.

It can therefore be stated that in the case of customers who consume high power with a considerable dynamic course of loading that causes flicker outside the permitted tolerance (and where production optimisation has already taken place, for instance, by the distribution of manufacturing over time in order to prevent the simultaneous running of those appliances that contribute to the flicker the most), the most suitable way to reduce the flicker is by increasing the short circuit power, for instance, by building a new transformer station, etc. For this type of power take-off, the authors recommend performing a connectivity study before connecting the industrial plant to the network, in which the elaborator should take into account a future increase in the reserved power input over the period of approximately the following 10 years.

Acknowledgment

This research was partially supported by the SGS grant from VSB–Technical University of Ostrava (No. SP2016/95).

References

[1] EN 50160 Ed. 3 Voltage characteristics of electricity supplied by public distribution systems. *European Committee for Electrotechnical Standardization.* 2010.

[2] KASPIREK, M., P. SANTARIUS, P. KREJCI and K. PROCHAZKA. Flicker vlivem prumyslovych technologii. In: *11th ERU conference.* Brno: ERU, 2014, pp. 1–44. ISBN 978-80-905933-0-5.

[3] TESAROVA, M. Methods for prediction of short voltage dips and interruptions and the possibility of their practical use. In: *12th International Scientific Conference on Electric Power Engineering.* Ostrava: VSB–Technical University of Ostrava, 2011, pp. 1–4. ISBN 978-80-248-2393-5.

[4] OTCENASOVA, A., P. BRANICIK and M. DUBOVSKY. Voltage dips in the distribution system and their spread. In: *7th International Scientific Symposium on Electrical Power Engineering.* Stara Lesna: Technical University of Kosice, 2013, pp. 280–283. ISBN 978-80-553-1441-9.

[5] KASPIREK, M. and D. MEZERA. Voltage quality parameters in LV distribution grids in dependence on short circuit impedance. In: *22nd International Conference and Exhibition on Electricity Distribution.* Stockholm: IET, 2013, pp. 1–3. ISBN 978-1-84919-732-8. DOI: 10.1049/cp.2013.0580.

[6] SANTARIUS, P., P. KREJCI, Z. CHMELIKOVA and J. CIGANEK. Long-term Monitoring of Power Quality Parameters in Regional Distribution Networks in Czech Republic. In: *13th international conference on Harmonics and Quality of Power.* Wollongong: IEEE, 2008, pp. 78–82. ISBN 978-1-4244-1770-4. DOI: 10.1109/ICHQP.2008.4668753.

About Authors

Martin KASPIREK was born in Pisek, Czech Republic. He received his M.Sc. in electrical power engineering in 2004. He is interested in problems of voltage quality and the operation of renewable energy sources.

Petr KREJCI was born in Karvina, Czech Republic. He received his M.Sc. in electrical power engineering in 1998. His research interests include Electrical Power Engineering of Industrial and Distribution Networks and Power Quality.

Pavel SANTARIUS was born in Bludovice, Czech Republic. He received his M.Sc. in the production and distribution of electrical energy in 1963. His research interests include EMC from the perspective of harmonics and the frequency characteristics of electrical networks.

Karel PROCHAZKA was born in Choustnik, Czech Republic. He received his M.Sc. in production and distribution of electrical energy in 1961. His research interests include the earthing of neutral points, the automation and protection of MV distribution networks, power quality and DER.

4

Sine Inverter Controller with 8 Bit Microcontroller

Radim KUNCICKY, Lacezar LICEV, Michal KRUMNIKL,
Karolina FEBEROVA, Jakub HENDRYCH

Department of Computer Science, Faculty of Electrical Engineering and Computer Science,
VSB–Technical University of Ostrava, 17. listopadu 15, 708 33 Ostrava, Czech Republic

radim.kuncicky@vsb.cz, lacezar.licev@vsb.cz, michal.krumnikl@vsb.cz,
karolina.feberova@vsb.cz, jakub.hendrych@vsb.cz

Abstract. *This article describes the design of a sine wave inverter control unit. We will define the basic principles and requirements to control and maintain a good quality of inverter's output. The solution that can achieve full 8 bit output resolution with 10 bit measurement and regulation time shorter than mains period is proposed. Furthermore, the possibilities and ways of implementing the control unit using the 8 bit microcontrollers are described, so it is possible to gain a complete understanding of this issue.*

Keywords

Inverter, microcontroller, PID, pulse width modulation, regulator, sine wave inverter.

1. Introduction

In the recent years, there has been a large expansion of the solar plants, mainly thanks to the huge subsidy provided by the Czech Republic government. It was followed by a rapid price fall of the solar panels, which started to appear on the roofs all over the country. Solar panels can serve as an independent autonomous power system. The power from the cells is stored in batteries and can be used later on, when needed. However, the batteries cannot be connected directly to the appliances, since they do not provide correct voltage. Therefore, it must be transformed via an inverter providing an output voltage of 230 V and frequency of 50 Hz.

Inverters are very well known since the time of first switching semiconductors. Due to the versatility, efficiency, size and costs, they became one of the best

choices in situations when there was a need to change the voltage, waveform or current.

The simplest constructions use a low frequency Pulse Width Modulation (PWM) to approximate a line sinusoidal waveform with a rectangular waveform. This type of converter is not very suitable to power many commonly used appliances and can also be dangerous. For example, let's mention devices based on a single phase asynchronous motor. The rectangular waveform is not able to create rotating magnetic field and the motor will not work correctly. These motors can be found in building's heating systems as a circulation pump and their failure may result in considerable damages. Another example is energy-saving lamps, which can be even destroyed by the sharp rising edge. Therefore, more advanced inverters must be used.

A review of common multilevel inverter topologies and control schemes has been covered in [1]. Multilevel inverters are typically used in medium and high power applications due to their lower dissipation, harmonic distortions and electromagnetic interference. However, the topic of this article is focused on smaller inverters, typical for small solar power plants. These inverters use Synchronous Pulse Width Modulation (SPWM) or simple PWM for driving output circuits. A comparative study of multiple inverters approaches, including SPWM and pure sine wave techniques, providing their main characteristics and advantages was publisher in [5]. The performance analysis of square wave and modified square wave inverters [6] showed significant performance drawbacks and higher harmonics. We can say that the only disadvantage of the sine wave inverters is the higher complexity compared to the simple PWM units. As the sine inverters need more complex drivers, several options were proposed ranging from general purpose microcontrollers, microcontrollers with DSP units, FPGA and single purpose (dedicated) integrated circuits. The inverter can be based even on multivibrator IC, but this is rather an awkward construction [10].

General purpose microcontrollers and FPGAs are popular choices for driving sine wave inverters. A single phase full bridge inverter using Field Programmable Logic Array (FPGA) was proposed in [2] and [4]. The designs based on microcontrollers differ mainly in the control software and microcontroller architecture, the hardware design remains almost the same, e.g. [7], [8] and [9]. Probably the most similar approach to ours is the work presented in [3]. However, proposed sine inverter based on 8 bit PIC microcontroller 16F84 provides only basic functions without voltage regulation.

The sine inverter we describe approximates the AC waveform using a high frequency PWM. Such inverter has the output waveform (after the filtering) much more similar to the true sine wave. Moreover, we have added additional features, such as frequency shifting, battery monitoring and computer control interface.

The operation of the proposed inverter is controlled by the control unit. The control unit is responsible for the correct pulse timing, pulse duration, measurement and maintaining the operating parameters within the required limits.

The remaining part of the article discusses the implementation of the converter control unit based on a modern 8 bit microcontroller. Special attention is given to the computational performance and optimization techniques.

2. Pulse Inverters

The selection of the suitable power source type is heavily affected by the requirements of the application. For the linear power sources speaks the simple design, silent operation and relatively low price. As the biggest disadvantage, we can consider low efficiency that ranges between $30 - 60$ %. Another problem is required power design factor, where every component needs to be designed according to the limits of possible operating load. These edge conditions are rarely met in everyday operation. It is important to mention a significant constraint of the linear power supplies - they are not able to operate in a step-up mode. The pulse inverters provide much higher efficiency, up to $68 - 90$ %, but they require more complex structural design. A substantial part of the structural complexity lies in the necessary control electronics. The following Section describes the basic principles of an increasing DC to AC inverter.

Understanding the basic principle is not hard. A major difference from the linear regulator is that the linear sources continuously regulate the amount of current to reach the constant output voltage. The switching power supplies regulate the flow of the current by chopping the same input voltage and maintaining the mean value of the output voltage by changing the duty cycle. If it is necessary to supply greater amount of the current to the load, the width of pulse is increased to compensate the change.

The DC input voltage can be converted to the AC by the semiconductor circuit operating in the switching mode. The AC current is transformed and the output voltage is, therefore, dependent on the conversion ratio. The operating frequency of such transducers is in the range above the audible range, at frequencies higher than 20 kHz. The transformer does not require a heavy iron core, but due to the higher operating frequency, it can use a lightweight ferrite core.

The transistors Q_A and Q_B in Fig. 1. are driven by the control unit. They supply the current to the different parts of the primary wiring. On the other hand, the current is rectified and smoothed by the output filter. The output voltage and current are sensed by the probes of the control unit. The control unit evaluates the deviation between the output voltage and the reference and uses the controlling algorithm to determine the width of the pulse width modulation. The unit calculates the duration of switching Q_A and Q_B transistors. The method is typical for the low-frequency modulation. For the high frequency PWM, the switching time of the transistor is based on the regulation algorithm and the current instantaneous voltage value.

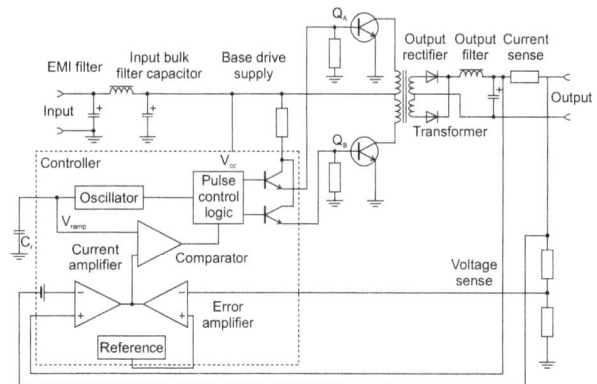

Fig. 1: A walk through a representative switching regulator circuit [12].

The control unit must provide the following functions:

- voltage measurement,

- current measurement,

- calculation of the regulation deviation,

- evaluation of the regulation algorithm,

- generation of the PWM carrier frequency,

- PWM coding.

The control unit may also include other parts like circuits for synchronization, soft start or remote control and monitoring.

For the testing purposes, we have created a simple push pull construction made of discrete components. The power of the prototype was limited to 100 W as it was not designed for deployment in the real-world conditions.

3.   Control Complexity and Microcontrollers Performance

In this Section will be discussed the performance requirements that are necessary to be considered for a suitable microcontroller core selection.

We have chosen Atmel RISC architecture (called AVR), with performance up to 20 MIPS at 20 MHz clock timing. The core is capable of processing single instruction per clock cycle and provides two cycles 8 bit hardware multiplier. An important advantage is that it is supported by the highly optimized C compiler. The developed control unit was based on the ATMega 644PA model.

The performance requirements for the inverter controller can be easily demonstrated as follows. In the previous Section, we have discussed that the switching frequency should be higher than 20 kHz. When using 16 MHz clock frequency and 8 bit counter in the Fast PWM mode, the switching frequency is given by the Eq. (1) [11]:

$$f_{\text{PWM}} = \frac{f_{osc}}{\text{prescaler} \cdot \text{counter}_{\text{max}}} = \\ = \frac{16}{1 \cdot 256} = 62.5 \text{ kHz}. \tag{1}$$

When using the phase-correct mode, the modulation frequency will be 31.3 kHz. From Eq. (1), it is obvious that the core of the microcontroller will be forced to handle the interrupt from counter every 255 machine cycles.

Handling the AVR architecture interrupt and the return jump require about 10 cycles on average. Therefore, only 245 cycles remain (until the arrival of the next interrupt) for calculating the regulation algorithm and processing any additional features.

Mathematical operations and especially multiplication can quickly become a bottleneck of the software

performance. However, they are necessary for computing the pulse width or measurement of the operating parameters. The Tab. 1 gives an idea of an AVR core performance, such as the machine cycles counts for different kinds of software multiplications with the optimized versions of the algorithm.

Tab. 1: AVR core performance [13].

	Machine cycles	Code size
8*8=16 bit unsigned	34	34
16*16=32 bit unsigned	105	105
8/8=8+8 bit unsigned	66	58
16/16=16+16 bit unsigned	196	173

In the worst case, the multiplication of two double data precision float types can take almost up to 1800 cycles. This is totally unacceptable for this application.

The preceding paragraphs assume that calculations are made by the software and processed by the binary adder. Most of the modern MCUs include a hardware multiplication unit. We provide the Tab. 2 for a quick comparison. The table shows performance for several families of microcontrollers. We have included different families of 8 to 32 bit microcontrollers. We have highlighted the number of cycles needed to execute unsigned 16×16 multiplications. It should be noted that not all architectures necessarily complete one instruction per one machine cycle. The instruction completion may consume more cycles.

Tab. 2: Machine cycles count to perform multiplication on different MCU architectures.

PIC 16F	PIC 18F	AVR 8	dsPIC 33F	Atmel xMega	Cortex M0
140	28	19	1	1	1

As you can see, the 32 bit MCUs are equipped with hardware multiplier of sufficient width. On the other hand, the 8 bit MCU multipliers are limited with the bus width, which is usually only 8 bit.

It was necessary to carefully consider the desired precision of each variable. For the development purposes, it was decided to keep the accuracy of calculations between 8 and 16 bits.

One of the basic techniques for limiting calculations is the usage of a look-up table. This technique can be demonstrated on the calculation of the sine function. If we know that the input data will always be in the range from 0 to 255 - i.e. with an accuracy of 8 bits and the desired result will be also in the same range, the simplest solution is to allocate an array of 256 elements to which the individual results will be pre-calculated as shown in following code example. The calculation functions can be then limited to a simple memory access, which is much faster.

```
void initSinusAprox() {
  for( uint8_t i=0;i<255;i++) {
    SinusAprox[i]
    = 255 * sin( i *(M_PI / 256));
  };
};
```

We can consider some loss of the legibility of the code as the biggest disadvantage. When we call the function sin(value), it is immediately apparent that the unit of value is radian. When accessing the SinusAprox[value] array the parameter is losing its dimension and obviously its accuracy as well.

If the situation requires the use of more tables for the calculation, it could be necessary to convert the dimensionless variable to some unit. This is the case of the creation of, so called, magic numbers. Their origin or purpose is not obvious at first glance. The limiting factor for this approach is the amount of available memory.

Another significant limitation factor is the very simple interruption unit, which does not allow the advanced techniques, such as the masking. Interrupts are processed according to the priorities that are set by the address of the interrupt - the lower address, the higher priority. It seems logical to divide software into individual layers according to the priority given to their functionality. The masking is a necessary precondition for this approach. The PWM signal generation is the most crucial function while for the communication, we can assign lower priority. Simple interrupt unit is a hardware limitation that is not possible to overcome by the software. The only option is to use the nested interrupts (but with the risk of stack overflow).

4. Generating PWM

The pulse width modulation is discrete in its value and the modulation variable is voltage or current. The value is transmitted and encoded by the impulse width during the constant period. Width can take a range of $0 - 100$ %. We need to know the range of modulation voltage in advance, because when the maximum value is exceeded, the information is lost.

The PWM has to fulfill the condition of the Shannon Nyquist theorem. The carrier frequency must be twice greater than the maximum frequency of the transferred signal.

The carrier and the modulation signal are compared in the comparator. When the value of the carrier signal is smaller than the modulation, the output of the comparator is set to logical one.

The microcontrollers offer the ability to generate the PWM with the usage of the integrated counters.

The counter operates with the constant frequency and the maximum is usually set to the maximum value of the counter. Usually, two interrupts are needed. When reaching the maximum value, the output is set to log 0. The current value of the modulation signal is stored in the comparator register. If the counter register value is equal to the comparator register, the output is flipped to log 1. Generating is implemented in the hardware and it does not require any additional cooperation with the core. The principle of the operation and usage of the registers is shown in Fig. 2.

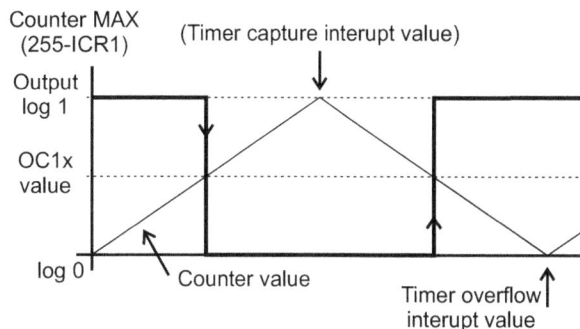

Fig. 2: Registers usage for PWM generating.

5. Measurement and Regulation

For the regulation purposes, it is necessary to know the regulation deviation, which is derived from the desired value and the real value of measured output. For the measurement we used an integrated 8-channel approximation converter with the 10 bit resolution.

The timing of the conversion was derived from the master clock of the microcontroller via the frequency divider. The suitable timing was determined to be in the range from 50 to 200 kHz. The conversion is finished in 13 cycles. Firmware was designed to measure three variables. From the above-mentioned parameters, we can derive the maximum sampling rate as in Eq. (2):

$$f_{\text{SAMPLE}} =$$
$$= \frac{f_{CPU}}{\text{Prescaler}_{\text{ADC}} \cdot \text{ConversionCycles} \cdot \text{Channels}} \quad (2)$$
$$= \frac{16}{128 \cdot 13 \cdot 3} = 3.2 \text{ kHz}.$$

The battery voltage, which is not galvanically separated from the control unit, is fed to the pins of the microcontroller via the voltage divider. The protection against the battery undervoltage has been implemented in the firmware to avoid the battery damage. The measuring range is $0 - 15$ V with an accuracy of 8 bits.

In the prototype, it was necessary to isolate the voltage and the current of the output. The output current is measured on the resistor using the SHF-615 optocoupler. When using the optocoupler, it is necessary to compensate its non-linear characteristics. So the conversion characteristics of the current and the ADC value had to be determined. Subsequently, the linearization using three lines with the least squares method was performed. The transfer function is stored in a look-up table. The measured current values are used in the overload protection.

The output voltage is measured with the voltage measurement transformer that reduces the amplitude for the ADC converter. The input has a floating center, which is equivalent to 511 after the conversion. With the 10 bit, it is possible to achieve a range of ±350 V with a resolution of ±512. That means that the accuracy of measurement is approximately 1 V.

The evaluation of the true Root Mean Square (RMS) of output voltage is performed according to the measured values. Each calculation is made from 256 samples according to the Eq. (3):

$$U_{\mathrm{RMS}} = \sqrt{\frac{1}{n} \sum x_n^2}. \tag{3}$$

The calculation of the true RMS is the most difficult operation, which is implemented in the control unit.

One of the considered options was to use the IC with the function of the true RMS to the DC conversion. The generally available ICs for this purpose are, e.g., AD 7360 or LTC 1966. However, the full integration was prioritized.

The converter, the output sensing and the control unit together constitute the control loop. The evaluation and the maintaining of a constant output are the main tasks of the controller.

The PWM width is changed based on sensing the output. The goal is to achieve a state where the actual value corresponds to the desired output value. The stability of the output can be disrupted by the interference sources, such as changing the load.

The control unit determines the control difference according to the Eq. (4):

$$e = w - x, \tag{4}$$

where e is regulation error, w is desired value, and x is measured value.

The deviation value itself is not enough to run the regulation. It is necessary to determine the y set value of the output. The procedure used for the determination of the value is called the regulation algorithm. In general, we can use two types of controllers. The adjustment algorithm calculates the adjustment function and the velocity algorithm calculates the additions to the y value. In our case we use the discrete (digital) PI velocity algorithm.

5.1.   PID Controller

The setting signal consists of a weighted sum of the three components. P part, the integral of P component (I part) and the derivation of P component (D part). The part P is proportional to the control difference. The P controller creates an immediate permanent lasting change of the adjustment signal, when the first control difference occurs. I part creates an active part of the adjustment signal even at small regulatory differences, which can remove the lasting P difference at the achievable time. D member is applied while changing the regulatory differences. It would predictable overshoot the adjustment of the signal and speed up the compensation of the difference.

The PID is difficult to setup. For this reason, P or PI controllers are used. The suitable controller is selected according to the regulatory dependences and the behavior of the object.

The Discrete adjustment algorithm is given by the following Eq. (5):

$$\begin{aligned}
\text{P part: } & y_{Pn} = K_p \cdot e_n, \\
\text{I part: } & y_{In} = K_p \cdot \frac{T_A}{T_I} \sum_{i=0}^{n} e_i, \\
\text{D part: } & y_{Dn} = K_p \cdot \frac{T_V}{T_A} \cdot (e_n - e_{n-1}),
\end{aligned} \tag{5}$$

where K_p is proportionality constant, T_I integral time constant, T_V is derivative time constant and T_A is sampling period time.

By using the modified Eq. (5), the Eq. (6) for the speed algorithm can be obtained. From this Eq. (6) we can obtain the equation of PI controller by omitting the derivative part.

$$\begin{aligned}
\Delta y_n = K_P \Big[e_n - e_{n-1} + \frac{T_A}{T_I} \cdot e_n + \dots \\
\dots + \frac{T_V}{T_A}(e_n - 2e_{n-1} + e_{n-2}) \Big].
\end{aligned} \tag{6}$$

It is obvious that the calculation is performed from control deviations and few constants that are the object of the optimization.

The controller is carefully set, so it quickly reaches the desired value without the unwanted overshooting. The value of regulatory response t_0 describes the time of reaching the set point.

The simplest method for achieving this is the Ziegler-Nichols method [14] of the critical gain. During the setting up, at first the integral and then the derivative part are eliminated. The constant of proportionality is incremented from zero until reaching KU, where the regulation loop starts to undamped oscillation with the constant amplitude and period t_p. Then the parameters of proportional and derivative component can be calculated according to the Tab. 3.

Tab. 3: Setting of controllers constants by Ziegler-Nichols method.

Type	K_P	K_I	K_D
P	$0.5K_P$	-	-
PI	$0.45K_U$	$\dfrac{1.2K_P}{T_P}$	-
PD	$0.8K_U$	-	$\dfrac{K_P T_P}{8}$
PID	$0.6K_U$	$\dfrac{2K_P}{T_P}$	$\dfrac{K_P T_P}{8}$

This optimization implementation is suitable for development and educational purposes, some advanced tuning methods based on fuzzy logic, artificial neural networks or genetically inspired algorithms achieve much more precise results. It is also possible to implement several advanced and modified PID algorithm versions for example predictive PID variant as supposed in [15].

6.   Communication

One of the design requirements was the integration of the communication and remote control of the unit.

The RS 232 interface has been selected due to its simplicity and widespread in personal computers, as well as in industrial applications.

Atmel AVR microcontrollers contain integrated, full-duplex, hardware-operated universal synchronous/asynchronous serial ports.

The clock generator is derived from the MCU clock. The port provides high precision and allows a large spectrum of modulation speeds. There are three interrupt vectors available for the communication control, receiving a byte, empty the output queue and complete a transmission. For the controlling purpose, a simple text-based protocol was implemented in the firmware. Characters are encoded in ASCII. The protocol operates in the request/response mode. The unit acts as the slave controlled by the master.

The beginning of the message is indicated with the Start of Text character (0×02). STX is followed by one character called the function code, which can be

followed by a numeric parameter. The message ends with the End of Transmission (0×04). In the firmware, we have implemented messages related to the operational status, monitoring of the input voltage, the output voltage and the current. It is possible to setup the control constants and the reference values as well. For example, the message for setting the reference output value in volts should look like this: STX, E230, EOT. If the completion of setup message is successful, the answer of the firmware is OK.

The only byte transfers from USART registers are operated within the interrupts. The actual processing is carried out in the main loop. The time used in the interrupt is minimized and the idle time in the main loop is used as much as possible.

7.   Functionality Testing

The development of both hardware and software parts coincided. The various implementation variants were considered in order to find the optimal solution.

In the first part, the basic principles of generating the sine wave using the PWM were tested. The output of this stage is shown in Fig. 3, showing the output sine wave filtered by the RC elements.

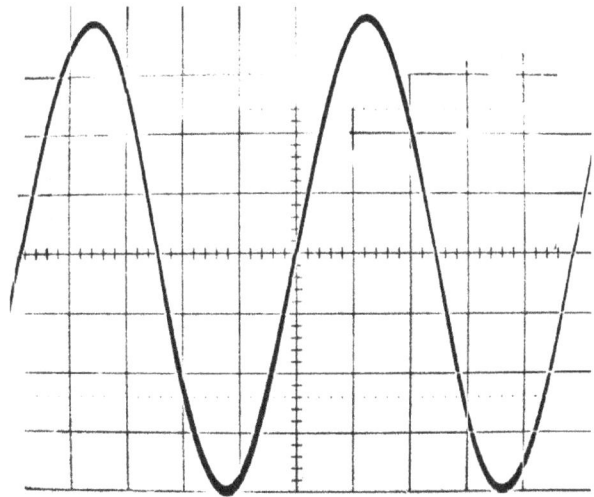

Fig. 3: Sine wave generated by MCU (H: 4 ms per div, V: 80 V per div).

The testing and the verification of the measurements followed. If the firmware is able to effectively measure the output values, the regulatory deviation can be calculated and the control algorithm checked. The P algorithm itself was not sufficient, because of its permanent deviation. The PI algorithm was sufficient enough. The output regulation worked relatively slowly due to the low computational power of the microcontroller. The convergence time t_0 was about 1.5 seconds. The integration constant was adjusted high enough to compensate this phenomenon. However, as the result, this causes a small overshoot of the desired value.

8.   Conclusion

The article describes the implementation of the controlling algorithm of the sine wave inverter for the appliances sensitive to the correct supply voltage waveform. The design used an 8 bit microcontroller. We have demonstrated the ability of the simple monolithic microcontrollers to handle this critical application. The price of needed optimization is a large number of man hours needed for coding.

The simple MCU did not offer priority interrupts; thereby it was not possible to divide the firmware into appropriate priority areas and implement all functionalities.

Today, the price of the ARM family is falling very fast and the developers do not have to be limited to 8 bit cores. More suitable architecture would be Atmel's XMEGA or Microchip's dsPIC [16].

This work provided a comprehensive understanding of the possibilities of the inverter control unit firmware development.

Acknowledgment

This work was partially supported by Grant of SGS No. SP2015/142, VSB–Technical University of Ostrava.

References

[1] COLAK, I., E. KABALCI and R. BAYINDIR. Review of multilevel voltage source inverter topologies and control schemes. *Energy Conversion and Management*. 2011, vol. 52, iss. 2, pp. 1114–1128. ISSN 0196-8904. DOI: 10.1016/j.enconman.2010.09.006.

[2] AFARULRAZI, A. B., M. ZARAFI, W. M. UTOMO and A. ZAR. FPGA implementation of Unipolar SPWM for single phase inverter. In: *International Conference on Computer Applications and Industrial Electronics*. Kuala Lumpur: IEEE, 2010, pp. 671–676. ISBN 978-1-4244-9054-7. DOI: 10.1109/ICCAIE.2010.5735019.

[3] MAMUM, A. A., M. F. ELAHI, M. QUAMRUZZAMAN and M. U. ZOMAL. Design and Implementation of Single Phase Inverter. *International Journal of Science and Research (IJSR)*. 2013, vol. 2, iss. 2, pp. 163–167. ISSN 2319-7064.

[4] NGALAMOU, L. and L. MYERS. Digital SPWM synthesis for the design of single phase inverters. *International Journal of Electronics*. 2008, vol. 95, iss. 5, pp. 489–503. ISSN 0020-7217. DOI: 10.1080/00207210801976420.

[5] CHEEMA, M. B., S. A. HASMAIN, M. M. AHSAN, M. UMER and G. AHMAD. Comparative analysis of SPWM and square wave output filtration based pure sine wave inverters. In: *15th IEEE International Conference on Environment and Electrical Engineering (EEEIC)*. Rome: IEEE, 2015, pp. 38–42. ISBN 978-1-4799-7992-9. DOI: 10.1109/EEEIC.2015.7165289.

[6] KUMAR, N., D. JOSHI and S. SINGHAL. Design, implementation and performance analysis of a single phase PWM Inverter. In: *6th IEEE India International Conference on Power Electronics (IICPE)*. Kurukshetra: IEEE, 2014, pp. 1–6. ISBN 978-1-4799-6045-3. DOI: 10.1109/IICPE.2014.7115742.

[7] QAZALBASH, A. A., A. AMIN, A. MANAN and M. KHALID. Design and implementation of microcontroller based PWM technique for sine wave inverter. In: *International Conference on Power Engineering, Energy and Electrical Drives*. Lisbon: IEEE, 2009, pp. 163–167. ISBN 978-1-4244-4611-7. DOI: 10.1109/POWERENG.2009.4915171.

[8] HAIDER, R., R. ALAM, N. B. YOUSUF and K. M. SALIM. Design and construction of single phase pure sine wave inverter for photovoltaic application. In: *International Conference on Informatics, Electronics & Vision (ICIEV)*. Dhaka: IEEE, 2012, pp. 190–194. ISBN 978-1-4673-1153-3. DOI: 10.1109/ICIEV.2012.6317332.

[9] CHOWDHURY, A. S. K., M. S. SHEHAB, M. A. AWAL and M. A. RAZZAK. Design and implementation of a highly efficient pure sine-wave inverter for photovoltaic applications. In: *International Conference on Informatics, Electronics & Vision (ICIEV)*. Dhaka: IEEE, 2013, pp. 1–6. ISBN 978-1-4799-0397-9. DOI: 10.1109/ICIEV.2013.6572634.

[10] HASAN, M., M. Q. BAIG, J. MAQSOOD, S. M. A. S. BUKHARI and S. AHMED. Design & Implementation of Single Phase Pure Sine Wave Inverter Using Multivibrator IC. In: *17th UKSIM-AMSS International Conference on Modelling and Simulation*. Cambridge: IEEE Computer Society, 2015, pp. 451–455. ISBN 978-1-4799-8713-9.

[11] AVR131: Using the AVR's High-speed PWM: AVR 8-bit Microcontrollers. In: *Atmel* [online]. 2016. Available at: `http://www.atmel.com/Images/Atmel-2542-Using-the-AVR-High-speed-PWM_ApplicationNote_AVR131.pdf`.

[12] BROWN, M. *Practical switching power supply design*. San Diego: Academic Press, 1990. ISBN 0121370305.

[13] AVR200: Multiply and Divide Routines: 8-bit Microcontrollers. In: *Atmel* [online]. 2009. Available at: `http://www.atmel.com/Images/doc0936.pdf`.

[14] ZIEGLER, J. G. and N. B. NICHOLS. Optimum settings for automatic controllers. *Journal of dynamic systems measurement and control*. 1993, vol. 115, iss. 2B, pp. 220–222. ISSN 0022-0434. DOI: 10.1115/1.2899060.

[15] AYLOR, J. H., R. L. RAMEY and G. COOK. Design and Application of a Microprocessor PID Predictor Controller. *IEEE Transactions on Industrial Electronics and Control Instrumentation*. 1980, vol. 27, no. 3, pp. 133–137. ISSN 0018-9421. DOI: 10.1109/TIECI.1980.351665.

[16] Microchip's Digital Pure Sine Wave Uninterruptible Power Supply (UPS) Reference Design. In: *Microchip Technology Inc.* [online]. 2016. Available at: `http://www.microchip.com/DevelopmentTools/ProductDetails.aspx?PartNO=Digital-Pure-Sine-Wave-UPS`.

About Authors

Radim KUNCICKY was born in Karvina, Czech Republic. He received his master's degree in informatics from VSB–Technical University of Ostrava, in 2014. He is Ph.D. student in informatics. His research interests include embedded systems development and the usage of unconventional computational methods.

Lacezar LICEV was born in Bulgaria in 1953. He is associate professor at VSB–Technical University of Ostrava. His research interests include computer architectures and digital image processing in biomedical applications.

Michal KRUMNIKL received the master's degree and the Ph.D. degree in computer science from VSB–Technical University of Ostrava in 2006 and 2014, respectively. From 2007 he works as an assistant professor at the Department of Computer Science and from 2014 as a junior researcher at IT4Innovations. His research interests include embedded systems and digital image processing - especially the 3D scene reconstruction.

Karolina FEBEROVA was born in Cesky Tesin, Czech Republic. In 2014 she received her Master's degree in biomedical engineering at VSB–Technical University of Ostrava. She is Ph.D. student in informatics. Her research is focused on analysis and processing of biomedical images.

Jakub HENDRYCH was born in Cesky Tesin, Czech Republic. He received his Master's degree in informatics from VSB–Technical University of Ostrava in 2014. Now he is a Ph.D. student in department of computer science. His research interests include image processing, embedded system development and the implementation of unconventional methods.

5

Experimental Enhancement for Electric Properties of Polyethylene Nanocomposites under Thermal Conditions

Ahmed THABET[1], Youssef MOBARAK[1, 2]

[1]Nanotechnology Research Center, Faculty of Energy Engineering, Aswan University, 81528 Aswan, Egypt
[2]Department of Electrical Engineering, Faculty of Engineering Rabigh, King Abdulaziz University, 21589 Jeddah, Kingdom of Saudi Arabia

athm@aswu.edu.eg, ysoliman@kau.edu.sa

Abstract. *Polymer properties can be experimentally tailored by adding small amounts of different nanoparticles for enhancing their mechanical, thermal and electrical properties. The work in this paper investigates enhancing the electric and dielectric properties of Low Density Polyethylene (LDPE), and High Density Polyethylene (HDPE) polymer materials with cheap nanoparticles. Certain percentages of clay and fumed silica nanoparticles are used to enhance electric and dielectric properties of polyethylene nanocomposites films. By using the Dielectric Spectroscopy; the electric and dielectric properties of each polyethylene nanocomposites have been measured with and without nanoparticles at various frequencies up to 1 kHz under different thermal conditions (20 °C and 60 °C). And so, we were successful in specifying the optimal nanoparticles types and their concentrations for the control of electric and dielectric characterization.*

Keywords

Dielectric properties, electric properties, nanocomposite, nanoparticles, polyethylene, polymers.

1. Introduction

Nanocomposites represent a very attractive route to upgrade and diversify properties of the polymers. Nano-filler-filled polymers might be differentiated from micro-filler-filled polymers and so the characteristics are reflected in their material properties [1]. In general, fillers are added to polymeric materials in order to enhance thermal and mechanical properties. Over the past few years there have been many researches on the effect of fillers on dielectric properties of polymers [2] and [3]. Polymer nanocomposites films have attracted wide interest for enhancing polymer properties and extending their utility in recent years. PolyEthylene (PE) is widely used as an insulating material for power cables. Electrical insulating polymers are usually modified with inorganic fillers to improve electrical, mechanical, thermal properties. Generally, inorganic fillers are dispersed non-uniformly in the polymer matrix, and the irregular interfaces are usually electrically weak spots. It is well known that the electrical properties of insulating polymer composites depend strongly on their microstructures. In particular, the size and shape of the fillers, the dispersion of the fillers, the filler-filler and filler-matrix interactions including interfacial strain, directly affect the electrical properties of composites [4], [5], [6], [7], [8], [9] and [10]. Nanoparticles/polymer composites are now interested for their specific electrical properties. It is recognized that the interfaces between the host dielectric and the nanometric particles can strongly influence the dielectric properties of the composite material as a whole. Since interfaces dominate dielectric situations at this level, nanodielectrics and interfaces become inextricable [11], [12], [13], [14], [15], [16] and [17].

As of now, work is underway to examine the physical properties of nanocomposites materials composed of nanoparticles of metals and their compounds stabilized within a polymeric dielectric matrix. It has been found that the dielectric properties have a close relationship with the interfacial behavior between the nanoparticles and the polymer matrix in such nanocomposites films [18], [19] and [20]. Nowadays, the effects of nanoparticles in many polymers have been enhanced electric and dielectric behaviour depending on the size, structure, and concentration of the nanoparticles, as well as the type of polymeric matrix [21], [22], [23], [24], [25], [26], [27], [28], [29] and [30]. With a continual progress in polymer nanocomposites films, the main objective of this paper is studying the effects of nanoparticles on conductance and susceptance of insulating polyethylene nanocomposites films to achieve more cost-effective, energy-effective and hence environmentally better materials for the electrical insulation technology. Also, this research depicts the effects of types and concentrations of cheap nanoparticles on electrical properties of industrial polymer material. Our experimental results show the effects of clay and fumed silica nanoparticles on electric and dielectric properties of polyethylene under thermal conditions.

2.　Experimental Setup

2.1.　Nanoparticles

Clay and fumed silica nanoparticles are cheap catalysts that change the properties of industrial materials with respect to physical manufacture process.

2.2.　Polyethylene Base Matrix Polymer

Polyethylene is a thermoplastic made from petroleum, unreactive at room temperatures, and with all but strong oxidizing agents, and some solvents causing swelling. It can withstand temperatures of 80 °C and 95 °C for a short time. This polymer is a commercial material that is used in the manufacturing of high-voltage industrial products. Polyethylene nanocomposites films have been manufactured by using melting polyethylene (LDPE and HDPE), then, mixing and penetrating nanoparticles inside the base matrix polyethylene by modern ultrasonic devices.

SEM images for polyethylene nanocomposites films illustrate the penetration of nanoparticles inside low-density polyethylene and high-density polyethylene as shown in Fig. 1. And so, Tab. 1 depicts the measured electric and dielectric properties of polyethylene nanocomposites materials.

2.3.　Electric Characterization Measurements

Figure 2 shows HIOKI 3522-50 LCR Hi-tester device that measured characterization of nanocomposites insulation industrial materials, it has been used for

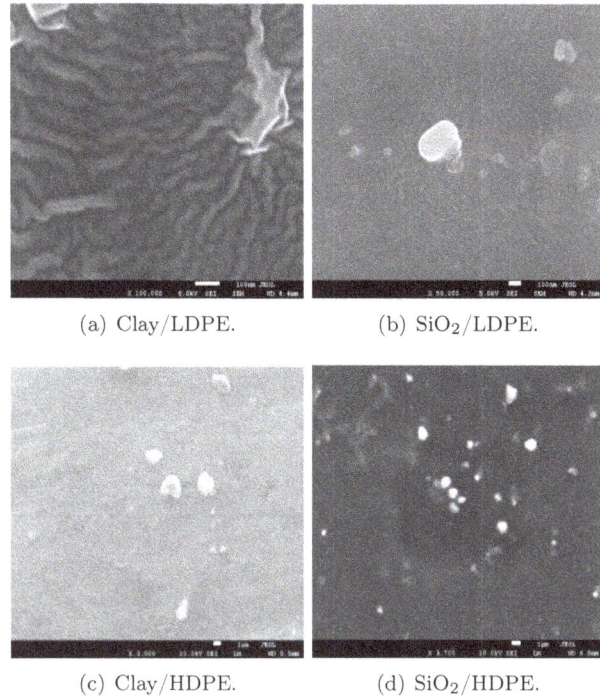

(a) Clay/LDPE.　　　　(b) SiO_2/LDPE.

(c) Clay/HDPE.　　　　(d) SiO_2/HDPE.

Fig. 1: SEM images for polyethylene nanocomposite films.

measuring electric and dielectric parameters of nanometric solid dielectric insulation specimens at various frequencies. Specification and accuracy of LCR Hi-tester device have been defined as follows, Power supply: 100, 120, 220 or 240 V (± 10 %) AC (selectable), 50/60 Hz, and Frequency: DC, 1 MHz to 100 kHz, Display Screen: LCD with backlight / 99999 (full 5 digits), Basic Accuracy: Z: ± 0.08 % rdg. θ: $\pm 0.05°$ and External DC bias ± 40 V max.(option) (3522-50 used alone ± 10 V max./ using 9268 ± 40 V max).

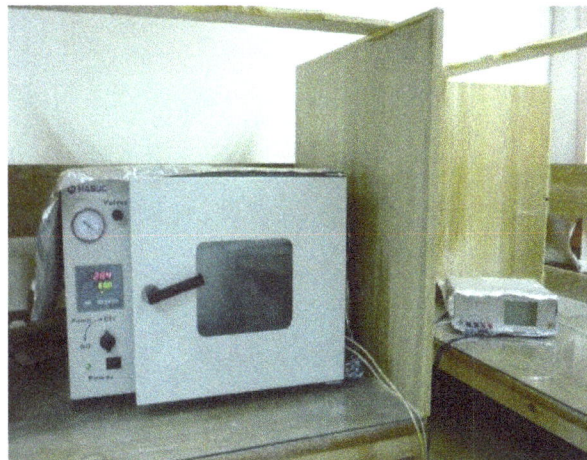

Fig. 2: HIOKI 3522-50 LCR Hi-tester device.

Tab. 1: Electric and dielectric properties of pure and nanocomposite materials.

Characteristics materials	Dielectric constant		Resistivity ($\omega \cdot$m)	
	LDPE	**HDPE**	**LDPE**	**HDPE**
PurePure	2.3	2.3	10^{14}	10^{15}
1 wt.% Clay	2.23	2.23	10^{15}	10^{16}
5 wt.% Clay	1.99	1.99	$10^{15} - 10^{18}$	$10^{16} - 10^{19}$
10 wt.% Clay	1.76	1.76	$10^{18} - 10^{20}$	$10^{19} - 10^{21}$
1 wt.% SiO$_2$	2.32	2.32	10^{13}	10^{14}
5 wt.% SiO$_2$	2.39	2.39	$10^{13} - 10^{11}$	$10^{14} - 10^{12}$
10 wt.% SiO$_2$	2.49	2.49	$10^{11} - 10^{9}$	$10^{12} - 10^{10}$

3. Results and Discussion

Dielectric Spectroscopy is a powerful experimental method to investigate the dynamical electric and dielectric behavior of the polymeric sample through frequency response analysis. This technique is based on the measurement of the resistance, conductance, and susceptance as a function of frequency for a sample sandwiched between pin-plate electrodes. Thus, the conductance and susceptance were measured for all samples as a function of frequency up to 1 kHz under variant temperatures of (20 °C and 60 °C).

3.1. Measurements on LDPE Nanocomposites Films

1) Effect of Nanoparticles on Conductance Property

Figure 3 depicts the conductance of clay/LDPE nanocomposites films that decreases with increasing concentration of clay nanoparticles in the nanocomposites up to 5 wt.% at room temperature (20 °C). However, at high temperature (60 °C), the conductance performance of clay/LDPE nanocomposites films is reversed within the same concentration range of nanoparticles.

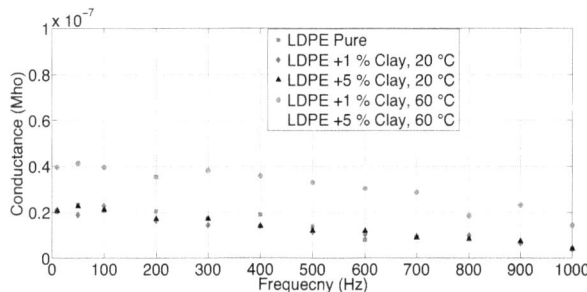

Fig. 3: Measured conductance of clay/LDPE nanocomposite films.

Therefore, increasing temperature of nanocomposites materials changes temperature degrees of nanoparticles that are changing the electric conductance behavior against normal conditions. On the other hand,

Fig. 4 shows the conductance of SiO$_2$/LDPE nanocomposites films as a function of frequency. Note that the measured conductance decreases with increasing concentration of fumed silica nanoparticles up to 1 wt.% but it increases with increasing concentration of fumed silica nanoparticles up to 5 wt.% without reaching to values of low-density polyethylene.

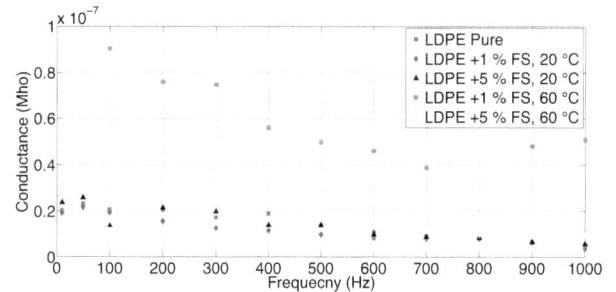

Fig. 4: Measured conductance of SiO$_2$/LDPE nanocomposite films.

Under high temperature (60 °C), the measured conductance of fumed silica/LDPE nanocomposites films increases with increasing concentration of fumed silica nanoparticles in the nanocomposites up to 1 wt.%, then, it decreases with increasing percentage of fumed silica nanoparticles in the nanocomposites up to 5 wt.%. Therefore, there is no stability in conductance property behavior for using fumed silica nanoparticles in low-density polyethylene that can reverse conductance property behavior under high temperature (60 °C).

2) Effect of Nanoparticles on Electric Susceptance Property

Figure 5 and Fig. 6 show the results of the measurements of susceptance as a function of frequency for clay/LDPE, and SiO$_2$/LDPE nanocomposites films samples under varying thermal temperatures. Note that, Fig. 5 shows that the susceptance of clay/LDPE nanocomposites films increases with the increasing concentration of clay nanoparticles in the nanocomposites up to 5 wt.% under varying thermal conditions (low and high). However, Fig. 6 shows the measured suscep-

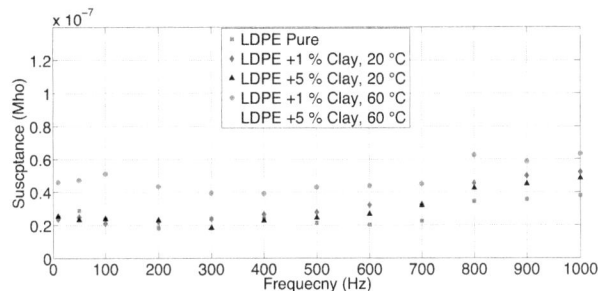

Fig. 5: Measured susceptance of cay/LDPE nanocomposite films.

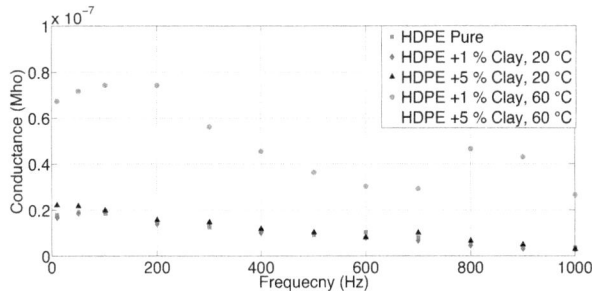

Fig. 7: Measured conductance of clay/HDPE nanocomposite films.

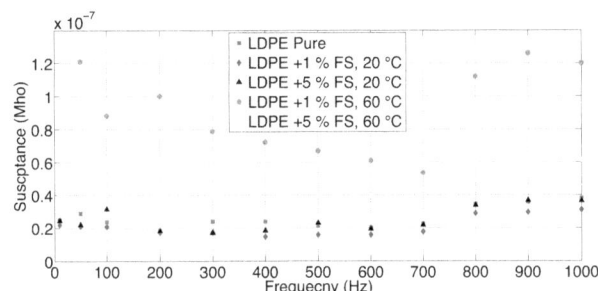

Fig. 6: Measured susceptance of SiO_2/LDPE nanocomposite films.

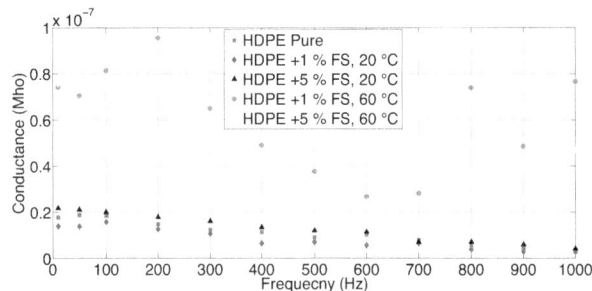

Fig. 8: Measured conductance of SiO_2/HDPE nanocomposite films.

tance of SiO_2/LDPE nanocomposites films that display the same performance of conductance with increasing fumed silica nanoparticles in low-density polyethylene under varying thermal conditions (low and high). Therefore, rising temperature of nanocomposites materials changes the temperature of nanoparticles that is changing the electric behavior against the normal conditions. Thus, presence of clay nanoparticles in low-density polyethylene causes instability of susceptance property behavior in case of high temperatures with respect to room temperature.

3.2. Measurements on HDPE Nanocomposites Films

1) Effect of Nanoparticles on Electric Conductance Property

In case of high density polyethylene, Fig. 7 shows the measured conductance of the tested samples of clay/HDPE nanocomposites films as a function of frequency at temperatures of (20 °C and 60 °C). It is obvious that the measured values of conductance are convergent and increases with the increase of the concentration of clay nanoparticles up to 5 wt.%. However, there is no convergence between the measured values of conductance of high-density polyethylene nanocomposites at high temperature (60 °C). On the other hand, Fig. 8 shows the convergence between the measured values of conductance for SiO_2/HDPE nanocompos-

ites films with increasing concentration of fumed silica nanoparticles up to 5 wt.% at room temperature (20 °C). Thus, the measured conductance increases with increasing concentration of fumed silica nanoparticles in the nanocomposites up to 5 wt.% gradually under high thermal conditions.

2) Effect of Nanoparticles on Electric Susceptance Property

Figure 9 and Fig. 10 give the results of the measurements of susceptance as a function of frequency for clay/HDPE, and SiO_2/HDPE nanocomposites films samples at temperatures of (20 °C and 60 °C). It is obvious that Fig. 9 focus on increasing susceptance with increasing concentration of clay nanoparticles in the nanocomposites up to 1 wt.%, then, the measured susceptance decreases with increasing concentration of clay nanoparticles up to 5 wt.%. On the other hand, the susceptance of clay/HDPE nanocomposites films increases with increasing concentration of clay nanoparticles up to 5 wt.% at high temperature (60 °C).

Figure 10 shows the measured susceptance of SiO_2/HDPE nanocomposites films samples versus frequency at temperatures of (20 °C and 60 °C), the susceptance of SiO_2/HDPE nanocomposites films increases with increasing concentration of fumed silica nanoparticles in high-density polyethylene nanocomposites up to 1 wt.% at room temperature (20 °C),

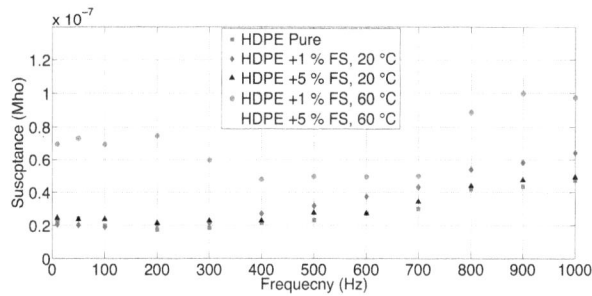

Fig. 9: Measured susceptance of clay/HDPE nanocomposite films.

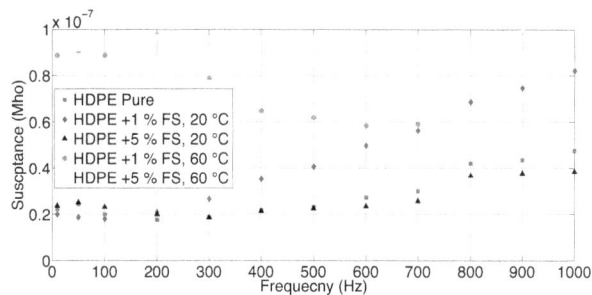

Fig. 10: Measured susceptance of SiO$_2$/HDPE nanocomposite films.

but it decreases with increasing concentration of fumed silica nanoparticles up to 5 wt.% at high temperature (60 °C).

It is clear that the dielectric properties of insulating polymer nanocomposites films have been investigated in the frequency domain from 0.1 Hz to 1 kHz and there is a convergence between the measured values of electric and dielectric polymer properties at room temperature (20 °C).

4. Trends of Nanoparticles on Polyethylene Under Thermal Conditions

The experimental results focused on effects of nanoparticles on electric characterization under variant thermal conditions. In the beginning, adding fumed silica nanoparticles increased permittivity of the fabricated polyethylene nanocomposites materials, however, adding clay has decreased permittivity of the new nanocomposites materials as shown in Tab. 1. Increasing concentration of clay and fumed silica nanoparticles at room temperature (20 °C) affects behavior of conductance and susceptance of polyethylene nanocomposites films and depends on changing the concentration of nanoparticles inside polyethylene materials under low and high frequencies. Types and concentrations of nanoparticles display the relationship between elec-

tric properties with interfacial medium behavior between the nanoparticles and the polymer matrix in nanocomposite thin films. The aim of adding nanoparticles of clay or fumed silica is controlling on the dielectric strength of commercial polyethylene by using nanotechnology techniques.

5. Conclusion

The variation of conductance and susceptance values in polyethylene nanocomposites films can be controlled by changing the types and concentrations of nanoparticles. Increasing concentration of clay nanoparticles in polyethylene decreases the effective permittivity. But, increasing concentration of fumed silica nanoparticles increases effective permittivity of polyethylene nanocomposites films.

Presence of special types of nanoparticles inside polyethylene will restrict the chain mobility, then, the result is increasing electric insulation and limiting the generation of mobile charge for the movement of charge carriers in polymer dielectrics. Therefore, the number of charge carriers and applied frequency become dominating factors of the electrical insulation of polyethylene nanocomposites films.

New fabricated polyethylene nanocomposites films have high thermal stability at small concentrations of clay or fumed silica nanoparticles. Adding large amounts of these nanoparticles to polyethylene may reverse electric and dielectric behavior characteristics gradually. In addition, rising thermal conditions of nanocomposites materials affect temperatures of nanoparticles and hence change the electric characterization.

Acknowledgment

The present work was supported by Nanotechnology Research Center at Aswan University that is established by aiding the Science and Technology Development Fund (STDF), Egypt, Grant No: Project ID 505, 2009–2011.

References

[1] TANIMOTO, G., M. OKASHITA, F. AIDA and Y. FUJIWARA. Temperature dependence of tan δ in polyethylene. In: *Proceedings of the 3rd International Conference on Properties and Applications of Dielectric Materials*. Tokyo: IEEE, 1991, pp. 1068–1071. ISBN 0-87942-568-7. DOI: 10.1109/ICPADM.1991.172259.

[2] TOKORO, T., M. NAGAO and M. KOSAKI. High-field dielectric properties and AC dissipation current waveforms of polyethylene film. *IEEE Transactions on Dielectrics and Electrical Insulation*. 2002, vol. 27, iss. 3, pp. 482–487. ISSN 0018-9367. DOI: 10.1109/14.142710.

[3] ARAOKA, M., H. YONEDA and Y. OHKI. Dielectric properties of new-type polyethylene polymerized using a single-site catalyst. In: *IEEE 6th International Conference on Conduction and Breakdown in Solid Dielectrics (ICSD)*. Vasteras: IEEE, 1998, pp. 493–497. ISBN 0-7803-4237-2. DOI: 10.1109/ICSD.1998.709332.

[4] VILCKAS, J. H., L. G. ALBIERO, S. A. CRUZ, M. M. UEKI and M. ZANIN. Study of electrical and mechanical properties of recycled polymer blends. In: *International Symposium on Electrical Insulating Materials*. Kitakyushu: IEEE, 2005, pp. 683–686. ISBN 4-88686-063-X. DOI: 10.1109/ISEIM.2005.193462.

[5] GUO, W. M., B. Z. HAN, H. ZHENG and Z. H. LI. The Effect of Basic Resins on Conductivity Properties of the Polyethylene and Carborundum Composite. In: *8th International Conference on Properties & applications of Dielectric Materials*. Bali: IEEE, 2006, pp. 747–750. ISBN 1-4244-0190-9. DOI: 10.1109/ICPADM.2006.284286.

[6] HINATA, K., A. FUJITA, K. TOHYAMA and Y. MURATA. Dielectric Properties of LDPE/MgO Nanocomposite Material under AC High Field. In: *2006 IEEE Conference on Electrical Insulation and Dielectric Phenomena*. Kansas City: IEEE, 2006, pp. 683–686. ISBN 4-88686-063-X. DOI: 10.1109/ISEIM.2005.193462.

[7] ULZUTUEV, A. N. and N. M. USHAKOV. Investigation of the charge localization processes in the metal polymeric materials based on the low density polyethylene matrix with stabilized nanoparticles. In: *4th International Conference on Advanced Optoelectronics and Lasers (CAOL 2008)*. Crimea: IEEE, 2008, pp. 435–437. ISBN 978-1-4244-1973-9. DOI: 10.1109/CAOL.2008.4671988.

[8] ISHIMOTO, K., T. TANAKA, Y. OHKI, Y. SEKIGUCHI, Y. MURATA and M. GOSYOWAKI. Comparison of Dielectric Properties of Low-density Polyethylene/MgO Composites with Different Size Fillers. In: *Annual Report Conference on Electrical Insulation and Dielectric Phenomena (CEIDP 2008)*. Quebec: IEEE, 2008, pp. 208–211. ISBN 978-1-4244-2548-8. DOI: 10.1109/CEIDP.2008.4772819.

[9] WANG, X., H. Q. HE, D. M. TU, C. LEI and Q. G. DU. Dielectric Properties and Crystalline Morphology of Low Density Polyethylene Blended with Metallocene Catalyzed Polyethylene. *IEEE Transactions on Dielectrics and Electrical Insulation*. 2008, vol. 15, iss. 2, pp. 319–326. ISSN 1070-9878. DOI: 10.1109/TDEI.2008.4483448.

[10] GREEN, C. D., A. S. VAUGHAN, G. R. MITCHELL and T. LIU. A Structure property relationships in polyethylene/montmorillonite nanodielectrics. *IEEE Transactions on Dielectrics and Electrical Insulation*. 2008, vol. 15, iss. 1, pp. 134–143. ISSN 1070-9878. DOI: 10.1109/TDEI.2008.4446744.

[11] LYNN, C., A. NEUBER, J. KRILE, J. DICKENS and M. KRISTIANSEN. Electrical conduction in select polymers under shock loading. In: *IEEE Pulsed Power Conference*. Washington: IEEE, 2009, pp. 171–174. ISBN 978-1-4244-4064-1. DOI: 10.1109/PPC.2009.5386199.

[12] SHAH, K. S., R. C. JAIN, V. SHRINET, A. K. SINGH and D. P. BHARAMBE. High Density Polyethylene (HDPE) Clay Nanocomposite for Dielectric Applications. *IEEE Transactions on Dielectrics and Electrical Insulation*. 2009, vol. 16, iss. 3, pp. 853–861. ISSN 1070-9878. DOI: 10.1109/TDEI.2009.5128526.

[13] SAMI, A., E. DAVID and M. FRECHETTE. Dielectric characterization of high density polyethylene/SiO_2 nanocomposites. In: *IEEE Conference on Electrical Insulation and Dielectric Phenomena*. Virginia Beach: IEEE, 2009, pp. 689–692. ISBN 978-1-4244-4557-8. DOI: 10.1109/CEIDP.2009.5377742.

[14] BOIS, L., F. CHASSAGNEUX, S. PAROLA, F. BESSUEILLE, Y. BATTIE, N. DESTOUCHES, A. BOUKENTER, N. MONCOFFRE and N. TOULHOAT. Growth of ordered silver nanoparticles in silica film mesostructured with a triblock copolymer PEO–PPO–PEO. *Journal of Solid State Chemistry*. 2009, vol. 182, iss. 7, pp. 1700–1707. ISSN 0022-4596. DOI: 10.1016/j.jssc.2009.01.044.

[15] DAO, N. L., P. L. LEWIN, I. L. HOSIER and S. G. SWINGLER. A comparison between LDPE and HDPE cable insulation properties following lightning impulse ageing. In: *IEEE International Conference on Solid Dielectrics*. Potsdam: IEEE, 2010, pp. 1–4. ISBN 978-1-4244-7945-0. DOI: 10.1109/ICSD.2010.5567944.

[16] KUZNETSOVA, I. E., B. D. ZAITSEV and A. M. SHIKHABUDINOV. Elastic and viscous properties of nanocomposite films based on low-density polyethylene. *IEEE Transactions on Dielectrics and Electrical Insulation*. 2010,

vol. 57, iss. 9, pp. 2099–2102. ISSN 0885-3010. DOI: 10.1109/TUFFC.2010.1658.

[17] SHENGTAO, L., Y. GUILAI, N. FENGYAN, B. SUNA, L. JIANYING and Z. TUO. Investigation on the dielectric properties of nano-titanium dioxide - low density polyethylene composites. In: *10th IEEE International Conference on Solid Dielectrics*. Potsdam: IEEE, 2010, pp. 1–4. ISBN 978-1-4244-7943-6. DOI: 10.1109/ICSD.2010.5568106.

[18] FANG, P., X. QIU, W. WIRGES, R. GERHARD and L. ZIRKEL. Polyethylene-naphthalate (PEN) ferroelectrets: cellular structure, piezoelectricity and thermal stability. *IEEE Transactions on Dielectrics and Electrical Insulation*. 2010, vol. 17, iss. 4, pp. 1079–1087. ISSN 1070-9878. DOI: 10.1109/TDEI.2010.5539678.

[19] AMAN, A., M. M. YAACOB, M. A. ALSAEDI and K. A. IBRAHIM. Polymeric composite based on waste material for high voltage outdoor application. *International Journal of Electrical Power & Energy Systems*. 2013, vol. 45, iss. 1, pp. 346–352. ISSN 0142-0615. DOI: 10.1016/j.ijepes.2012.09.004.

[20] DA SILVA, D. A., E. C. M. DA COSTA, J. L. DE FRANCO, M. ANTONIONNI, R. C. DE JESUS, S. R. ABREU, K. LAHTI, L. H. I. MEI and J. PISSOLATO. Reliability of Directly-Molded Polymer Surge Arresters: Degradation by Immersion Test Versus Electrical Performance. *International Journal of Electrical Power & Energy Systems*. 2013, vol. 53, iss. 1, pp. 488–498. ISSN 0142-0615. DOI: 10.1016/j.ijepes.2013.05.023.

[21] SARATHI, R. and R. UMAMAHESWARI. Understanding the Partial Discharge Activity Generated Due to Particle Movement in a Composite Insulation Under AC Voltages. *International Journal of Electrical Power & Energy Systems*. 2013, vol. 48, iss. 1, pp. 1–9. ISSN 0142-0615. DOI: 10.1016/j.ijepes.2012.11.017.

[22] MCCALLEY, J. D. and V. KRISHNAN. A Survey of Transmission Technologies for Planning Long Distance Bulk Transmission Overlay in US. *International Journal of Electrical Power & Energy Systems*. 2014, vol. 54, iss. 1, pp. 559–568. ISSN 0142-0615. DOI: 10.1016/j.ijepes.2013.08.008.

[23] GOUDA, O., A. THABET, Y. A. MOBARAK, M. SAMIR. Nanotechnology Effects on Space Charge Relaxation Measurements for Polyvinyl Chloride Thin Films. *International Journal on Electrical Engineering and Informatics*. 2014, vol. 6, iss. 1, pp. 1–12. ISSN 2085-5830. DOI: 10.15676/ijeei.2014.6.1.1.

[24] THABET, A. Experimental Enhancement for Dielectric Strength of Polyethylene Insulation Materials Using Cost-fewer Nanoparticles. *International Journal of Electrical Power & Energy Systems*. 2015, vol. 64, no. 1, pp. 469–475. ISSN 0142-0615. DOI: 10.1016/j.ijepes.2014.06.075.

[25] THABET, A. and Y. A. MOBARAK. Experimental Dielectric Measurements for Cost-fewer Polyvinyl Chloride Nanocomposites. *International Journal of Electrical and Computer Engineering*. 2015, vol. 5, no. 1, pp. 13–22. ISSN 2088-8708. DOI: 10.11591/IJECE.V5I1.6743.

[26] THABET, A. Experimental Verification for Improving dielectric strength of polymers by using clay nanoparticles. *Advances in Electrical and Electronic Engineering*. 2015, vol. 13, no. 2, pp. 182–190. ISSN 1336-1376. DOI: 10.15598/aeee.v13i2.1249.

[27] THABET, A. Thermal experimental verification on effects of nanoparticles for enhancing electric and dielectric performance of polyvinyl chloride. *Measurement*. 2016, vol. 89, no. 1, pp. 28–33. ISSN 0263-2241. DOI: 10.1016/j.measurement.2016.04.002.

[28] THABET, A. Theoretical analysis for effects of nanoparticles on dielectric characterization of electrical industrial materials. *Electrical Engineering*. 2016, vol. 98, iss. 2, pp. 1–7. ISSN 1432-0487. DOI: 10.1007/s00202-016-0375-4.

[29] EBNALWALED, A. A. and A. THABET. Controlling the optical constants of PVC nanocomposite films for optoelectronic applications. *Synthetic Metals*. 2016, vol. 220, iss. 1, pp. 374–383. ISSN 0379-6779. DOI: 10.1016/j.synthmet.2016.07.006.

[30] THABET, A and Y. A. MUBARAK. The effect of cost-fewer nanoparticles on the electrical properties of polyvinyl chloride. *Electrical Engineering*. 2016, vol. 98, iss. 2, pp. 1–7. ISSN 1432-0487. DOI: 10.1007/s00202-016-0392-3.

About Authors

Ahmed THABET was born in Aswan, Egypt in 1974. He received the B.Sc. (FEE) Electrical Engineering degree in 1997 and M.Sc. (FEE) Electrical Engineering degree in 2002 both from Faculty of Energy Engineering, Aswan, Egypt. Ph.D. degree had been received in Electrical Engineering in 2006 from El-Minia University, Minia, Egypt. He joined with Electrical Power Engineering Group of Faculty of Energy Engineering in Aswan University as a Demonstrator at July 1999, until; he held Associate Professor Position at October 2011 up to date. His research interests lie in the areas of analysis and developing electrical engineering models and applications, investigating novel nano-technology materials via addition nano-scale particles and additives for usage in industrial branch, electromagnetic materials, electroluminescence and the relationship with electrical and thermal ageing of industrial polymers. On 2009, he had been a Principle Investigator of a funded project from Science and Technology Development Fund "STDF" for developing industrial materials of ac and dc applications by nano-technology techniques. He has been established first Nano-Technology Research Centre in the Upper Egypt. He has many of publications which have been published and under published in national, international journals and conferences and held in Nano-Technology Research Centre website.

Youssef MOBARAK was born in Luxor, Egypt in 1971. He received his B.Sc. and M.Sc. degrees in Electrical Engineering from Faculty of Energy Engineering, Aswan University, Egypt, in 1997 and 2001 respectively and Ph.D. from Faculty of Engineering, Cairo University, Egypt, in 2005. He joined Electrical Engineering Department, Faculty of Energy Engineering, Aswan University as a Demonstrator, as an Assistant Lecturer, and as an Assistant Professor during the periods of 1998–2001, 2001–2005, and 2005–2009 respectively. He joined Artificial Complex Systems, Hiroshima University, Japan as a Researcher 2007–2008. Also, he joined King Abdulaziz University, Rabigh, Faculty of Engineering 2010 to present. His research interests are power system planning, operation, and optimization techniques applied to power systems. Also, his research interests are Nanotechnology materials via addition nano-scale particles and additives for usage in industrial field.

6

Driving Pressure Influence in Voltage Maps Measurement Process Using Advanced Pneumatic Mapping Probe

Marek KUKUCKA[1], Andreas WEISZE[2], Daniela DURACKOVA[3], Zuzana KRAJCUSKOVA[3], Viera STOPJAKOVA[3]

[1]Institute of Automotive Mechatronics, Faculty of Electrical Engineering and Information Technology, Slovak University of Technology, Ilkovicova 3, 81219 Bratislava, Slovak Republic
[2]Dr. Ing. h.c. Porsche AG, Porscheplatz 1, 70435 Stuttgart, Germany
[3]Institute of Electronics and Photonics, Faculty of Electrical Engineering and Information Technology, Slovak University of Technology, Ilkovicova 3, 81219 Bratislava, Slovak Republic

marek.kukucka@stuba.sk, post@weisze.de, daniela.durackova@stuba.sk, zuzana.krajcuskova@stuba.sk, viera.stopjakova@stuba.sk

Abstract. *Our paper deals with the method of the voltage-impedance map measurement process as a method useful for the electric mapping of human skin. The area of research extends from the basic research to its practical application in acupuncture skin mapping and acupuncture point localization and visualization. The problem of sufficient skin coverage and electrical contact with measuring electrodes is solved by the conventional mechanical telescopic electrodes and by the pneumatic matrix electrode probe. A 2D or 3D voltage-impedance map of skin is an output of the measuring, interpretation and evaluation process. New pneumatic construction of measuring probe was implemented to achieve a better coverage of specified skin area and get a reduced force range of the touching electrodes allowing the steady contact of the skin-electrode. A skin contact is related to the driving pressure of touching electrodes. Our paper offers experimentally measured results, voltage maps of skin on specific areas, selected measured and described acupuncture points and their applications in electro-acupuncture.*

Keywords

Acupuncture point, driving pressure, human skin, measuring probe, pneumatic electrodes.

1. Introduction

Scientific research in physics, medicine and in electrotechnics consider a human skin as a subject of interesting and perspective research area. Skin behaviour from the electro-technical point of view is influenced by its structure and material. Specific parts of human or animal skin are called acupuncture points or active points; they are also depicted in various research publications. These points can be located by the results of the impedance-voltage map measurement. Acupuncture points (ACU-points) have been used in Traditional Chinese Medicine for more than few thousand years. Not only nowadays researchers try to describe them properly, to measure their electrical properties and to publish their results [1]. ACU-points are used in electro-acupuncture and in new non-standard diagnostic and therapeutic methods in medical devices [2]. Researchers consider these channels (meridians) as areas in the extracellular space where the electric charges flow through. A deviation of energy flow in comparison with normal values causes a misbalance of the body and can cause illness. Measuring of voltage-impedance minimums and reduced impedance on some areas of skin is useful for the unfolding of ACU-points. Acupuncture points are specific dynamic objects. They are changing their shape and intensity according to the circadian biorhythms [3] as various living systems [4]. During the specific hours of the day and also according to the measured person they are larger and deeper and more distinguishable but in some hours of the day (low meridian activity) they are scarcely visible and measurable. Their shape and properties are also influenced by the health conditions of the examined person and the corresponding organs. The maximal dynamic range of measurement is not the only considering factor for skin mapping and the active point recognition process. The state of skin, wet or dry surface, particular skin-electrode connection quality are changing and influence the measurement and recognition process.

2.   Mapping Method and Skin Voltage Measuring Device

An experimental acupuncture measuring and mapping device developed at Faculty of Electrical Engineering and Information Technology, Slovak University of Technology in Bratislava was used for experimental measurements and skin mapping [5]. Correct operation and function of experimental skin voltage mapping device are controlled by processor ATmega16. For the communication with the control Personal Computer (PC), the USB to UART converter is used as a virtual serial port. The processor is equipped with a serial port, 10-bit A/D converter and Serial Peripheral Interface (SPI). Four analog multiplexers were used for the controlled reconnection of measuring electrodes touching measured object. The device utilizes a modified version of a peak detector, the DDS synthesis generator AD9833 (Fig. 1).

The control computer and the measuring mapping device are galvanically separated by a two channel insulator circuits, because of the safety of the patient (Fig. 2). There is also a different way to protect the operator or the measured human object from potential electric injury. It is fully guaranteed by supplying the device from accumulators.

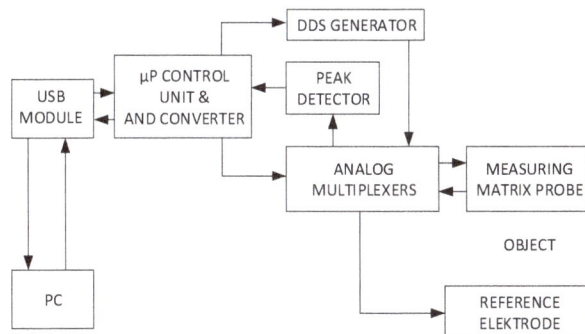

Fig. 1: Functional block diagram of mapping device.

The skin interaction and surface contact are provided by the electrodes of the measuring probe. All the electrodes measure the change of voltage with regard to a reference electrode [5] - a drop of potential on the skin induced by driving measuring electrical current

from the generator of the device, flowing through the unknown skin or body impedance.

A measuring device is allowed to apply driving voltage from 0 V to maximally 5 V. An electrode with a larger contact area was used as reference. Mechanical or pneumatic telescopic matrix probes were used for measuring. Similar mapping method of acupuncture point measurement was introduced also in [6], where a larger and less sensitive non-telescopic kind of measuring probe was used.

Fig. 2: Realized experimental mapping device with connected matrix probe, the reference electrode and practical supporting devices with the control computer.

Figure 2 shows a look into the laboratory with pneumatic-mechanic-electronic measuring system with connected sensing probe, mechanical holding and driving system and mapping device connected to control computer. For software design of the measuring system in MATLAB software were designed scripts and control GUI. MATLAB environment offers a wide area for real-time processing and file post-processing of clinical data and graphical interpretation (similar to the system in [4]).

3.   Pneumatic Matrix Electrode Probe Construction

There were designed, developed and realized several constructions of non-invasive sensing electrode probe, in our Research Laboratory of Biomechatronics. The whole probe consists of 64 electrodes placed into an 8×8 matrix on an isolative holding construction. Non-planar, bended and side areas of skin on the human body are not easily accessible and during the measurement exhibit non-stable and non-proper contact electrode-skin. In specific non-planar and complicated skin areas as fingers, ears and parts of face the problem with skin irritation occurs. The experimental laboratory system contains pneumatic sensing electrode probe that uses pneumatic telescopic electrodes with the complex driving mechanism (Fig. 3), externally controlled air compressor and device of pressed air pipe distribution subsystem (Fig. 4) with pressure distributing driving elements and measurement [7].

Detail of the connected pneumatic probe with air pipes and manometer in driving, holding and positioning mechanism can be seen in Fig. 4. In realized pneumatic probe are used electrodes with pistons and extended contact peaks of 0.5 mm tip diameter on a substrate square of 2.5×2.5 cm (Fig. 3).

Fig. 3: Detail of pneumatic sensing probe in initial electrode position.

Fig. 4: Detail of the pneumatic probe with the air pipes and manometer in driving and holding mechanism.

4. Measurement of ACU-Points with Pneumatic Electrode Probe in Driving Pressure Progression

Most of the acupuncture points are situated on flat and good accessible skin areas. However, specific ACU-points are in hardly accessible parts of a human body. A skin area or acupuncture point is difficult to map, to visualize or to measure in the case it is situated on non-planar parts of human body or the acupuncture point is during the time of measurement smaller than the electrode probe resolution. Problems are also caused by the uneven skin areas of certain parts on human body, e.g. lateral side of fingers on hand or leg where specific active points of several meridians are situated. Performing a mapping skin measurement on the non-planar skin of lateral side of hand finger, the tips of all the electrodes have been touching the skin and pressing on it by the same lower and comfortable force, because of the same controlled constant air pressure in the system [8], [9]. Air in system is pressing on pistons of all the electrodes and invokes the same force pressing on the skin under the single electrodes (Fig. 5). That is why the skin irritation, perspiration and temporary change of electrical contact quality for all the measuring electrodes are the constant [8], [9]. Specific experimental measurements with pneu-

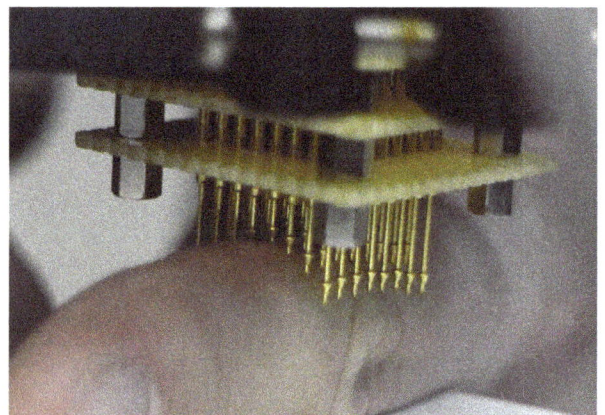

Fig. 5: Electrodes-skin detail of touching area in measurement by the pneumatic probe.

matic sensing probe were performed in our laboratory. We offer a representative set of measurements on an area described as "B" for this paper (see Fig. 6). It is significant mainly due to practical documentation of skin mapping measurement on complicated and uneven skin area. This is an operating space for a new pneumatic probe. Mapping skin measurements have been performed in specified parts on the lateral side of hand finger area.

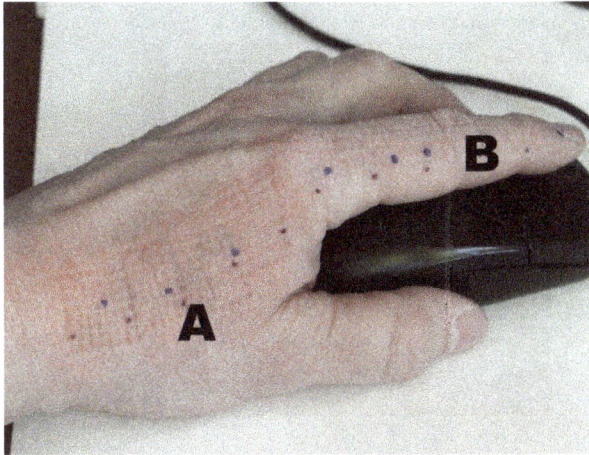

Fig. 6: Experimental object for measurement - hand and finger.

5. Experimental Results and Discussion

The first set of measurements was performed on the place of LI1b ACU-point (according to Voll description [10], between classical LI1 and LI2 ACU-points positions). This diagnostic point is situated on the area "B" (Fig. 6), respectively according to displayed location (Fig. 7).

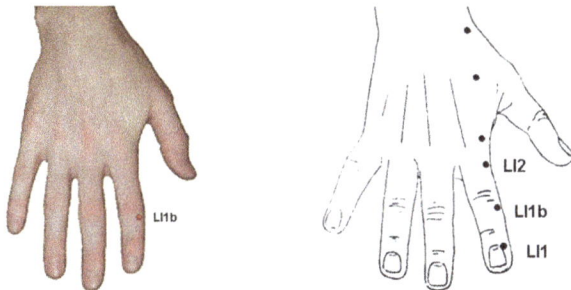

Fig. 7: LI1b ACU-point on LI meridian and next diagnostic points LI1 and LI2 according to Voll [10].

Measured graphical results are displayed on voltage-impedance maps below. In general, they contain the blue parts of map which are places on the skin with lower voltage/impedance drop (ACU-points), yellow or red places are surroundings with higher measured voltage or impedance, the voltage scale bar is situated on the right side of single maps.

5.1. Measurements of LI1b ACU-Point

Results are presented in both tabular and graphical form. Both forms contain voltage drops values of measured voltage Ux induced by harmonic driving current ($I = 1$ µA and $f = 1$ kHz) flowing through the skin.

The voltage drop is the measure of skin impedance under the measuring electrode. Measured maps of relatively uneven skin area "B" in the position of LI2 ACU-point obtained in four steps of driving pressure are shown in Fig. 8. Common consequence and shape relation of them all is apparent. Primarily the force of probe application in measurement process has been considered. The driving pressure p_D of application was varied by choosing the pressure range between 350 kPa and 500 kPa in 50 kPa steps. According to the technical point of view, the air pressure 400 kPa and higher, ejects all the pistons in measuring electrodes (Fig. 3 and Fig. 5) in correct working position. The mechanical construction of some electrodes allows to push out them with higher driving pressure. The higher pressure, the better transition contact, and the lower transition impedance.

According to the measured human object, the lower pressure causes higher comfort of measurement and also lower skin irritation. While the maximal voltage drop measured on skin for driving pressure $p_D = 350$ kPa was $Ux_{max} = 2.31$ V and $Ux_{min} = 1.21$ V, for $p_D = 400$ kPa it was $Ux_{max} = 2.29$ V and $Ux_{min} = 0.98$ V, for $p_D = 450$ kPa it was $Ux_{max} = 2.29$ V and $Ux_{min} = 0.93$ V and for the driving pressure $p_D = 500$ kPa it was $Ux_{max} = 1.36$ V and $Ux_{min} = 0.96$ V. Maximal dynamic range of measurement was obtained with $p_D = 450$ kPa, $(Ux_{max} - Ux_{min}) = (2.29$ V $- 0.93$ V$) = 1.36$ V.

Considering all the influences and parameters, the optimum pressure for the recognition results in these measurements appears to be 450 kPa, as the compromise value because of the higher dynamic range of measurement (see also surrounded shape in the middle of single maps on Fig. 8).

5.2. Measurements of LI2 ACU-Point

During the measurement on selected area of LI2 ACU-point position the maximal voltage drop measured on the skin for driving pressure $p_D = 350$ kPa was $Ux_{max} = 2.28$ V and minimal $Ux_{min} = 1.26$ V, for $p_D = 400$ kPa it was $Ux_{max} = 2.18$ V and $Ux_{min} = 0.65$ V, for $p_D = 450$ kPa it was $Ux_{max} = 2.09$ V and $Ux_{min} = 0.72$ V and for the driving pressure $p_D = 500$ kPa it was $Ux_{max} = 2.00$ V and $Ux_{min} = 0.53$ V. Maximal dynamic range of measurement was obtained with $p_D = 400$ kPa and it was $(Ux_{max} - Ux_{min}) = (2.18$ V $- 0.65$ V$) = 1.53$ V (Fig. 9).

(a) $p_D = 350$ kPa.

(b) $p_D = 400$ kPa.

(c) $p_D = 450$ kPa.

(d) $p_D = 500$ kPa.

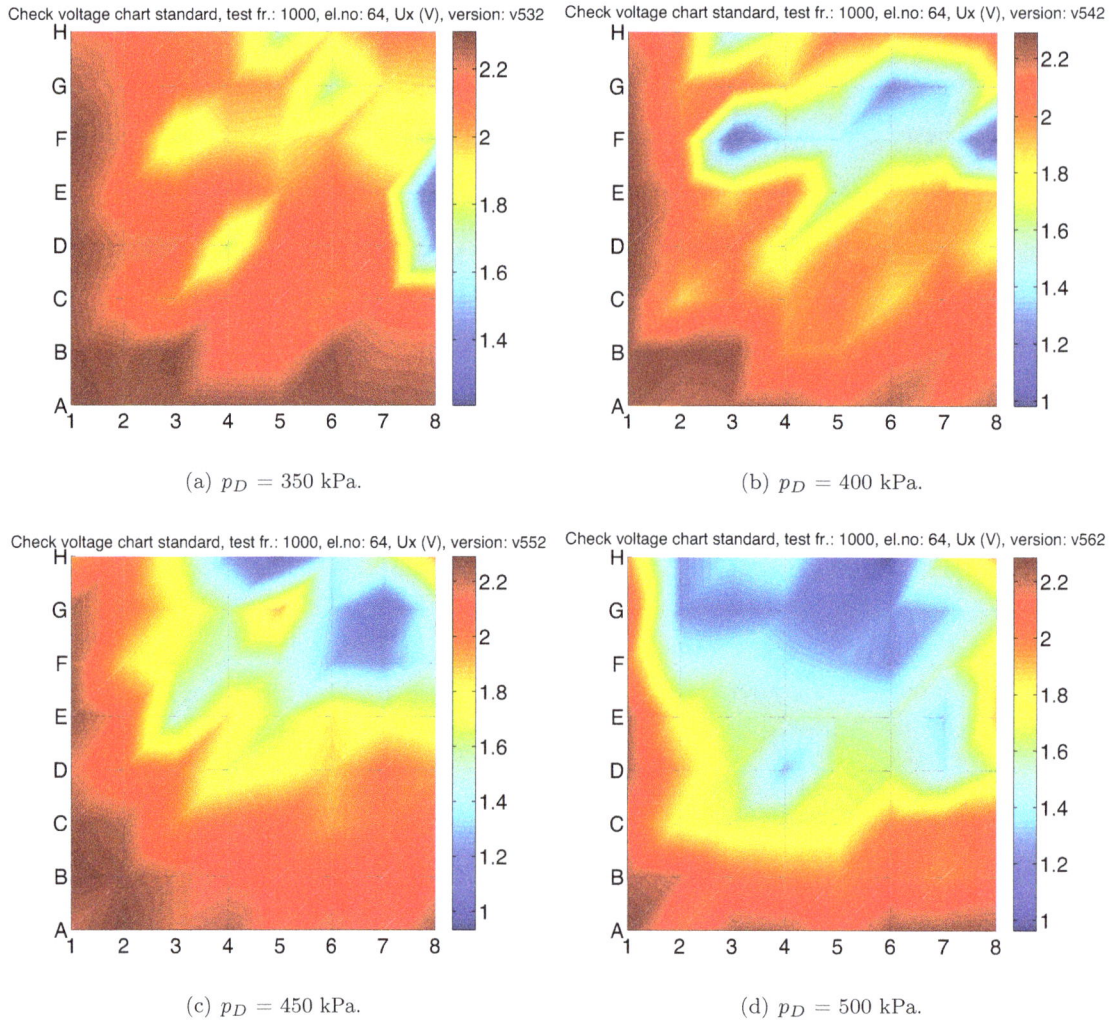

Fig. 8: Voltage map of skin on selected of LI1b ACU-point position (uneven skin area "B") measured with driving pressure p_D.

Tab. 1: Measured voltage map of skin on LI1b ACU-point position for driving pressure $p_D = 350$ kPa, $p_D = 400$ kPa, $p_D = 450$ kPa and $p_D = 500$ kPa in tabular form.

Ux (V)	$p_D = 350$ kPa								$p_D = 400$ kPa							
	1	**2**	**3**	**4**	**5**	**6**	**7**	**8**	**1**	**2**	**3**	**4**	**5**	**6**	**7**	**8**
A	2.31	2.28	2.29	2.28	2.28	2.29	2.28	2.27	2.26	2.21	2.22	2.23	2.14	2.23	2.11	2.24
B	2.30	2.27	2.14	2.30	2.13	2.13	2.22	2.22	2.25	2.26	2.27	1.98	1.98	2.11	2.15	2.08
C	2.28	2.15	2.04	2.13	2.15	2.08	2.15	2.14	2.27	1.88	2.03	1.94	2.02	1.86	2.08	2.10
D	2.30	2.21	2.20	1.79	2.16	2.13	2.15	1.44	2.29	2.07	2.06	1.71	1.98	1.96	1.96	1.78
E	2.29	2.14	2.10	2.14	2.02	2.13	2.02	1.21	2.28	2.12	1.86	2.08	1.58	1.88	1.98	1.90
F	2.31	2.20	1.81	1.94	1.92	2.00	1.85	1.78	2.27	2.02	1.10	1.41	1.43	1.52	1.52	0.98
G	2.26	2.19	2.11	2.04	2.03	1.72	1.94	1.95	2.24	1.95	2.05	1.88	1.69	1.24	1.35	1.79
H	2.20	2.17	2.08	2.04	1.71	1.95	2.09	2.07	2.15	2.02	1.45	1.73	2.05	1.96	2.03	2.14
Ux (V)	$p_D = 450$ kPa								$p_D = 500$ kPa							
	1	**2**	**3**	**4**	**5**	**6**	**7**	**8**	**1**	**2**	**3**	**4**	**5**	**6**	**7**	**8**
A	2.20	2.29	2.17	2.22	2.10	2.17	2.08	2.25	2.29	2.00	2.22	2.21	2.09	2.24	2.24	2.25
B	2.29	2.26	2.18	2.13	2.09	2.10	2.07	2.09	2.22	2.17	2.03	1.98	1.97	2.05	1.97	2.05
C	2.27	2.22	1.96	2.03	2.06	1.93	2.01	2.06	2.10	1.78	1.70	1.70	1.72	1.99	2.01	2.07
D	2.17	2.04	2.01	1.68	1.77	1.87	2.03	1.94	2.23	1.92	1.66	1.31	1.67	1.62	1.40	1.67
E	2.27	2.11	1.48	1.84	1.60	1.83	1.74	1.69	2.23	1.77	1.59	1.56	1.57	1.55	1.33	1.90
F	2.25	1.86	1.67	1.50	1.52	1.28	1.23	1.46	2.01	1.46	1.49	1.38	1.31	1.24	1.56	1.85
G	2.24	2.06	1.81	1.59	1.91	1.32	1.04	1.47	2.14	1.25	1.21	1.25	1.16	1.25	1.31	1.56
H	1.93	2.08	1.70	1.25	0.93	1.38	1.60	1.94	1.95	1.25	1.40	1.47	1.13	0.96	1.67	1.91

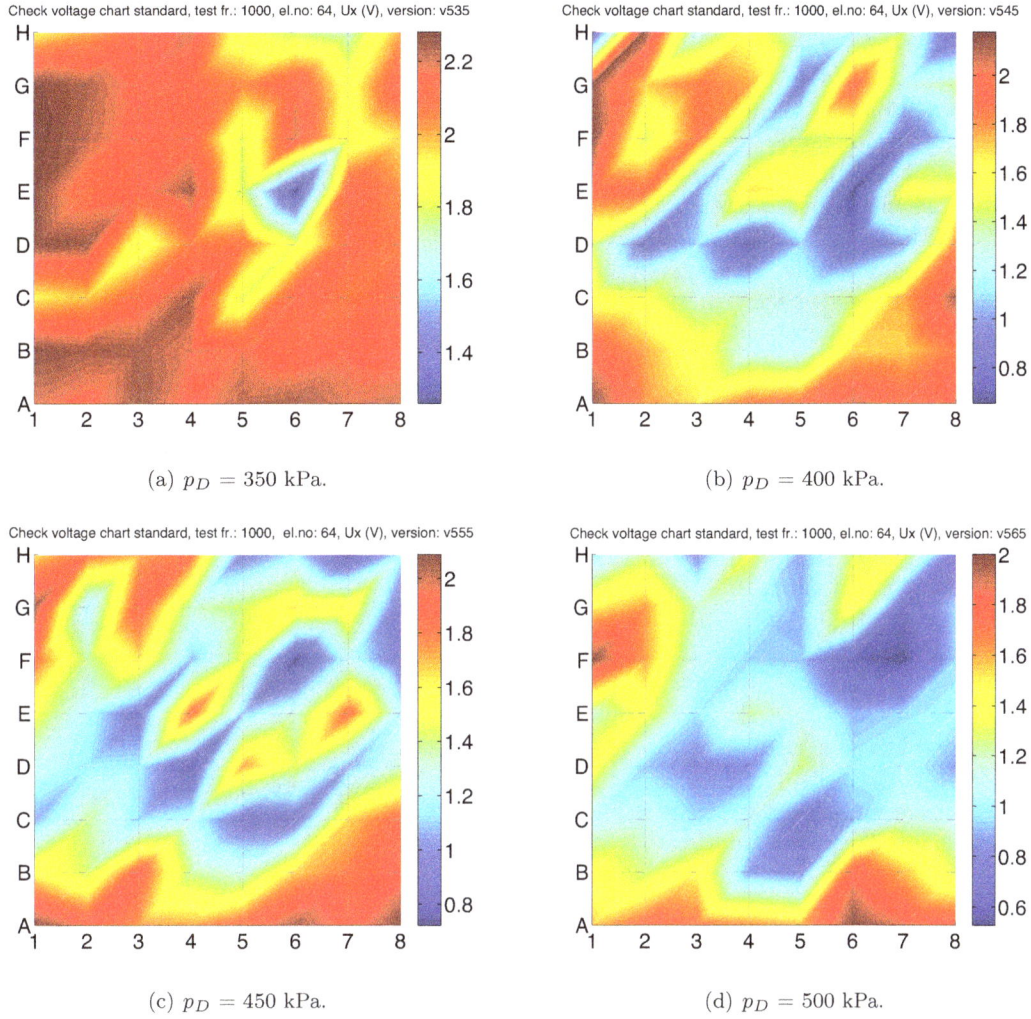

(a) $p_D = 350$ kPa.

(b) $p_D = 400$ kPa.

(c) $p_D = 450$ kPa.

(d) $p_D = 500$ kPa.

Fig. 9: Voltage map of skin on selected area of LI2 ACU-point position (left side of uneven skin area "B") measured with driving pressure p_D.

Tab. 2: Measured voltage map of skin on LI2 ACU-point position for driving pressure $p_D = 350$ kPa, $p_D = 400$ kPa, $p_D = 450$ kPa and $p_D = 500$ kPa in tabular form.

Ux	\multicolumn{8}{c}{$p_D = 350$ kPa}	\multicolumn{8}{c}{$p_D = 400$ kPa}
(V)	**1**	**2**	**3**	**4**	**5**	**6**	**7**	**8**	**1**	**2**	**3**	**4**	**5**	**6**	**7**	**8**
A	2.20	2.14	2.25	2.15	2.23	2.25	2.20	2.21	2.17	1.80	1.68	1.87	1.90	2.06	1.95	2.00
B	2.27	2.25	2.25	2.19	2.18	2.12	2.16	2.14	1.96	1.76	1.58	1.24	1.20	1.70	1.70	1.86
C	1.93	1.96	2.11	2.25	1.80	2.17	2.06	2.15	1.58	1.53	1.30	1.33	1.25	1.41	1.77	2.13
D	2.27	2.26	1.85	2.06	2.09	1.82	2.12	2.07	1.44	0.86	1.10	0.69	1.08	0.79	1.01	1.84
E	2.27	2.11	2.15	2.25	1.78	1.26	2.14	2.04	1.95	1.91	1.00	1.68	1.63	0.65	1.58	1.52
F	2.28	2.22	2.14	2.13	1.80	2.21	1.84	1.95	2.18	1.45	1.88	1.16	1.43	1.48	0.83	1.20
G	2.27	2.24	2.15	2.12	2.05	2.16	1.88	2.17	2.15	1.70	1.87	1.77	0.89	1.95	1.26	1.40
H	1.92	1.99	2.08	2.20	1.85	1.77	1.71	2.01	0.72	2.16	1.24	1.59	1.28	1.19	1.40	0.76
Ux	\multicolumn{8}{c}{$p_D = 450$ kPa}	\multicolumn{8}{c}{$p_D = 500$ kPa}
(V)	**1**	**2**	**3**	**4**	**5**	**6**	**7**	**8**	**1**	**2**	**3**	**4**	**5**	**6**	**7**	**8**
A	2.08	1.92	1.73	1.73	1.86	2.05	1.89	2.09	1.57	1.75	1.70	1.65	1.60	2.00	1.87	1.80
B	1.60	1.41	1.89	1.37	1.28	1.67	1.79	1.90	1.42	1.26	1.48	0.86	0.82	1.81	1.58	1.30
C	1.05	1.38	1.12	1.21	0.90	0.92	1.66	1.87	1.06	1.12	0.96	1.11	0.74	0.92	1.15	1.10
D	1.26	1.14	1.15	0.80	1.73	1.46	1.10	1.33	1.41	0.92	0.80	0.81	1.27	0.99	1.03	0.86
E	1.49	1.18	0.92	1.84	1.09	1.30	1.85	1.19	1.18	1.38	0.97	1.23	1.11	0.89	0.95	0.99
F	1.82	1.37	1.64	1.08	1.48	0.72	1.24	0.91	2.00	1.62	1.08	0.94	0.92	0.58	0.53	0.95
G	2.05	1.17	1.90	1.33	1.43	1.64	1.62	0.87	1.49	1.62	1.19	1.05	0.90	1.47	0.65	0.96
H	1.65	1.82	1.72	1.84	0.95	1.13	0.73	1.15	0.95	1.24	0.80	1.35	0.87	1.29	1.35	0.69

5.3. Measurements of LI1 ACU-Point

Very specific position for measurement and skin mapping is a position of acupuncture point LI1 at a lateral side of the distal phalanx of the forefinger. That position is very difficult to access (constantly by all the electrodes) and if we had a mechanical version of measuring electrode probe only, it would by practically impossible. The press of mechanical electrodes on the central part of measuring area would cause a strong skin irritation and false voltage drop. A pneumatic electrode probe construction allowed the uniform and non disturbing electrode contact with skin and the performance of the mentioned mapping measurement. The measurement of LI1 ACU-point position (Fig. 7) showed the maximal voltage drop measured on skin for driving pressure $p_D = 450$ kPa, it was $Ux_{max} = 2.30$ V and minimal $Ux_{min} = 1.20$ V (measured voltage map of skin on Fig. 10) and for the driving pressure $p_D = 500$ kPa it was $Ux_{max} = 2.32$ V and $Ux_{min} = 1.00$ V. Maximal dynamic range was obtained with $p_D = 500$ kPa and it was $(Ux_{max} - Ux_{min}) = (2.32$ V $- 1.00$ V$) = 1.32$ V. A significant bordered blue area of the measured acupuncture point is visible in upper part of the voltage map on Fig. 10, also in Tab. 3, similarly on Fig. 11, also in Tab. 4.

Check voltage chart standard, test fr.: 1000, el.no: 64, Ux (V), version: v551

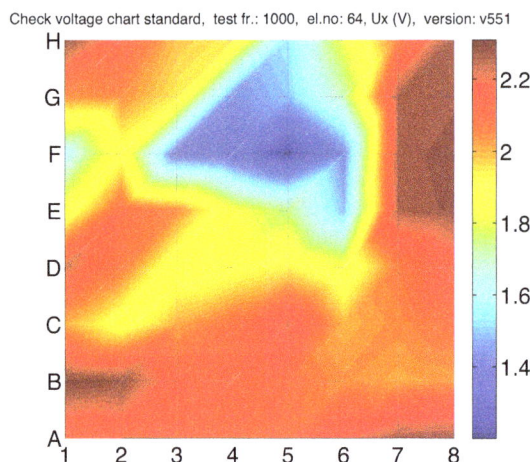

Fig. 10: Voltage map of skin on selected area of LI1 ACU-point position in 2D view (right side of uneven skin area "B") measured with driving pressure $p_D = 450$ kPa.

The measured, described and recognized position and the shape of acupuncture point LI1 (Fig. 10) is possible to characterize numerically by the table of voltage drops values (Tab. 3) scanned on considered skin surface.

The position and the shape of acupuncture point LI1 (Fig. 11) are also possible to numerically characterize by the table of voltage drops values (Tab. 4) scanned

Tab. 3: Measured voltage map of skin on LI1 ACU-point position in tabular form measured with driving pressure $p_D = 450$ kPa.

Ux (V)	1	2	3	4	5	6	7	8
A	2.11	2.14	2.11	2.11	2.08	2.19	2.27	2.28
B	2.31	2.29	2.14	2.12	2.08	2.04	2.01	2.02
C	2.01	1.87	2.02	2.10	2.21	2.00	2.04	2.08
D	2.24	2.08	1.85	1.89	1.94	1.84	2.07	2.15
E	1.85	2.13	2.06	1.84	1.74	1.49	2.26	2.28
F	1.60	1.89	1.39	1.34	1.20	1.42	2.27	2.29
G	2.06	2.06	1.95	1.44	1.50	1.71	2.26	2.27
H	2.29	2.15	2.00	1.95	1.53	1.81	2.07	2.28

Check voltage chart standard, test fr.: 1000, el.no: 64, Ux (V), version: v561

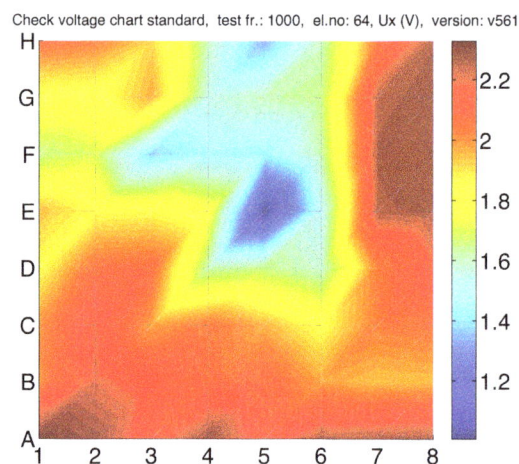

Fig. 11: Voltage map of skin on selected area of LI1 ACU-point position (right side of uneven skin area "B") measured with driving pressure $p_D = 500$ kPa.

on considered skin surface. An increased driving pressure on pneumatic electrodes in regard to the previous measurement (Fig. 10) caused higher skin irritation, a projection of the acupuncture point lost its clear contour (Fig. 11).

Tab. 4: Measured voltage map of skin on LI1 ACU-point position in tabular form measured with driving pressure $p_D = 500$ kPa.

Ux (V)	1	2	3	4	5	6	7	8
A	2.33	2.28	2.12	2.29	2.13	2.26	2.27	2.27
B	2.01	2.20	2.14	2.08	2.09	1.97	1.93	1.94
C	2.00	2.19	1.97	2.06	1.96	1.85	2.03	2.05
D	1.86	2.07	2.08	1.60	1.65	1.57	2.03	2.18
E	1.96	1.86	1.87	1.80	1.01	1.51	2.26	2.28
F	1.64	1.69	1.37	1.45	1.43	1.62	2.27	2.29
G	1.84	1.83	1.98	1.56	1.62	1.65	2.26	2.28
H	2.07	2.04	1.93	1.60	1.32	1.60	2.05	2.27

Measured and described structure of skin surface voltage and impedance distribution obtained by the advanced pneumatic electrode matrix probe (controlled by mapping measuring device) [9] offers a relatively precious, rich and compact look on the skin surface, its structure and the position, the size and the shape of certain acupuncture points. Obtained density of measured voltage map realized by our pneumatic probe has greater resolution than results of authors [6], see Fig. 12 - the realized measuring probe of authors [6]. German researchers team led by Sybille Kramer published their research results in 2008 [6]. They performed their measurements and in the area of chosen acupuncture point GB34. They measured signals from the skin surface by they own designed and constructed flexible sensor, which consisted of he isolative plastic foil with regularly placed metal electrodes in array of 8×8 of the complex dimension 6×6 cm (Fig. 12).

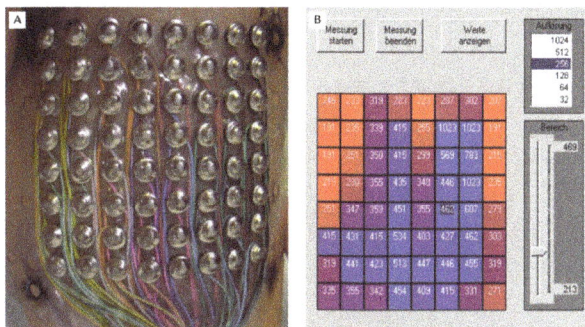

Fig. 12: Array of 64 metal touching electrodes on plastic foil (A) and graphic output of analyzing software with numerical and color-map outputs (B) [6].

A mapping resolution density of our pneumatic electrode probe is 20.24 electrodes per cm^2 in compare with the resolution of only 1.78 electrodes per cm^2 achieved by the realized probe of mentioned authors [6]. See Fig. 12 and Fig. 13.

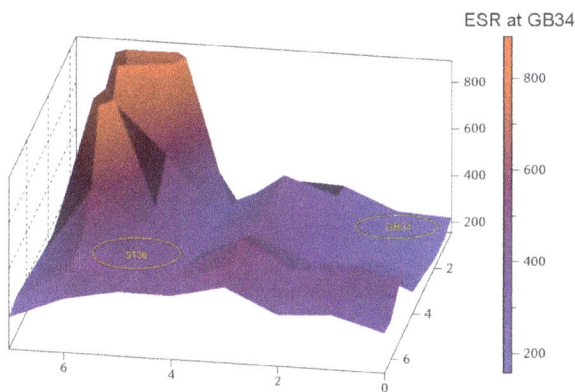

Fig. 13: Graphical 3D representation of the skin conductivity values measured on skin surface on acupuncture points GB34 and ST36 [6].

An ability to distinguish the position and the shape of certain acupuncture point on skin by measuring system of authors [6], (see Fig. 13) is smaller and not so effective as using our system Our effectivity is in the ability of the constructed electrode probe to cover chosen skin area and scan the parameters of skin in sufficient resolution. The driving pressure on measuring probe of authors [6] was guaranteed by mechanical

fixation of plastic foil probe on skin or by the weight pressing on the probe in stable or lying position.

Specific parts of uneven skin area or small sections on certain parts of human body surface were immeasurable. Using a mechanical telescopic matrix probe or the probe with pneumatically driven electrodes from our laboratory team could also be an answer for these complicated areas. See an example of the especially difficult skin mapping the area between fingers realized by our mechanical matrix probe [5] and by our pneumatically driven electrode matrix probe [8], [9] (see Fig. 14 and Fig. 16).

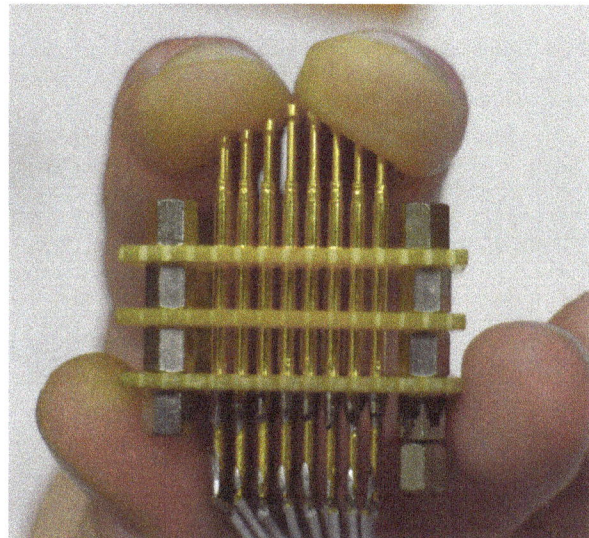

Fig. 14: Example of difficult skin mapping measurement realized by mechanical matrix probe - detail.

A driving pressure influence in mapping measurement realized by mechanical matrix probe cannot be additionally controlled. The impress of the mechanical telescopic electrodes is done by construction. Pressing force per one electrode is maximally 0.6 N, in maximal impress of the electrode. The quality of electric contact of touching electrodes differs by the depth of impress each of electrode. More pressed side electrodes of the probe (Fig. 14) achieve a better electric contact then less pressed central electrodes of the probe, also the level of skin disturbance is varying in the area of the matrix probe contact square. An ACU-point situated in the side area of the probe (measured blue parts on map, see Fig. 14) is then difficult to distinguish from the irritated skin on laterally from the measured square of skin (see Fig. 15).

A red like mountain chain part on map (Fig. 15) is the finger-side skin area between the fingers. A numerical representation of the voltage map on complicated skin area of LI-PC meridians measured by mechanical matrix probe can be seen in Tab. 6.

LI–PC voltage chart, test fr.: 1000, el.no: 64, Ux (V)

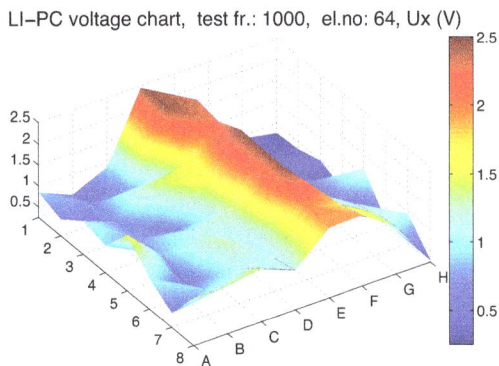

Fig. 15: 3D voltage map measured on area in between LI and PC meridians on fingers by mechanical matrix probe.

Tab. 5: Tabular form of voltage map of skin on area in between LI and PC meridians on fingers measured by mechanical matrix probe.

Ux (V)	1	2	3	4	5	6	7	8
A	0.85	0.65	1.04	0.89	1.57	0.97	0.60	1.39
B	0.52	0.73	0.73	0.76	0.83	0.60	0.88	1.38
C	0.90	0.79	1.36	0.67	1.13	1.32	1.07	1.52
D	2.46	2.46	1.54	1.61	1.30	1.06	1.05	1.19
E	1.62	2.33	2.25	2.50	2.48	2.49	1.96	1.91
F	0.54	0.82	1.15	1.25	1.26	1.28	1.74	1.92
G	0.45	0.46	0.53	0.70	0.45	1.07	1.28	1.38
H	0.29	0.33	0.50	0.25	1.19	0.37	0.97	0.30

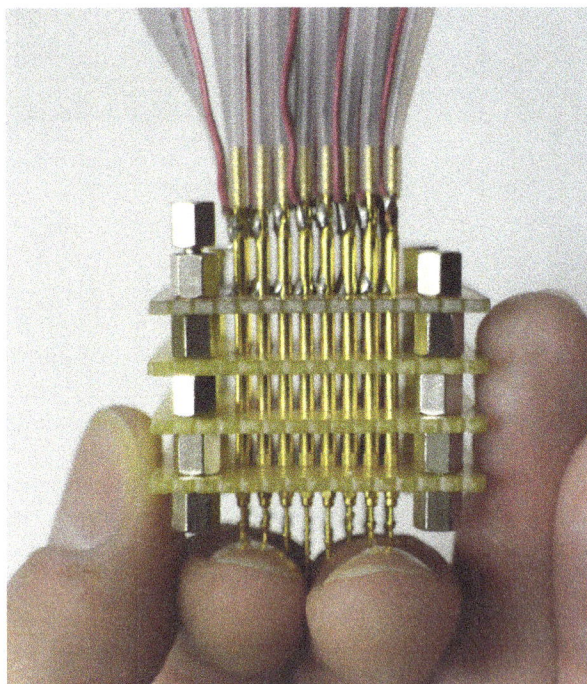

Fig. 16: Example of difficult skin mapping measurement realized by pneumatic matrix probe - detail.

A pressure controlled influence in mapping measurement realized by the pneumatic way causes the balanced impress of the pneumatic telescopic electrodes. A pressing force per one electrode is equal to pressing force in all the electrodes of the probe. The force pressing on the skin surface is regulated by the air pressure in driving pipes. The aim of regulation is to achieve the regular and constant electric contact for all the electrodes regardless the difficult and non planar surface of measured object.

The contact is not influenced by the depth of impress each of electrode. More pressed side electrodes of probe (Fig. 16) achieve the same electric contact as less pressed central electrodes of the probe, also the level of skin disturbance is the same in all the area of the matrix probe contact square

LI–PC voltage chart, test fr.: 1000, el.no: 64, Ux (V)

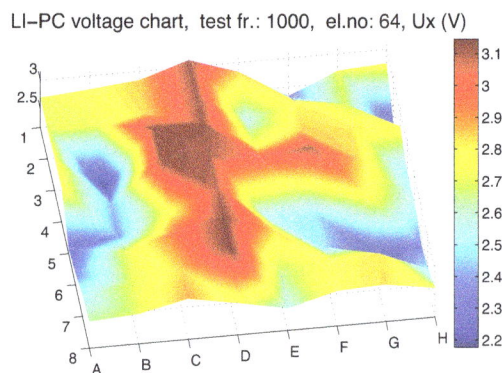

Fig. 17: 3D voltage map measured on area in between LI and PC meridians on fingers by sensitive pneumatic matrix probe.

A central red like mountain chain part on measured map (Fig. 17) is the lateral skin area between the fingers (Fig. 16). An ACU-point situated in side area of the measurement (blue parts of map, Fig. 17) becomes more significant and distinguishable from the rest of skin on side parts of the measured square of skin (see Fig. 15 and compare with Fig. 17).

Tab. 6: Tabular form of voltage map of skin on area in between LI and PC meridians on fingers measured by pneumatic matrix probe.

Ux (V)	1	2	3	4	5	6	7	8
A	0.68	0.61	0.88	0.75	1.00	0.58	0.46	2.36
B	0.61	1.29	1.40	1.03	1.06	1.32	0.44	2.13
C	1.70	1.65	1.60	1.84	1.60	1.04	1.01	2.01
D	2.41	2.28	1.61	2.21	2.07	2.27	2.22	2.29
E	1.88	1.55	1.86	1.93	2.03	1.92	2.00	1.93
F	1.50	0.88	1.39	1.72	1.58	1.74	1.68	0.74
G	1.51	0.71	0.52	1.48	1.52	1.79	1.86	0.65
H	1.25	1.19	1.76	1.71	1.91	1.63	0.92	1.26

A numerical representation of voltage map on complicated skin area of LI-PC meridians, mentioned above, measured by pneumatic matrix probe can be seen in Tab. 6.

5.4.   Conclusion

Achieved mapping results measured using the pneumatic probe construction driven by various pressure range for the piston-electrode movement, impress on the skin surface and the constant electrode-skin electrical contact proved the hopeful and perspective results. Voltage-impedance maps show electrical skin structure, positions and shape of acupuncture points and can serve acupuncture physician as a useful tool for medical diagnostics and therapy. As we have mentioned above, the skin conditions are changing in time and weather conditions, but these changes are not contraindication of successful measurement and an acupuncture point successful recognition. Thereforeour mapping measurement is based on relative and not on absolute values of the voltage-impedance measured results. For active point on skin position recognition is not crucial the final map offset or lower dynamic range if the surrounded position of the measured active point on mapped skin is explicitly visible and recognizable. The acupuncture voltage/impedance mapping of the skin, unfolding the position and the shape of certain active points on the human skin surface, basic gradual research and unfolding of their properties has a tradition in our Research Laboratory of Biomechatronics at the Institute of Automotive Mechatronics and at the Institute of Electronics and Photonics FEI SUT (Faculty of Electrical Engineering and Information Technology of Slovak University of Technology) in Bratislava.

Acknowledgment

The paper has been created as a part of a research and education process at the Institute of Automotive Mechatronics and the Institute of Electronics and Photonics, FEI SUT in Bratislava, Slovak Republic and is supported by the Grant VEGA 1/0987/12, VEGA 1/0823/13, VEGA 1/0937/14, KEGA 011STU-4/2015 and VEGA 2/0138/16.

References

[1] ZHANG, C. L. SKIN RESISTANCE vs. BODY CONDUCTMTY: On the Background of Electronic Measurement on Skin. *Subtle Energies and Energy Medicine.* 2003, vol. 14, no. 2, pp. 151–174. ISSN 1099-6591.

[2] BECKER, R. O., M. REICHMANIS, A. MARINO and J. A. SPADARO. Electrophysiological Correlates of Acupuncture Points and Meridians. *Psychoenergetic Systems.* 1979, vol. 1, iss. 1, pp. 105–112. ISSN 0305-7224.

[3] CERNY, M. and M. PENHAKER. The Home-Care and Circadian rhythm. In: *Information Technology and Applications in Biomedicine.* Shenzhen: IEEE, 2008, pp. 245–248. ISBN 978-1-4244-2255-5. DOI: 10.1109/ITAB.2008.4570546.

[4] AUGUSTYNEK, M., J. PINDOR, M. PENHAKER and D. KORPAS. Detection of ECG Significant Waves for Biventricular Pacing Treatment. In: *Second International Conference on Computer Engineering and Applications.* Bali Island: IEEE, 2008, pp. 164–167. ISBN 978-1-4244-6080-9. DOI: 10.1109/ICCEA.2010.186

[5] KUKUCKA, M. and Z. KRAJCUSKOVA. Automatized multi-electrode voltage map measurement of active points on skin. *Communications: Scientific Letters of the University of Zilina.* 2011, vol. 13, no. 1, pp. 51–55. ISSN 1335-4205

[6] KRAMER, S., D. ZAPS, B. WIEGELE and D. IRNICH. Changes in electrical skin resistance at gallbladder 34 (GB34). *Journal of Acupuncture and Meridian Studies.* 2008, vol. 1, iss. 2, pp. 91–96. ISSN 2005-2901. DOI: 10.1016/S2005-2901(09)60028-5.

[7] WEISZE, A., M. KUKUCKA, Z. KRAJCUSKOVA, D. DURACKOVA and V. STOPJAKOVA. Changes in electrical skin resistance at gallbladder 34 (GB34). In: *25th International Conference Radioelektronika.* Pardubice: IEEE, 2015, pp. 364–368. ISBN 978-1-4799-8117-5. DOI: 10.1109/RADIOELEK.2015.7129033

[8] KUKUCKA, M., S. KOZAK, P. DRAHOS and A. WEISZE. Pneumatic Model of Measuring Probe in Voltage Acupuncture Skin Mapping with Influence of Frequency of Driving Signal. *International Review of Automatic Control.* 2015, vol. 8, no. 4, pp. 307–314. ISSN 1974-6059. DOI: 10.15866/ireaco.v8i4.6971.

[9] KUKUCKA, M., S. KOZAK, A. WEISZE, D. DURACKOVA, V. STOPJKAOVA, Z. KRAJCUSKOVA and M. TEPLAN. Mechanical vs. Pneumatic Model of Measuring Probe in Voltage Acupuncture Skin Mapping. *International Review of Automatic Control.* 2015, vol. 8, no. 6, pp. 434–441. ISSN 1974-6059. DOI: 10.15866/ireaco.v8i6.7455.

[10] LEONHARDT, H. *Grundlagen der Elektroakupunktur nach Voll - Ein Leitfaden zur Einfuehrung in die Elektroakupunktur nach Voll (EAV).* 2nd ed. Uelzen: Medizinisch-literarische Verlagsgesellschaft, 1986. ISBN 978-3-881-36113-2.

About Authors

Marek KUKUCKA was born in Bratislava, Slovakia, November 8th 1975. In 2000 he finished masters' study at Faculty of Electrical Engineering and Information Technology at Slovak University of Technology and became M.Sc. He worked as researcher at Department of Radio and Electronics. In 2005 defended dissertation thesis "Diagnostic processing of medical signals" and obtained Ph.D. in electronics. He has been working as assistant lecturer at Slovak University of Technology, Faculty of Electrical Engineering and Information Technology, Institute of Automotive Mechatronics. Before that he worked as researcher. He is interested in biomechatronics, medical signal processing, measurement, biomedical sensors, diagnostic methods and systems and research in bio-medical mechatronics. Dr. Kukucka is scientific secretary of Society for Biomedical Engineering and Medical Informatics of Slovak Medical Association. He is member of International Federation for Medical and Biological Engineering. Dr. Kukucka was also solutionist of many scientific projects and coorganiser of various scientific conferences.

Andreas WEISZE has been working at Dr. Ing. h.c. F. Porsche AG, Stuttgart, in Germany and is an external postgraduate student at Institute of Electronics and Photonics, Faculty of Electrical Engineering and Information Technology, Slovak University of Technology in Bratislava.

Daniela DURACKOVA has been working as professor specialized in design of integrated circuits at Institute of Electronics and Photonics, Faculty of Electrical Engineering and Information Technology, Slovak University of Technology in Bratislava.

Zuzana KRAJCUSKOVA has been working as a university teacher at the Institute of Electronics and Photonics, Faculty of Electrical Engineering and Information Technology, Slovak University of Technology in Bratislava. Dr. Krajcuskova is member of Society for Biomedical Engineering and Medical Informatics of Slovak Medical Association, member of Society for Medical Informatics of Slovak Medical Association and member of International Federation for Medical and Biological Engineering.

Viera STOPJAKOVA has been working as professor specialized in design of integrated circuits at Institute of Electronics and Photonics and as a vice dean of Faculty of Electrical Engineering and Information Technology, Slovak University of Technology in Bratislava.

7

A Novel Method for Detection and Classification of Covered Conductor Faults

Stanislav MISAK[1], Michal KRATKY[2], Lukas PROKOP[1]

[1]Centre ENET, VSB–Technical University of Ostrava, 17. listopadu 15, 708 33 Ostrava, Czech Republic
[2]Department of Computer Science, Faculty of Electrical Engineering and Computer Science,
VSB–Technical University of Ostrava, 17. listopadu 15, 708 33 Ostrava, Czech Republic

stanislav.misak@vsb.cz, michal.kratky@vsb.cz, lukas.prokop@vsb.cz

Abstract. *Medium-Voltage (MV) overhead lines with Covered Conductors (CCs) are increasingly being used around the world primarily in forested or dissected terrain areas or in urban areas where it is not possible to utilize MV cable lines. The CC is specific in high operational reliability provided by the conductor core insulation compared to Aluminium-Conductor Steel-Reinforced (ACSR) overhead lines. The only disadvantage of the CC is rather the problematic detection of faults compared to the ACSR. In this work, we consider the following faults: the contact of a tree branch with a CC and the fall of a conductor on the ground. The standard protection relays are unable to detect the faults and so the faults pose a risk for individuals in the vicinity of the conductor as well as it compromises the overall safety and reliability of the MV distribution system. In this article, we continue with our previous work aimed at the method enabling detection of the faults and we introduce a method enabling a classification of the fault type. Such a classification is especially important for an operator of an MV distribution system to plan the optimal maintenance or repair the faulty conductors since the fall of a tree branch can be solved later whereas the breakdown of a conductor means an immediate action of the operator.*

Keywords

Covered conductor faults, fault type classification, medium-voltage, overhead lines with covered conductors, Partial Discharges, PD-pattern.

1. Introduction

The Covered Conductor (CC) have been primarily utilized for their high operation reliability guaranteed by the insulation compared to the ACSR. Consequently, the CC have been often used in forested or dissected terrain areas, in general, in all areas with extreme climatic conditions [1], [2] and [3]. Under these conditions, we can identify a high number of faults of the ACSR: the fall of a tree branch on the conductors, a phase-to-phase contact due to strong winds resulting in the phase-to-phase fault or, even, the breakdown of a conductor.

Therefore, the CC have been used for MV distribution lines for several decades. The major advantages of the conductor include:

- A high operational reliability since no short-circuit appears at the moment of the phase-to-phase contact.

- The individual phases can be located closer to one another and a demand on the protective zone of the overhead lines is lower compared to the ACSR.

- No immediate phase-to-phase contact appears when a tree branch falls on the conductors, and finally.

- The lines are not dangerous for birds sitting on the conductors.

The only disadvantage of using the CC is a difficult detection of faults when the conductor falls down on the ground. In this case, there is no single-phase-to-earth fault due to the insulation of the CC (let us note that it does not matter whether the conductor is broken). The standard protection relays work on the principle of an evaluation of the current or the voltage at the point of the single-phase-to-earth fault. When the CC fault is not detected, the live CC lying on the ground can endanger individuals being close to the conductor or it can threat the reliability and safety of the distribution system. At the contact point, low-power capacitive current flows are generated by Partial Discharges (PDs). These PDs gradually degrade the CC insulation over time until the insulation system is locally destroyed. At this moment, the PDs are transformed into an arcing phase-to-ground fault and this fault is only detected by standard protection relays.

In this article, we continue with our previous work aimed at an electric voltage-based method enabling a detection of the CC faults [4] and [5] and we introduce a method enabling a classification of the fault type. Such a classification is especially important for an operator of an MV distribution system to plan the optimal maintenance or repair the faulty conductors since the contact of a tree branch can be solved later whereas the breakdown of a conductor means an immediate action of the operator.

In Section 2. , we categorize types of CC faults, in Section 3. , related works are described. In Section 4. , we introduce new indicators, and, in Section 5. , we put forward experiments in a climatic cabinet showing an ability of the indicators to classify the CC fault type. In Section 6. , we put forward an algorithm for the classification of the fault type. In Section 7. , the indicators are verified on a real 22 kV overhead line. In the last section, we conclude this article and we outline possibilities of our future work.

2. Problem Formulation - CC Fault Types

Following the previous section, the main problem of CC operating is a difficult detection of faults. The CC fault types can be divided into two basic categories differing from one another in terms of the time in which they are required to be eliminated.

Category I: It includes the most frequent faults where the climatic conditions in forested areas cause the contact of a tree branch with one or more phases of a CC overhead line or the fall of a tree on two or three phases of a CC overhead line. Category I faults represent the less dangerous incidents since a low degradation of the insulation system; these faults can be eliminated later [1] and [6].

Category II: It includes more dangerous faults which should be eliminated as soon as possible. In this case, a CC is exposed with an atmospheric overvoltage or an extensive degradation of the CC insulation leading to its breakdown and fall on the ground.

Both of the fault categories are characterized by PDs at the point where the CC insulation touches the ground, a tree branch, and so on. At this point, a non-homogeneous electric stray field appears and various types of PDs arise (inner, outer or surface [7], [8] and [9]); we talk about a PDs activity. It is important that these PDs types and corona discharges can be measured at the point of the fault [1], [6] and [10].

3. Research Background

The most widely discussed detection of the CC faults by PDs activity monitoring is a method using an analysis of the current signal flowing through the insulation system of a CC by means of a Rogowski coil [3], [11], [12], [13] and [14].

The electric current signal has two components, the carrier component with the frequency of the power supply (50 or 60 Hz) and the impulse component characterised by a PDs activity with a typical frequency range (e.g. 1–20 MHz). All the methods evaluate the impulse component. The main advantages of these methods are the good selectivity and sensitivity for CC faults of Category II. However, the good selectivity and sensitivity of these methods are required by the relative high cost of measuring apparatus. The most important parts of this measuring apparatus are the Rogowski coil and an A/D converter to impose the most demanding requirements in terms of (i) the accuracy of measuring the amplitude and phase in a wide frequency range, (ii) the resistance against external interferences, and (iii) the stability of a measurement for various climatic conditions. Evaluation of the actual state of the insulation system in the case of a fault of Category I is very difficult because of a small value of the impulse component [1], [3], [11] and [12]. Moreover, the small value of the impulse component is influenced by the load of CC overhead lines and also it has direct influence on the sensitivity of the impulse component evaluation.

In [4] and [5], we introduce a methodology for an evaluation of the PDs activity in the CC insulation system based on the electric voltage principle; we call it the CC fault detector. The main principles of the methodology are as follows: the electric stray field is measured using a sensor in the vicinity of the CC as a voltage signal (see Fig. 1). The sensor can be a single layer coil or a metal ring wrapped on the CC. The single layer coil with a turn-to-turn capacity forms one electrode of a compound dielectric medium, the second electrode is formed by the ground potential. The measured voltage signal is modified to the low level signal by a capacitive divider with a fixed ratio (CA in Fig. 1) whose primary side is connected with the terminal of the sensor. The voltage signal of the electric stray field clearly reveals the dominance of the carrier component

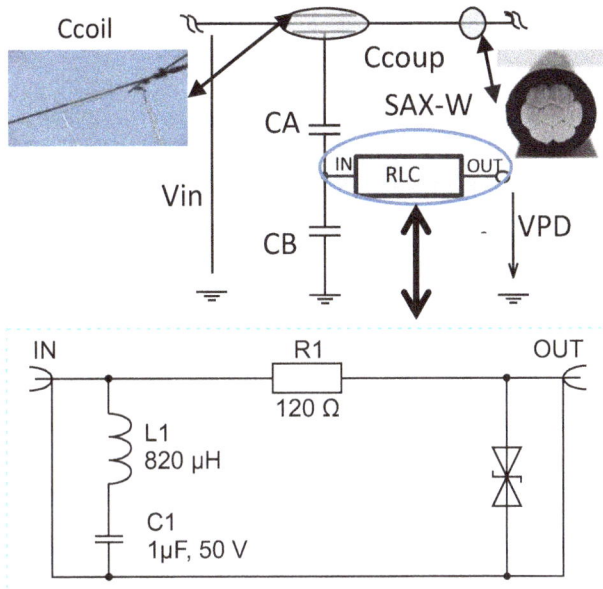

Fig. 1: Principal measurement diagram of the electric voltage-based method.

with the frequency of the power-supply unit, i.e. 50 Hz (the red line, the left axis in Fig. 2).

To evaluate the PDs activity, the impulse component is necessary (VPD in Fig. 1). The carrier component of the voltage signal has been therefore eliminated using the RC derivation block (the RLC block in Fig. 1). In this way, a time pattern of the impulse component is obtained. An example of the time pattern can be seen in Fig. 2 (the green line). We call the time pattern the PD-pattern in our article. An evaluation of a PD-pattern can provide an information about the actual state of the CC insulation system [4], [5], [15] and [16]. In [4] and [5] we also show that TRMS (True Root Mean Square) can be used for the evaluation of a PD-pattern.

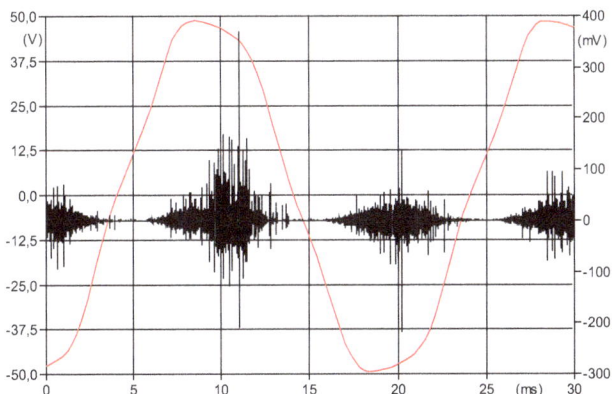

Fig. 2: Voltage signal of the electric stray field (the carrier component - the red line, the left axis; the impulse component - the green line, the right axis).

4. Indicator of CC Faults

The PD-pattern is the time pattern of the PDs activity which is specific for each CC fault type and climatic conditions at the point of a fault. Since a PD-pattern can provide an information about the actual state of the CC insulation system, in this paper, we utilize it for the classification.

The process of developing a fault indicator is divided into the following steps: 1) An analysis of PD-patterns for various types of CC faults and climatic conditions using the climatic cabinet in a framework of long-term testing (see Section 5.). 2) The development and implementation of the fault indicator and its integration in the CC fault detector (see Section 6.). 3) The verification of the fault indicator integrated in the CC fault detector in a real MV distribution system with the CC (see Section 7.).

We define three basic requirements on a fault indicator: (1) the low computational complexity, (2) the sensitivity for various CC fault types, (3) the capability to classify the CC fault type during a long-term time period. Let us describe these requirements in more details. (1): there is a need to find an indicator or indicators requiring the low computational performance in order not to increase the overall consumption of the fault detector. Moreover, the detector can be installed in inaccessible terrain and it can be supplied from, for example, a photovoltaic power station combined with a storage device. (2): the fault type has to be recognized during a short-term as well as 3): a long-term time period when a fault arises.

5. PD Pattern Analysis

As mentioned in the previous section, one aspect of the development of a fault indicator is an analysis of PD-patterns for various fault types and climatic conditions during a long-term measurement. For this purpose, an MV climatic cabinet was built, and three CC fault types of Categories I and II (see Tab. 1) in various climatic conditions (see Tab. 2) was simulated. During the tests, three different climatic conditions using the temperature and relative humidity were set.

The CC insulation system is naturally exposed with many other climatic influences [1] and [12], such as the atmospheric pressure or ultraviolet radiation, however these influences have not been simulated since it is difficult to keep the same conditions during the whole experiment. Similarly, it is not possible to simulate another fault type such as the fall of a tree on a CC.

Another factor important for a description of a PD-pattern is the time interval of a fault under various cli-

Tab. 1: CC faults simulated in the MV climatic cabinet.

Category	CC fault type	
I	The contact of a tree branch with the CC	
II	The fall of a CC on the ground	The soil
		The water

Tab. 2: Climatic conditions in the MV climatic cabinet.

Tempe-rature	Relative humidity	Abbr.	Type of day (The temperate zone)
(°C)	%		
−20	5	−20/5	Winter day
20	70	20/70	Spring or autumn day
60	15	60/15	Summer day(extremely sunny day)

matic conditions. Considering Indicator Requirement 3 (see Section 4.), the climatic cabinet enables us to test faults in a long period with a step size. For example, it enables us to test a fault for several days with the step size of 5 minutes. The following section puts forward the results of the tests for the selected fault types and climatic conditions. Since the PDs activity comes out as peeks in the impulse component of the voltage signal, we select the frequency of the PDs activity n (s^{-1}), i.e. the number of peaks per second, as a new indicator.

5.1. Category I - Contact of the Tree Branch with CC

The contact of a tree branch with a CC is one of the most frequent fault types. In Fig. 3, a PD-pattern of this fault is shown for the 20/70 climatic condition. In this figure, the PD-pattern is visualized as two lines; the red line represents the positive semi-period, and the green line represents the negative semi-period.

This fault type has two characteristic features. The first one is that peaks arise in both semi-periods of the PD-pattern (n is approximately 50 s^{-1}) and the

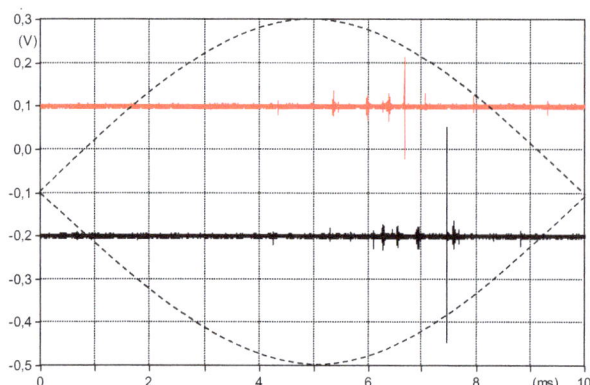

Fig. 3: PD-pattern - The contact of a tree branch with a CC, the climatic condition 20/70.

average amplitudes of the peaks are approximately the same in both semi-periods (70 mV). The second feature is the fact that the peaks occur when the supply voltage is going through the maximum values to 0.

In Fig. 4, the values of the indicator n during 75 hours after the fault arises can be seen. A long time interval of the contact of a tree branch with the CC means increased degradation of the insulation with the PDs activity. As a result, a diffusion of carbon particles into the contact point appears and the PDs activity is changed according to a climatic condition.

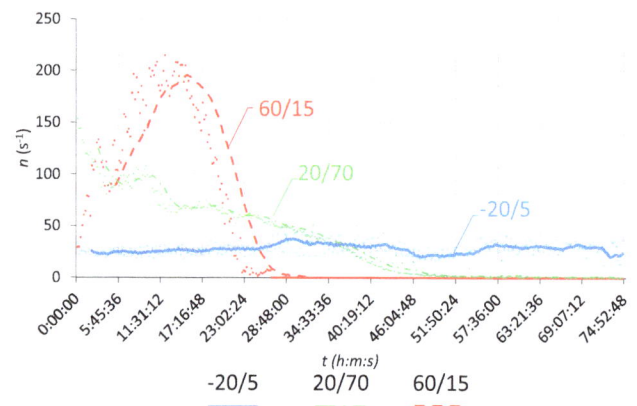

Fig. 4: Indicator n for the contact of a tree branch with CC.

In the case of the 20/70 climatic condition (see the green line in Fig. 4), a relatively intensive PDs activity ($n = 150$ s^{-1}) is caused by the high relative humidity. After some time (5 hours), a tree branch becomes dry at the contact point. As a result, the PDs activity is gradually reduced to $n = 80$ s^{-1}. Since the branch can have more contact points with the CC, the above mentioned process can appear again; however, the PDs activity is lower since electrical conductivity of the branch is lower.

When a time interval of the fault exceeds 46 hours, the branch becomes completely dry and the PDs activity is decreased to a negligible level.

In the second test, the 60/15 climatic condition (see the red line in Fig. 4), the PDs activity is low ($n = 25$ s^{-1}) when the fault appears due to the low relative humidity. Due to the high temperature, the insulation becomes soft and the number of contact points is increased. As a result, the PDs activity is gradually increased and it reaches the maximum $n = 200$ s^{-1} after 17 hours. After this point, the PDs activity rapidly decreases as the branch becomes dry; the PDs activity reaches a negligible level after 29 hours. In the third test, the −20/5 climatic condition (see the blue line in Fig. 4), the PDs activity is low ($n = 25$ s^{-1}) during the whole test since the low relative humidity.

Let us summarize the results; $n \in [25; 200]$ s^{-1} during 29 hours after the fault appears for all defined climatic conditions and the frequency of peaks is approximately the same in both semi-periods.

5.2. Category II - Fall of CC on Ground - Soil

Although the fall of a CC on the ground does not belong to the most frequent faults, it is extremely dangerous. In this case, a CC falls on a solid surface, e.g. the soil or the clay, PDs occur in cavities of the conductor insulation. In Fig. 5, a PD-pattern of this fault is shown for the 20/70 climatic condition.

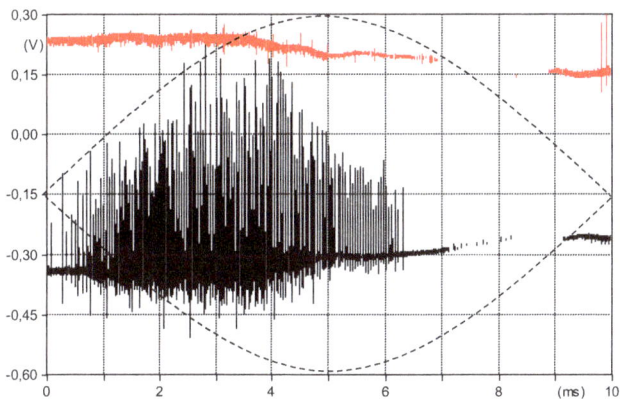

Fig. 5: PD-pattern - The fall of a CC on the ground - the soil, the climatic condition 20/70.

In this case, we distinguish three basic features. The first one, peaks mainly appear before the amplitudes in both semi-periods. The second one, n is up-to $13\,000$ s^{-1} compared to maximally $n = 200$ s^{-1} proposed for the previous fault type. The third one, there are less peaks with the low amplitude in the positive semi-period ($n = 150$ s^{-1}, the amplitudes is approx. 10 mV), on the other hand, there are more peaks with the high amplitude in the negative semi-period (n is up-to 10–100 higher, the amplitude is approx. 500 mV).

In the first test, the climatic condition 60/15 (see the red line in Fig. 6), a solid surface becomes dry and particles of isinglass create a glaze on the solid surface. It means a reduction of the number of contact points, and, consequently, the PDs activity is reduced ($n = 8\,000$ s^{-1} when the fault appears); the PDs activity is decreased bellow a measurable level after 50 hours.

In the second test, the climatic condition $-20/5$ (see the blue line in Fig. 6), we can observe a similar situation as in the previous section in the case of the low relative humidity; the PDs activity is relatively low (compared to results of other climatic conditions) during the whole test because of low relative humidity. However, now $n = 1.500$ s^{-1} compared to $n = 25$ s^{-1} of the previous fault type.

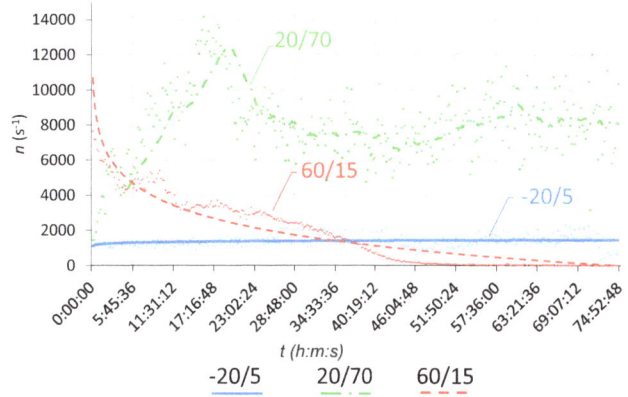

Fig. 6: Indicator n for the fall of a CC on the ground - the soil.

In the third test, the climatic condition 20/70 (see the green line in Fig. 6), the PDs activity gradually increases as the CC insulation degrades until a maximum is reached (approximately $n = 13\,000$ s^{-1} after approx. 16 hours). After the maximum, the insulation becomes dry and particles of isinglass create a glaze on the solid surface and the PDs activity begins to decrease.

However, this process repeatedly appears since the electrical stability is exceeded in other parts of the insulation and we can observer other local maxima, e.g. after 60 hours in this case, and n therefore oscillates around $8\,000$ s^{-1}.

5.3. Category II - Fall of CC on Ground - Water

A PD-pattern of the fault type and the climatic condition 20/70 is shown in Fig. 7; the peaks appear close to the amplitudes of both semi-periods. We also see that the amplitudes of peaks are approx. 2 times higher in the positive semi-period, however, n is approx. 2 times higher in the negative semi-period.

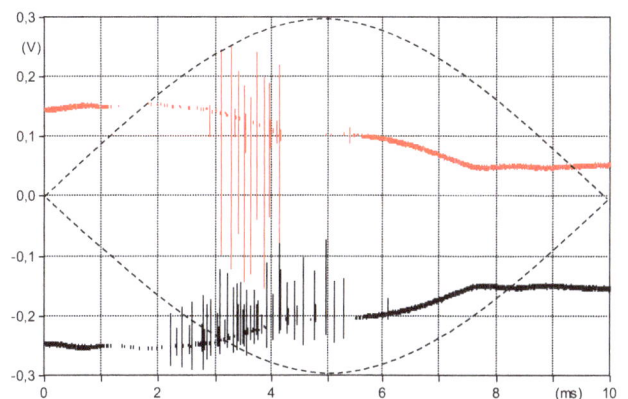

Fig. 7: PD-pattern - The fall of a CC on the ground - the water, the climatic condition 20/70.

In the first test, the climatic condition 20/70 (see the green line in Fig. 8), the long-term period of the measurement has no significant influence on the PDs activity; n oscillates around 300 s^{-1}.

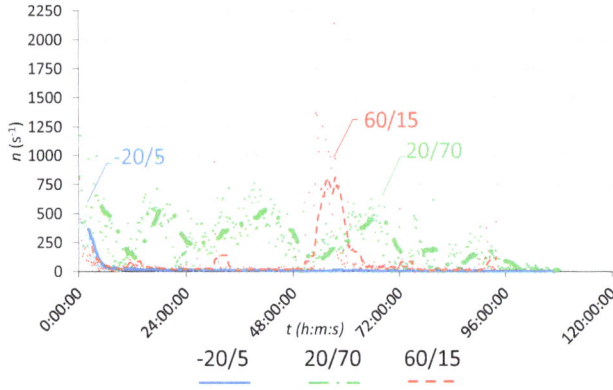

Fig. 8: Indicator n for the fall of a CC on the ground - the water.

In the second test, the climatic condition $-20/5$ (see the blue line in Fig. 8), the PDs activity is initially the same as in case of the first test ($n = 500$ s^{-1}). However, after a small time period (10 minutes), the PDs activity is minimal since the relative humidity is low. As a result, the insulation becomes dry and the number of contact points is reduced since the water surface is frozen.

In the third test, the climatic condition 60/15 (see the red line in Fig. 8), the PDs activity is initially low ($n = 50$ s^{-1}) since the number of contact points of a CC with the water surface is high (it is not necessary to reach so high energy to excite the PDs activity). After a number of hours, a condensation of water steam arises and the relative humidity is shortly increased. As a result, the PDs activity is also increased in this moment; $n = 1\,300$ s^{-1}, after 60 hours.

6. Fault Indicator Development and Implementation

In the previous section, we can see that the frequency of peaks, i.e. the indicator n, characterizes well different fault types; however, there are some cases where it is necessary to compute the frequency of peaks in both semi-periods. Therefore, we define n_+ (s^{-1}) for the frequency of peaks in the positive semi-period and n_- (s^{-1}) for the frequency of peaks in the negative semi-period. We define the following indicator:

$$k = \frac{n_+}{n_-}(-). \qquad (1)$$

Let us note that both indicators, n and k, satisfies three basic requirements proposed in Section 6.

In Tab. 3, the values of the indicators n and k for three fault types considered in a long-term period (after the 1st day, 3rd day, and 5th day) can be seen. We have to distinguish the fall of a CC on the ground (Category II) from other fault types (Category I) since it is the most dangerous fault and it has to be immediately repaired.

In Tab. 3, we can see that k is two orders of magnitude lower and n is one order of magnitude higher for the fall of a CC on the ground (the soil) compared to other fault types. As a result, if $k < 0.1$ it means this fault type occurs.

When we compare n for the contact of a tree branch with a CC and the fall of a CC on the ground, the water, n is often higher in the former case. Evidently, we need to utilize k to precisely distinguish both fault types; $k > 0.95$ for the contact and $k < 0.95$ for the fall. After the analysis, we can create a simple algorithm classifying the fault type (see Diagram on Fig. 9). To identify any fault type this algorithm utilizes a threshold, e.g. $n = 10$ s^{-1}.

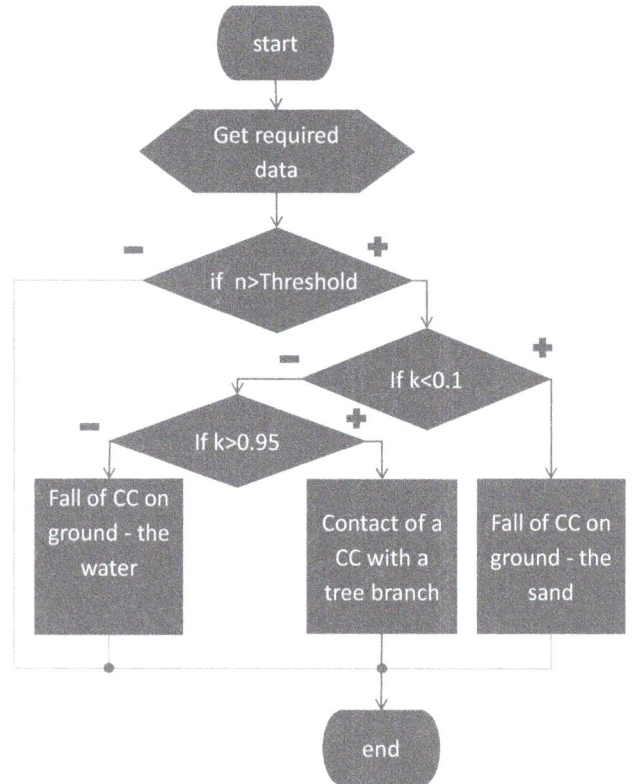

Fig. 9: Algorithm for a classification of the CC fault type.

In Tab. 3, we see that TRMS introduced in our previous work [4] and [5] is not appropriate for such a fault type classification. For example, we get the range $[1.1, 8.9]$ for the fall of a CC on the ground (the

Tab. 3: Evaluation on indicators n and k for three fault types.

Tempe-rature (°C)	Time period	The contact of a tree branch with a CC			The fall of a CC on the ground - the soil			The fall of a CC on the ground - the water		
		k (-)	n (s^{-1})	TRMS (V)	k (-)	n (s^{-1})	TRMS (V)	k (-)	n (s^{-1})	TRMS (V)
−20	1st day	1.23	52	0.10	0.03	1 670	1.46	0.75	158	3.57
	3rd day	1.00	54	0.15	0.04	2 009	1.53	0.67	28	0.79
	5th day	1.11	48	0.19	0.05	2 192	1.85	0.68	52	0.66
20	1st day	1.46	152	0.10	0.02	1 361	1.97	0.41	826	2.10
	3rd day	1.11	31	0.17	0.02	8 032	7.34	0.86	661	1.40
	5th day	1.00	18	0.12	0.04	8 992	8.48	0.83	17	0.41
60	1st day	1.00	15	0.16	0.07	4 664	8.85	0.81	77	4.62
	3rd day	1.11	13	0.08	0.08	329	2.10	0.52	88	1.17
	5th day	1.23	12	0.02	0.09	57	1.14	0.40	141	1.09

soil) and the range $[0.4, 4.7]$ for the fall of a CC on the ground (the water). These ranges are intersected and it means that we cannot distinguish these fault types using TRMS. Moreover, this value depends on the distance to the location of a fault.

7. Verification of Fault Detector

To verify the indicators n and k, we have installed the fault detector on a real 22 kV overhead line with the CC of the length 15 km. The overhead line is located in a forested area with a high number of faults; it is therefore appropriate for testing of the fault detector.

This detector was installed in September 2014 (see Fig. 10). PD-patterns together with the meteorological data have been measured one time per hour and they have been stored in a local disk of the fault detector. To analyse the data, we periodically download them and run a computation of the indicators n and k. To detect a fault, we set up the threshold of n to 10 s^{-1}. After the threshold is applied, we can classify the fault type using k.

Fig. 11: PD-pattern of the fault.

Tab. 4), we distinguish the fault as the contact of a tree branch with a CC. The contact of a tree branch and the extent of damage are shown in Fig. 12. We measure the PDs activity during 48 hours after the fault appeared. This fault is repaired after 48 hours by an operator of the overhead line; we see that n is below the threshold. As a result, we verify the functionality of the fault detector in this way.

Tab. 4: Analysis of the PD-Pattern after the fault appears.

	2 hours	24 hours	48 hours
k	1.085	1.213	1.008
n	75	56	6

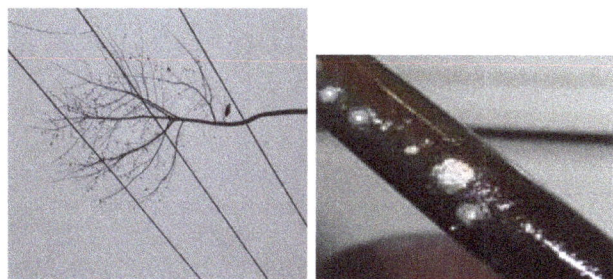

Fig. 10: Fault detector and a detail of a sensor (a single layer coil with 75 turns).

Using the algorithm in Diagram in Fig. 9, we detect a fault after 2 months after the installation; the PD-pattern of the fault is shown in Fig. 11. The fault has been located 2.5 km from the detector. Since we measure and compute $n = 75 \text{ s}^{-1}$ and $k = 1.085$ (see

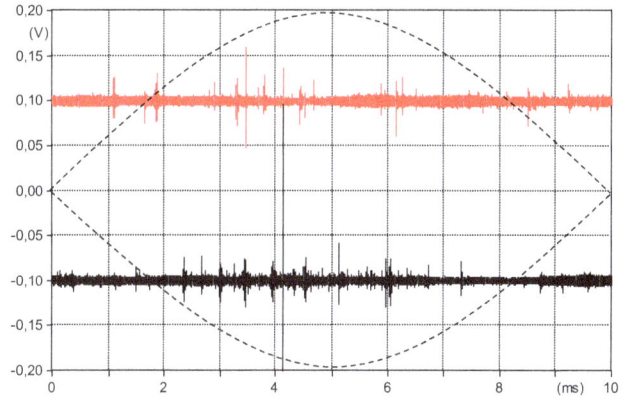

Fig. 12: Detected fault - the contact of a tree branch with a CC.

8.   Conclusion

The main goal of this paper is to introduce new indicators for a classification of the CC fault type. We consider two categories of the fault type: Category I (the contact of a tree branch with a CC) and Category II (the fall of a CC on the ground). In our previous work, we utilize TRMS for a detection of a fault using an electric voltage-based method. In this paper, we show that it is not possible to use TRMS for the classification of the fault type, therefore, in this article we introduce two novel indicators n and k and we show that it is possible to use the indicators for the classification of the fault type using a simple algorithm. This algorithm has been designed after an analysis of experimental results in the climatic cabinet. It is important that the indicators fulfil three requirements defined in this paper. Compared to the other existing methods, electric current-based methods, the cost of our detector is lower and it enables the detection of a fault as well as the classification of the fault type. The fault detector utilizing the novel indicators has been installed on a real 22 kV overhead line with the CC and we have verified the fault detector in this way. In the near future, we will focus our research on development of new tools and methods for detecting long-distance faults and also we focus on improving sensitivity of developed method.

Acknowledgment

This paper is supported by the following projects: LO1404: Sustainable development of ENET Centre; SP2016/177, SP2016/128 (Students Grant Competition); LE13011: Creation of PROGRES 3 Consortium Office to Support Cross-Border Cooperation (CZ.1.07/2.3.00/20.0075) and the project of TACR TH01020426, Czech Republic.

References

[1] PAKONEN, P. *Detection of Incipient Tree Faults on High Voltage Covered Conductor Lines*. Tampere, 2007. Dissertation thesis. Tampere University of Technology.

[2] ZHANG, W., Z. HOU, H.-J. LI, C. LIU and N. MA. An improved technique for online PD detection on covered conductor lines. *IEEE Transactions on Power Delivery*. 2014, vol. 29, iss. 2, pp. 972–973. ISSN 0885-8977. DOI: 10.1109/TPWRD.2013.2288008.

[3] HASHMI, G., M. LEHTONEN and M. NORDMAN. Calibration of on-line partial discharge measuring system using rogowski coil in covered-conductor overhead distribution networks. *IET Science, Measurement & Technology*. 2011, vol. 5, iss. 1, pp. 5–13. ISSN 1751-8822. DOI: 10.1049/iet-smt.2009.0124.

[4] MISAK, S. and V. POKORNY. Testing of a covered conductor's fault detectors. *IEEE Transactions on Power Delivery*. 2015, vol. 30, iss. 3, pp. 1096–1103. ISSN 0885-8977. DOI: 10.1109/TPWRD.2014.2357072.

[5] HAMACEK, S. and S. MISAK. Detector of covered conductor fault. *Advances in Electrical and Electronic Engineering*. 2012, vol. 10, no. 1, pp. 7–12. ISSN 1804-3119. DOI: 10.15598/aeee.v10i1.567.

[6] AGARWAL, H., K. MUKHERJEE and P. BARNA. Partially and fully insulated conductor systems for low and medium voltage overhead distribution lines. In: *IEEE 1st International Conference on Condition Assessment Techniques in Electrical Systems (CATCON)*. Kolkata: IEEE, 2013, pp. 100–104. ISBN 978-1-4799-0082-4. DOI: 10.1109/CATCON.2013.6737537.

[7] BARTNIKAS, R. Partial discharges. Their mechanism, detection and measurement. *IEEE Transactions on Dielectrics and Electrical Insulation*. 2002, vol. 9, iss. 5, pp. 763–808. ISSN 1070-9878. DOI: 10.1109/TDEI.2002.1038663.

[8] DABBAK, S., H. ILLIAS, B. CHIN and A. TUNIO. Surface discharge characteristics on HDPE, LDPE and PP. *Applied Mechanics and Materials*. 2015, vol. 785, no. 5, pp. 383–387. ISSN 1662-7482. DOI: 10.4028/www.scientific.net/AMM.785.383.

[9] FORSSEN, C. and H. EDIN. Partial discharges in a cavity at variable applied frequency part 1: measurements. *IEEE Transactions on Dielectrics and Electrical Insulation*. 2008, vol. 15, iss. 6, pp. 1601–1609. ISSN 1070-9878. DOI: 10.1109/TDEI.2008.4712663.

[10] ACHILIDES, Z., E. KYRIAKIDES and G. GEORGHIOU. Partial discharge modelling: an improved capacitive model and associated transients along medium voltage distribution cables. *IEEE Transactions on Dielectrics and Electrical Insulation*. 2013, vol. 20, no. 3, pp. 770–781. ISSN 1070-9878. DOI: 10.1109/TDEI.2013.6518947.

[11] HASHMI, G. and M. LEHTONEN. Effects of rogowski coil and covered-conductor parameters on the performance of PD measurements in overhead distribution networks. *International Journal of Innovations in Energy Systems and Power*. 2010, vol. 4, no. 2, pp. 14–20. ISSN 0142-0615.

[12] HASHMI, G., M. LEHTONEN and M. NORD-MAN. Modelling and experimental verification of on-line PD detection in MV covered-conductor overhead networks. *IEEE Transactions on Dielectrics and Electrical Insulation* 2010, vol. 17, iss. 1, pp. 167–180. ISSN 1070-9878. DOI: 10.1109/TDEI.2010.5412015.

[13] HEMMATI, E. and S. SHAHRTASH. Evaluation of unshielded rogowski coil for measuring partial discharge signals. In: *11th International Conference on Environment and Electrical Engineering (EEEIC)*. Venice: IEEE, 2012, pp. 434–439. ISBN 978-1-4577-1830-4. DOI: 10.1109/EEEIC.2012.6221417.

[14] SHAFIG, M., G. A. HUSSAIN, L. KUTT and M. LEHTONEN. Effect of geometrical parameters on high frequency performance of rogowski coil for partial discharge measurements. *Measurement*. 2013, vol. 49, iss. 1, pp. 126-137. ISSN 0263-2241. DOI: 10.1016/j.measurement.2013.11.048.

[15] CAVALLINI, A. and G. MONTANARI. Effect of supply voltage frequency on testing of insulation system. *IEEE Transactions on Dielectrics and Electrical Insulation*. 2006, vol. 13, iss. 1, pp. 111–121. ISSN 1070-9878. DOI: 10.1109/TDEI.2006.247849.

[16] STONE, G. C. Partial discharge diagnostics and electrical equipment insulation condition assessment. *IEEE Transactions on Dielectrics and Electrical Insulation*. 2005, vol. 12, iss. 5, pp. 891–904. ISSN 1070-9878. DOI: 10.1109/TDEI.2005.1522184.

About Authors

Stanislav MISAK was born in Slavicin in 1978. He received his Ing. and Ph.D. degrees in Electrical power engineering in 2003 and 2007, respectively, from the Department of Electrical Power Engineering, VSB–Technical University of Ostrava, Czech Republic, where he is currently an Associate Professor. Dr. Misak has published a number of articles in peer-reviewed journals and conference proceedings. He also holds a patent for a fault detector for medium-voltage lines. In 2012 he was appointed a delegate to the European Union Commission for the strategic management of renewable energy sources. He has been successful in obtaining a number of research contracts and grants from industry and government agencies for projects related to the areas of power systems and renewable energy integration. His current work includes the implementation of smart grid technologies using prediction models and bio-inspired methods and diagnostic of insulation systems.

Michal KRATKY was born in Olomouc in 1978. He is an associate professor at Department of Computer Science (http://www.cs.vsb.cz/), VSB–Technical University of Ostrava (http://www.vsb.cz/), from 2007. He received the Ph.D. degree in Computer Science in 2004. He has published more than 80 papers in the field of database systems (7 articles in journals with impact factor, e.g. VLDB Journal, Information Systems). He has been a head or a team member of 10 research projects (funded, e.g., by The Grant Agency of the Czech Republic, The Technology Agency of the Czech Republic). He has served as a member of 40 program and review committees of international conferences (e.g. DEXA 2007–2016, IDEAS 2015) and he has reviewed several journal and conference papers (Information Sciences, PODS 2015, DKE, IEEE Transactions on Information Theory and so on). There are 116 citations of his articles on SCOPUS, 390 on Google Scholar. He is the head of the Database Research Group (http://db.cs.vsb.cz/). He is a supervisor of 5 Ph.D. students, 2 students successfully defended their Ph.D. theses.

Lukas PROKOP was born in 1978 in Karvina. He received his Ing. degree in Electrical power engineering in 2002 from Brno University of Technology, Czech Republic, and his Ph.D. in Electrical power engineering in 2006 from VSB–Technical University of Ostrava, Czech Republic. In 2005 he joined the Department of Electrical and Computer Engineering at the VSB–Technical University of Ostrava, Czech Republic, where he is currently a research scientist. He is involved in a number of research contracts and grants from industry and government agencies, in the areas of power systems and renewable energy integration. His current work is focused on smart grid technologies, renewable energy sources, off-grid systems, power system reliability and power quality.

8

Comparison of VSC and Z-Source Converter: Power System Application Approach

Masoud Jokar KOUHANJANI, Ali Reza SEIFI

Department of Power and Control Engineering, School of Electrical and Computer Engineering,
Shiraz University, Zand Street, Shiraz, Iran

masoudjokar@hotmail.com, seifi@shirazu.ac.ir

Abstract. *Application of equipment with power electronic converter interface such as distributed generation, FACTS and HVDC, is growing up intensively. On the other hand, various types of topologies have been proposed and each of them has some advantages. Therefore, appropriateness of each converter regarding to the application is a main question for designers and engineers. In this paper, a part of this challenge is responded by comparing a typical Voltage-Source Converter (VSC) and Z-Source Converter (ZSC), through high power electronic-based equipment used in power systems. Dynamic response, stability margin, Total Harmonic Distortion (THD) of grid current and fault tolerant are considered as assessment criteria. In order to meet this evaluation, dynamic models of two converters are presented, a proper control system is designed, a small signal stability method is applied and responses of converters to small and large perturbations are obtained and analysed by PSCAD/EMTDC.*

Keywords

Small signal analysis, state space model, Total Harmonic Distortion, Voltage-Source Converter, Z-Source Converter.

1. Introduction

Nowadays, applications of power electronic-based equipment are predominantly widespread from high power converter usage in power system such as Distributed Generation (DG), HVDC and FACTS, until electric machinery drive circuits and low power converters in communication industry. The chief parts of them are converters classified into two main categories relevant to high power applications in power system, according to terminal voltage and current waveforms at the DC port:

- Current-Source Converter (CSC) is a converter in which the DC-side current maintains the same polarity; therefore, the direction of power flow is adjusted by the DC-side voltage polarity.

- Voltage-Source Converter (VSC), in contrast, keeps the same voltage polarity. The direction of power flow is determined by the polarity of DC-side current.

The Z-Source Converter proposed by [1] has a specific characteristic that provides the capability to buck or boost the output voltage when in rectification or inversion mode. It has a valuable feature due to two particular ideas added to the basic VSC.

- Turning on at least one leg of switches set, a shoot-through mode, which is forbidden in conventional converters.

- X-shaped impedance network includes two pairs of capacitors and inductances, maintains energy in the shoot-through state.

As shown in Fig. 1(b), a typical ZSC includes an impedance network. Its duty is storing energy in the shoot-through state when all six switches are on. In the shoot-through state all three parts, impedance network DC and AC-sides, are isolated and the inductances of network are charged by capacitors. While the non-shoot-through state, the energy of DC-source is transferred to capacitors and the absorbed energy of inductances is delivered to AC-grid at the point of common coupling. The shoot-through state is implemented by turning on all the switches. It is forbidden in VSC, shown in Fig. 1(a), because the capacitor is a short circuit and discharged in switches, leading to intensive damage.

(a) Voltage-Source Converter.

(b) Z-Source Converter.

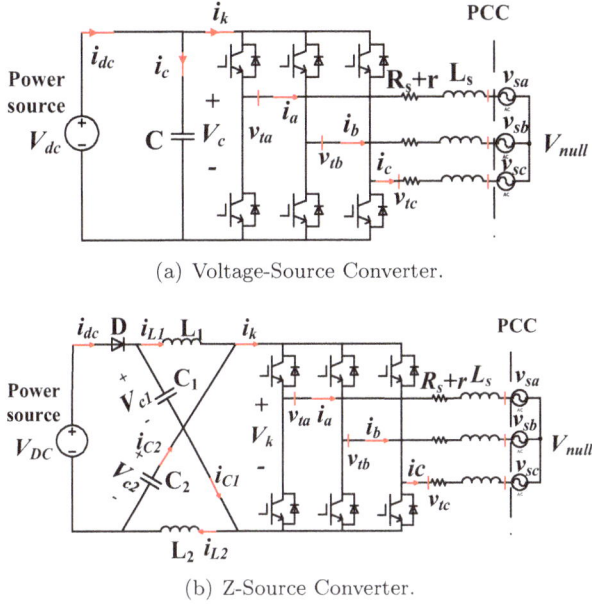

Fig. 1: Circuit of 3-phase.

A lot of literature has been devoted to various usage of ZSC. There is a clear evidence of the ascendance in doing research on finding various suitable applications of ZSC For instance, distributed generation [2], [3], [4], [5], [6], [7], [8] and [9], uninterruptable power supply [10], electric vehicles [11] and [12] adjustable speed drive [13] and [14] and so on. A dynamic model that describes the behavior of VSC with appropriate DC-side voltage control and active and reactive power controllers are provided in [15]. In [16], [17] and [18] a dynamic model and a small signal model of Z-Source Converter are proposed; however, Z-Source Converter is considered as a basic DC/DC-power conversion application. Therefore, a section of this study is allocated to expand the proposed dynamic model to become appropriate for comprehensive assessment of its response.

In terms of comparison, a comprehensive work is done by [19]. It concludes that, ZSC has lower average switching device power and provides higher efficiencies in most operation ranges. Constant power ratio is raised over conventional PWM converter notably. Moreover, significant reliability development of ZSC is very important advantage. Due to the fact that ZSC has potential advantages in contrast with traditional converters, it is expected to be applied in vast range of power electronic-based equipment used in power system and motor drive circuits. In order to realize whether it is a suitable substitution for conventional converters in high power applications, this paper mainly allocated to compare dynamic responses of VSC and ZSC when they are encountered to changes of active and reactive demands of AC-side. Stability margin and THD which are other important criteria in

power system point of view are taken into account and contribute to make the study more precise.

This paper is organized as follows. The mathematical models describing dynamic behaviour of VSC and ZSC are presented in Section 2. The comprehensive control system based on current control method which obtains AC-side currents and the voltage at PCC in rotary reference frame as input signals and deliver well-suited modulation signals to the converter are considered in Section 3. A comparison between VSC and ZSC regarding to stability and output harmonics of two converters is done in Section 4. Time domain results that illustrate the response of indicated converters to small and large changes in working point based on various scenarios are provided in Section 5. Conclusion commonly is delivered in Section 6.

2.   Dynamic Model of VSC and ZSC

Figure 1 illustrates the generic electrical circuit model of typical VSC and ZSC. As illustrated, DC-circuit which can be DC-source of energy such as a renewable energy source or a part of DC-line is considered by constant DC-power supply that is also capable to absorb the energy in rectification mode of operation. In this paper, without loss of generality, some simplifications are assumed:

- Switching losses are neglected, resulting from both shoot-through and active states operation. Moreover, all switches are considered fast sufficient that has no time constant.

- Conduction loss which is a consequence of interior resistance of switches and passive components in impedance network is ignored. If more accuracy is desired, the converter losses can be represented by small resistance in series with R_s.

2.1.   Voltage-Source Converter

The fundamental frequency component of terminal voltages for both converters in Fig. 1 is [17]:

$$v_t(t) = \frac{mV_{dc}(t)}{2}, \qquad (1)$$

where $v_t(t)$ is the AC-terminal voltage and m is the converter modulation index which is M (arbitrary constant modulation index) for VSC and $M/(1-2d(t))$ for ZSC [1] and [17]. A dynamic model of AC-circuit in abc reference frame is [17]:

$$v_{tabc}(t) = L_s \frac{di_{sabc}}{dt} + R_s i_{sabc} + v_{sabc}, \qquad (2)$$

where $v_{sabc}(t) = [v_{sa}(t) \cdot v_{sb}(t) \cdot v_{sc}(t)]^T$ and $i_{sabc}(t) = [i_{sa}(t) \cdot i_{sb}(t) \cdot i_{sc}(t)]^T$. R_s and L_s are equivalent series resistance and inductance of the RL filter and transformer, between the converter terminal and PCC. By applying transformation matrix $\mathbf{K}_s$ [17] in Eq. (2), differential equations describing dynamic behavior of AC-circuit in dq reference frame are obtained as follows:

$$\frac{di_{sdq}(t)}{dt} = \frac{R_s}{L_s}i_{sdq}(t) \pm \omega_e i_{sqd}(t)$$
$$+\frac{1}{L_s}(v_{tdg}(t) - v_{sdq}(t)), \tag{3}$$

where $\theta = \int_0^t \omega_e(\zeta) + \theta_0$ and ω_e is the power system frequency. The angle is $\omega_e(t)$ provided by the Phase-Locked Loop (PLL), arbitrary assumed here to align system voltage with the d-axis leading to $v_{sq}(t) = 0$. The DC-voltage dynamic expression is deduced based on energy balance between the AC and DC-side of the converter.

$$V_c(t)i_{dc}(t) = p(t) - P_L(t), \tag{4}$$

where $p(t)$ is instantaneous real power at AC-terminal of converter which is defined in dq reference frame as [17]:

$$p(t) = \frac{3}{2}(v_{td}(t)i_{sd}(t) + v_{tq}(t)i_{sq}(t)+$$
$$+2v_{t0}(t)i_{t0}(t). \tag{5}$$

It should be noted that the last term of Eq. (5) is omitted due to the fact that $i_{ta}(t) + i_{tb}(t) + i_{tc}(t) = 0$. $P_L(t)$ includes both conduction and switching losses that are ignored in this study based on [15].

In order to present all dynamic equations in the form of differential equations, capacitor voltage considered as a state variable and Eq. (5) is rewritten as:

$$\frac{dV_c(t)}{dt} = \frac{1}{CV_c(t)}\left(\frac{3}{2} \cdot (v_{td}(t)i_{sd}(t) +\right.$$
$$\left.+v_{tq}(t)i_{sq}(t))\right) + \frac{i_{dc}(t)}{C}. \tag{6}$$

dq-frame components of converter terminal voltage, $v_{td}(t)$ and $v_{tq}(t)$, will be determined based on control system equations.

2.2.   Z-Source Converter

As opposed to VSCs, Z-Source Converters have the shoot-through state, leading to different circuit topology from the non-shoot-through state illustrated in Fig. 1. Therefore, two operation modes in the switching period should be taken into account and average procedure will be held consequently. In mode 1, real

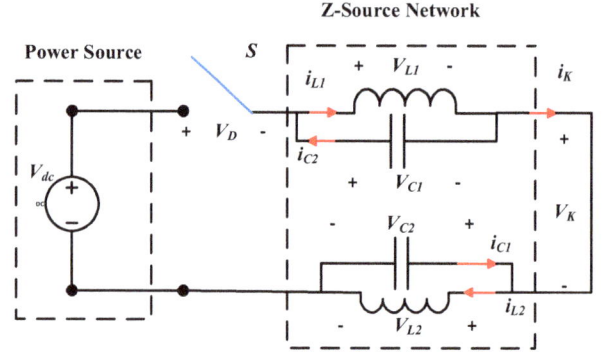

Fig. 2: ZSC in shoot through state.

energy transferred between AC and DC-sides; however, in mode 2, the path of transferring energy is blocked by diode D in DC-side and set of IGBTs in AC-side, so DC-power source, impedance network and AC-grid are decoupled as shown in Fig. 2. The duty ratio of switches, $d(t)$, which is generally not constant, is defined as the shoot-through duty ratio. Correspondingly, equation Eq. (3) is eligible for describing dynamic behavior of AC-circuit of ZSC; however, all terms should be multiplied to $1 - d(t)$, non-shoot through factor, in operation mode 1 and $d(t)$ for mode 2, distinctly. In order to decrease the number of equations, the impedance network is considered symmetric, which is $L_1 = L_2 = L$ and $C_1 = C_2 = C$.

$$(1 - d(t))L \cdot \frac{di_L(t)}{dt} =$$
$$= (V_{dc}(t) - V_c(t)) \cdot (1 - d(t)), \tag{7}$$

$$(1 - d(t))C \cdot \frac{dV_c(t)}{dt} =$$
$$= (i_L(t) - i_k(t)) \cdot (1 - d(t)). \tag{8}$$

As shown in Fig. 2, DC-side equations for the shoot-through state are modified into:

$$d(t)L \cdot \frac{di_L(t)}{dt} = V_c(t)d(t), \tag{9}$$

$$d(t)C \cdot \frac{dV_c(t)}{d(t)} = -i_L(t)d(t). \tag{10}$$

So far, electrical circuits of VSC and ZSC are modeled in the form of differential equations. By averaging on a switching period, adding equations for shoot-through and non-shoot-through states, state space close form of equations is obtained completely.

3.   Active and Reactive Power Controllers

The main purpose of the control system is to achieve appropriate output-active and output–reactive power

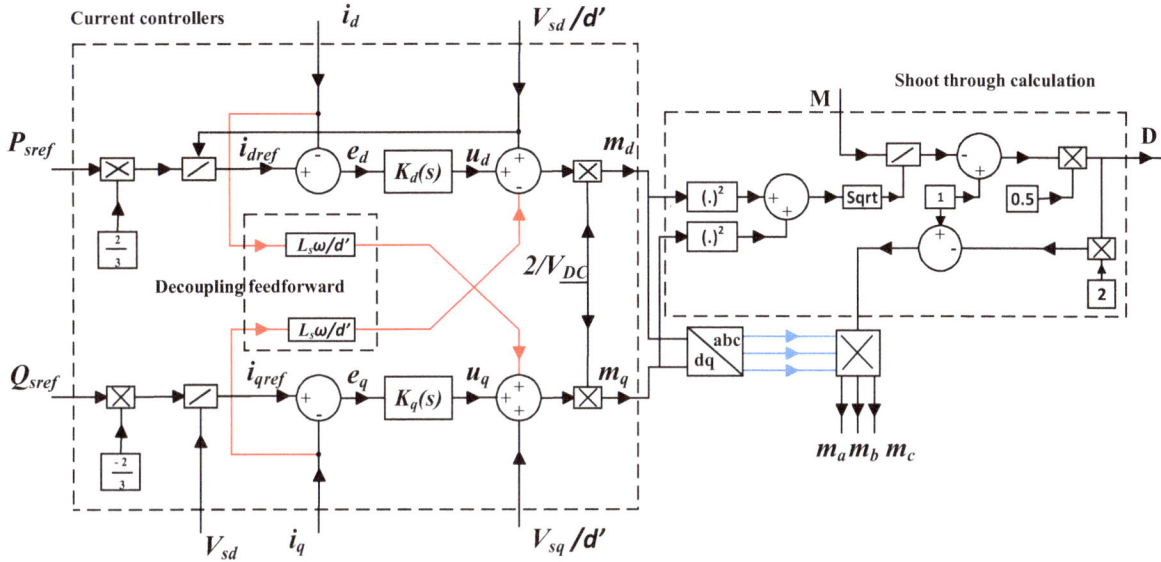

Fig. 3: Control system.

tracking exchanged with AC-grid at PCC. PI controller based on adjusting AC-current is taken into account for this work [15]. As illustrated in Fig. 3, active and reactive demand of AC-side are considered as reference input signals:

$$P_s(t) = \frac{3}{2} V_{sd}(t) i_d(t), \qquad (11)$$

$$Q_s(t) = -\frac{3}{2} V_{sq}(t) i_d(t). \qquad (12)$$

The reference value of dq-components of current, i_{dref} and i_{qref} can be obtained from a desired value of active and reactive powers by means of Eq. (11) and Eq. (12), respectively. A proportional-integral compensator in the form of $K(s) = k_p + \frac{k_i}{s}$ is sufficient for implementing the control process, due to the fact that all required signals are completely DC-quantities. The presence of $L_s \omega_e$ in Eq. (3) results in coupling two mentioned equations. Thus, $m_d(t)$ and $m_q(t)$ are identified as follows that are accomplished by cross coupling terms as shown in Fig. 3 so as to decouple remarked equations:

$$m_{dq}(t) = \frac{2\left(u_{dq}(t) + V_{sdq} \mp \frac{L_s \omega_e i_{qd}(t)}{d'}\right)}{V_{dc}(t)}, \qquad (13)$$

where $u_d(t)$ and u_q are:

$$u_{dq}(t) = (k_p) + \frac{k_i}{p}(i_{dqref} - i_{dq}(t)), \qquad (14)$$

$$\frac{dz_{dq}(t)}{dt} = i_{dqref} - i_{dq}(t). \qquad (15)$$

The output signals of current controllers are dq-component of modulation signal, $m_d(t)$ and $m_q(t)$,

which include only buck factor for VSC. In the other words, when they are transferred to abc-frame the amplitudes of sinusoidal signals are less than unity except in third harmonic injection condition [15]. However, by transforming modulation signals from dq-frame to abc-frame, the amplitude of AC-port voltage is acquired according to trigonometric equations:

$$\frac{M}{1 - 2d(t)} = \sqrt{m_d^2(t) + m_q^2(t)}. \qquad (16)$$

On the other hand, similar abc-component of modulation signals could be more than one for ZSC, because of the fact that $m_d(t)$ and $m_q(t)$ contain buck and boost factors simultaneously. In addition, the function of current controllers is to find out appropriate value of shoot through $d(t)$ in order to provide desired value of instantaneous active and reactive powers. Therefore, shoot-through calculation box is added to control system that get buck index M as the arbitrary factor and produce suitable shoot through index. It should be noted that consequently the amplitude of abc-frame of modulation signals is equal to M. In order to complete dynamic formulations, dq-components of AC-terminal voltage are delivered based on the state variables and inputs:

$$v_{tdq}(t) = m_{dq}(t) \frac{V_{dc}(t)}{2}. \qquad (17)$$

Finally, after mathematical calculations differential equations describing dynamic behavior of ZSC connected to AC-grid can be presented as follows:

$$L_s \frac{di_d(t)}{dt} = -(R_s + d'K_p)i_d(t) + \\ + d'K_i z_d(t) + d' I_{dref} K_p, \qquad (18)$$

$$L_s \frac{\mathrm{d}i_q(t)}{\mathrm{d}t} = -(R_s + d'K_p)i_q(t)+$$
$$+d'K_i z_q(t) + d'I_{qref}K_p, \tag{19}$$

$$L\frac{\mathrm{d}i_L(t)}{\mathrm{d}t} = (d-d)v_c(t) + d'V_{dc}(t) - R_L i_L(t), \tag{20}$$

$$C\frac{\mathrm{d}v_c(t)}{\mathrm{d}t} = (d'-d)i_L(t) - \frac{3d'(1-2d)}{2V_{dc}(t)}\cdot$$
$$\cdot [i_d(t)K_p(I_{dref} - i_d(t)) + i_d(t)z_d(t)K_i +$$
$$+\frac{V_{sd}i_d(t)}{d'} + i_q(t)z_q(t)K_i+ \tag{21}$$
$$+i_q(t)K_q(I_{qref} - i_q(t))]\,,$$

$$\frac{\mathrm{d}z_d(t)}{\mathrm{d}t} = I_{dref} - i_d(t), \tag{22}$$

$$\frac{\mathrm{d}z_q(t)}{\mathrm{d}t} = I_{qref} - i_q(t). \tag{23}$$

4. Stability and THD Comparison

In this section, a comparison between ZSC and conventional VSC according to first stability margin region and secondly harmonic of AC-terminal voltages are evaluated.

4.1. Small Signal Stability

The main method that attempts to identify stability of a nonlinear system like both ZSC and VSC directly is Lyapunov's method based on the sign of Lyapunov's function and its time derivative. However, the stability of a nonlinear system can also be given by the roots of characteristic equation of the system of first approximation obtained by linearizing equations around an equilibrium point [20].

The studied case is applied to analyze the stability of the converters after encountering small disturbances based on the parameters shown in Tab. 1. For improving dynamic response of the converters, selection of appropriate parameters of controllers can be done based on traces of eigenvalues. The eigenvalue spectrum of VSC is shown in Fig. 4 as a function of K_p and K_i respectively. It can be seen that increase of K_p causes oscillation reduction, however, after $K_p = 0.04$ the stability margin of the system is decreased due to real part reduction of eigenvalues. On the other hand, K_i increasing can improve stability, but after $K_i = 2$ all modes become oscillating. Similar study has been done for ZSC, however, in addition to controller's parameters the impact of impedance network components are taken into account. A base-case scenario and participation factors, indicating the relation between states and modes, are provided in Tab. 2.

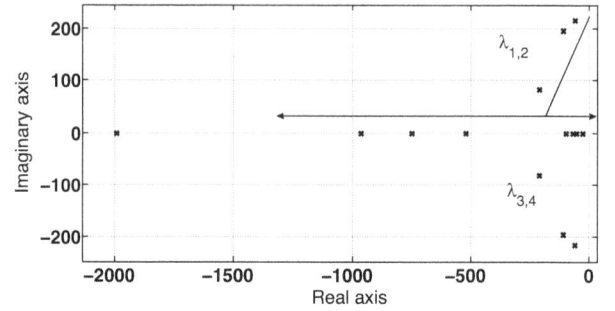

(a) Increase of k_p, $0.01 < k_p < 0.2$.

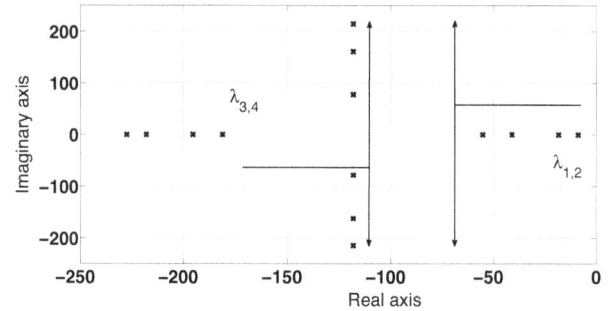

(b) Increase of k_i, $0.1 < k_i < 6$.

Fig. 4: Traces of VSC's eigenvalues for changes of controller's parameters.

Tab. 1: Parameters of case study.

Parameters	Value
Rated active power P_s	2.5 MW
Rated reactive power Q_s	1 MVar
AC system nominal voltage (L−L)	480 V
AC equivalent Inductance L_s	100 μH
AC equivalent Resistance R_s	1.63 mΩ
Nominal frequency	60 Hz
DC capacitance C	1000 μF
DC inductance L	1 mH
Nominal DC voltage V_{dc}	1250 V
Current controller K_p	0.05 Ω
Current controller K_i	5 Ω·s^{-1}
Carrier frequency	1.5 kHz

Tab. 2: Parameters of case study.

	$\lambda_{1,2} = -5 \pm j840$	$\lambda_{3,4} = -2017$	$\lambda_{5,6} = -22.8$
i_d	0	1.011	0.011
i_q	0	1.001	0.011
i_L	0.5	0	0
v_C	0.5	0	0
z_d	0	0.011	1.011
z_q	0	0.011	1.011

All participation factors are normalized into infinite norm. There are four evanescent modes and two oscillation modes. As shown in Tab. 2, $\lambda_{3,4}$ are highly sensitive to state variables i_d and i_q; therefore, they are chiefly under influence of k_p and k_i. Likewise, $\lambda_{5,6}$ are mainly affected by k_p and k_i. $\lambda_{1,2}$ are only sensitive to state variables i_L and v_C, so they are affected mostly by the value of impedance network elements. As

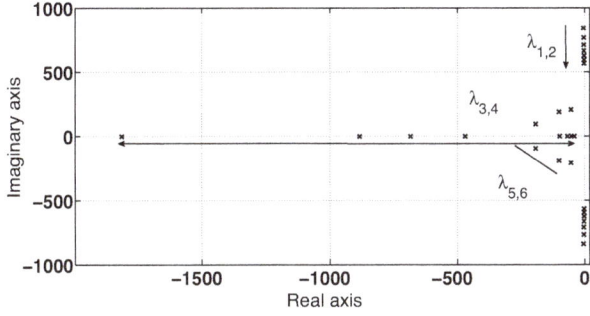

(a) Increase of k_p & L, $0.01 < k_p < 0.2$, 1 mH $< L < 2.2$ mH.

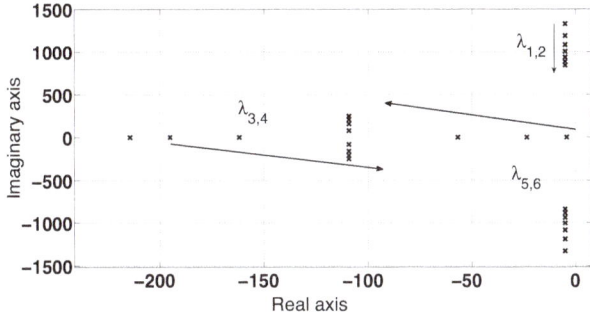

(b) Increase of k_i & C, $0.1 < k_i < 6$, 400 μF $< C < 1000$ μF.

Fig. 5: Traces of ZSC's eigenvalues for changes of controllers' and impedance network's parameters individually.

predicted from participation factors and illustrated in Fig. 5, $\lambda_{1,2}$ is dependent on DC-side parameters, while controllers' parameters effects are restricted to eigenvalues related to AC-side states. In Fig. 5(a), traces of eigenvalues are sketched as a function of k_p and L individually. By increasing L, real part of $\lambda_{1,2}$ is decreased seldom; however, it causes stability margin reduction. But the imaginary part or oscillation reduction is significant, resulting from the fact that larger inductance is more reluctant against changes. It can be seen that as k_p is changed, only $\lambda_{3,4}$ and $\lambda_{5,6}$ are changed, when k_p is small they are conjugated eigenvalues, as k_p is increased, imaginary parts are decreased and they move toward stable region. Finally, they become four evanescent modes that two moves toward stable region and the other ones move toward instable region. Figure 5(b) shows the traces of eigenvalues as function of k_i and C.

Correspondingly, $\lambda_{1,2}$ are affected by C and the other eigenvalues are moved by k_i changing. The oscillation of $\lambda_{1,2}$ is reduced as C is increased. Although C is not in direct path of active power flow, at the early moments of changing the level of transferred energy, L is reluctant and C delivers a part of demanding energy. Therefore, larger capacitance leads to smaller oscillations. In contrast with the effect of k_p increasing, larger k_i can overall improve the disturbance suppression of AC-side dynamic response.

4.2. Total Harmonic Distortion

Both VSC and ZSC have an adverse relation between their output harmonics and the modulation index. In VSCs, modulation index is designed near 0.8 in order to improve the voltage profile during voltage sag occurrence. However for ZSCs, due to boosting capability, modulation index can be considered about 0.95. Therefore, ZSC potentially has lower THD. In spite of this, as shown in Fig. 6, the number of harmonics and THD are larger for VSC at the same modulation index, since impedance network prevents engendering subharmonics.

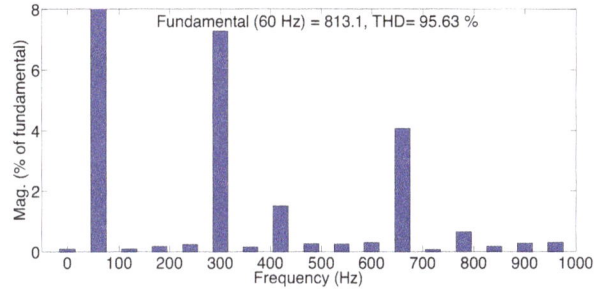

(a) VSC.

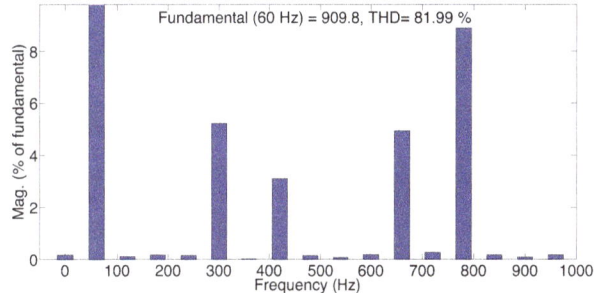

(b) ZSC.

Fig. 6: Total harmonic distortion.

5. Simulation and Results

A simulation model of Fig. 1 and controllers of Fig. 3 are implemented in PSCAD/EMTDC to verify the correctness of developed model of ZSC and accuracy of controllers, also compare the dynamic response and short current capacity of two converters. Small disturbance in active and reactive power demands for VSC are considered and the results are shown in Fig. 7.

As shown in Fig. 7, the active and reactive power signals are implemented at $t = 0.6$ s and $t = 1.1$ s, respectively in order to analyze the responses of converters more accurately. The ZSC is similarly encountered this scenario and results are shown in Fig. 8.

The proposed control system is working accurately and controlling the active and reactive power decou-

(a) I_d.

(b) I_q.

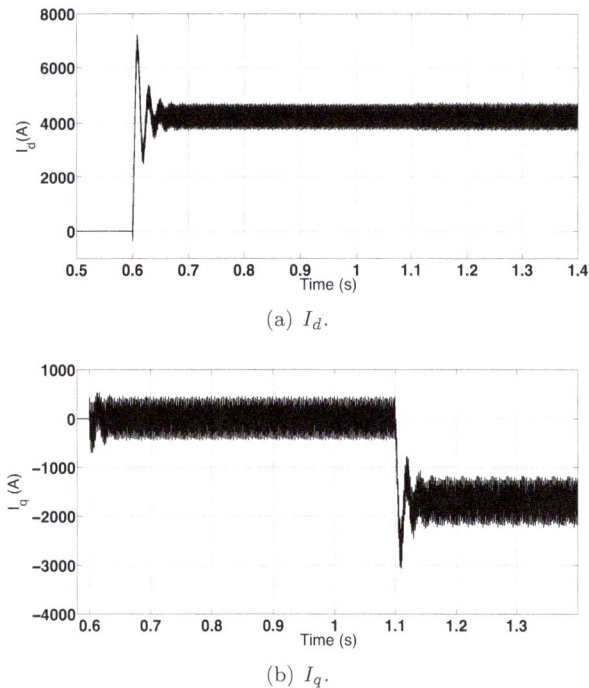

Fig. 7: Dynamic response of VSC to step change in active and reactive power.

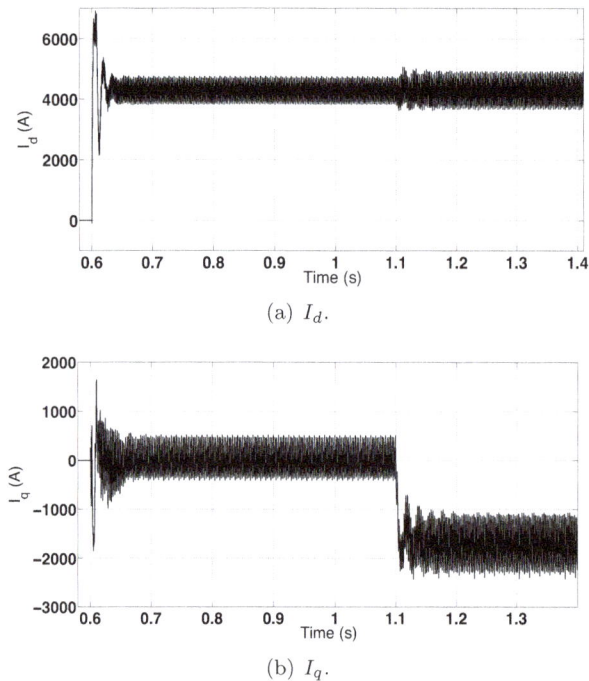

(a) I_d.

(b) I_q.

Fig. 8: Dynamic response of ZSC to step change in active and reactive power.

pled enough. It could be concluded from Fig. 7 and Fig. 8 that the control paths of I_d and I_q are separated for both converters. Because the active power changes have no effect on the value of reactive power and vice versa. By contrasting Fig. 7 and Fig. 8, ZSC has faster and smoother response to the reference sig-

nals; however, controller parameters are adjusted similarly. Presence of L in impedance network of ZSC justifies better response of I_d, since it resists versus rapid changes of energy by means of current. In another words, changing set point of active power by demand signal from the grid is realized by changing DC-current as DC-voltage is constant. For this purpose, DC-current should be passed through the inductances of impedance network which are reluctant against current changes.

Correspondingly, in Fig. 1(b), the voltage of capacitor is in relation with $V_k(t)$ accordingly reactive power demand of grid. Because the reactive power of grid could be adjusted by changing the amplitude of $v_t(t)$ that is in direct relation with $V_k(t)$ and $V_c(t)$. Its voltage is boosted during increment reactive power reference signal, so it allows increasing the voltage of AC-port. As a result, presence of C in the ZSC's circuit leads to better dynamic response of I_q than VSC one which is shown in Fig. 8(b).

In Fig. 9 a comparison according to fault tolerant criterion is held. ZSC restricts the fault current more than VSC, because of inductance L in the impedance network. As a result, this advantage leads to protecting accessories and switches.

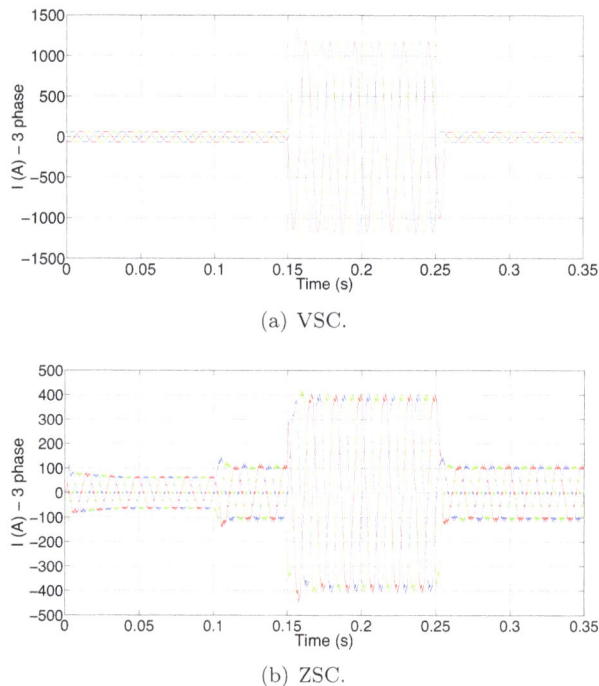

(a) VSC.

(b) ZSC.

Fig. 9: Three phase short circuit current.

6.   Conclusion

In this paper, small signal model of a ZSC suitable for connecting to AC-grid is presented, and the effect of each parameter on the behavior of the system is evaluated. In power electronic-based equipment such as DGs, dynamic response of the converter for frequency and voltage stability is definitely important. In addition to significant advantages of ZSC mentioned in previous works, it has also faster and smoother dynamic response, lower order of harmonics at AC-port and ability to limit the short circuit current in contrast with VSC that are significant criteria for designers and engineers in the process of selecting appropriate converter. The stability margin of this converter is fairly acceptable. Therefore, it could be a proper alternative for high power applications in near future.

References

[1] PENG, F. Z. Z-source inverter. *IEEE Transactions on Industry Applications*. 2003, vol. 39, iss. 2, pp. 504–510. ISSN 0093-9994. DOI: 10.1109/TIA.2003.808920.

[2] LI, Y., S. JIANG, J. G. CINTRON-RIVERA and F. Z. PENG. Modeling and Control of Quasi-Z-Source Inverter for Distributed Generation Applications. *IEEE Transactions on Industrial Electronics*. 2013, vol. 60, iss. 4, pp. 1532–1541. ISSN 0278-0046. DOI: 10.1109/TIE.2012.2213551.

[3] SIWAKOTI, Y. P. and G. E. TOWN. Performance of distributed DC power system using quasi Z-Source Inverter based DC/DC converters. In: *Twenty-Eighth Annual IEEE Applied Power Electronics Conference and Exposition (APEC)*. Orlando: IEEE, 2013, pp. 1946–1953. ISBN 978-147992407-3. DOI: 10.1109/APEC.2013.6520561.

[4] GE, B., H. ABU-RUB, F. Z. PENG, Q. LEI, A. T. DE ALMEIDA, F. J. T. E. FERREIRA, D. SUN and Y. LIU. An Energy-Stored Quasi-Z-Source Inverter for Application to Photovoltaic Power System. *IEEE Transactions on Industrial Electronics*. 2012, vol. 60, iss. 10, pp. 1532–1541. ISSN 0278-0046. DOI: 10.1109/TIE.2012.2217711.

[5] GAJANAYAKE, C. J., D. M. VILATHGAMUWA, P. C. LOH, R. TEODORESCO and F. BLAABJERG. Z-Source-Inverter-Based Flexible Distributed Generation System Solution for Grid Power Quality Improvement. *IEEE Transactions on Energy Conversion*. 2009, vol. 24, iss. 3, pp. 695–704. ISSN 1558-0059. DOI: 10.1109/TEC.2009.2025318.

[6] GAJANAYAKE, C. J., D. M. VILATHGAMUWA and P. C. LOH. Development of a Comprehensive Model and a Multiloop Controller for Z-Source Inverter

DG Systems. *IEEE Transactions on Industrial Electronics*. 2007, vol. 54, iss. 4, pp. 2352–2359. ISSN 1557-9948. DOI: 10.1109/TIE.2007.894772.

[7] DEHGHAN, S. M., M. MOHAMADIAN and A. Y. VARJANI. A New Variable-Speed Wind Energy Conversion System Using Permanent-Magnet Synchronous Generator and Z-Source Inverter. *IEEE Transactions on Energy Conversion*. 2009, vol. 24, iss. 3, pp. 714–724. ISSN 1558-0059. DOI: 10.1109/TEC.2009.2016022.

[8] JUNG, J. W. and A. KEYHANI. Control of a Fuel Cell Based Z-Source Converter. *IEEE Transactions on Energy Conversion*. 2007, vol. 22, iss. 2, pp. 467–476. ISSN 1558-0059. DOI: 10.1109/TEC.2006.874232.

[9] TANG, Y., J. WEI and S. XIE. Grid-tied photovoltaic system with series Z-source inverter. *IET Renewable Power Generation*. 2013, vol. 7, iss. 3, pp. 275–283. ISSN 1752-1416. DOI: 10.1049/iet-rpg.2012.0335.

[10] HANIF, M., M. BASU and K. GAUGHAN. Understanding the operation of a Z-source inverter for photovoltaic application with a design example. *IET Power Electronics*. 2011, vol. 4, iss. 3, pp. 278–287. ISSN 1755-4535. DOI: 10.1049/iet-pel.2009.0176.

[11] ZHOU, Z. J., X. ZHANG, P. XU and W. X. SHEN. Single-Phase Uninterruptible Power Supply Based on Z-Source Inverter. *IEEE Transactions on Industrial Electronics*. 2008, vol. 55, iss. 8, pp. 2997–3004. ISSN 1557-9948. DOI: 10.1109/TIE.2008.924202.

[12] KULKA, A. and T. UNDELAND. Voltage harmonic control of Z-source inverter for UPS applications. In: *13th Power Electronics and Motion Control Conference (EPE-PEMC)*. Poznan: IEEE, 2008, pp. 657–662. ISBN 978-1-4244-1741-4. DOI: 10.1109/EPEPEMC.2008.4635339.

[13] GUO, F., L. FU, C.-H. LIN, C. LI, W. CHOI and J. WANG. Development of an 85-kW Bidirectional Quasi-Z-Source Inverter With DC-Link Feed-Forward Compensation for Electric Vehicle Applications. *IEEE Transactions on Power Electronics*. 2013, vol. 28, iss. 12, pp. 5477–5488. ISSN 1941-0107. DOI: 10.1109/TPEL.2012.2237523.

[14] PENG, F. Z., M. SHEN and K. HOLLAND. Application of Z-Source Inverter for Traction Drive of Fuel Cell-Battery Hybrid Electric Vehicles. *IEEE Transactions on Power Electronics*. 2007, vol. 22, iss. 3, pp. 1054–1061. ISSN 1941-0107. DOI: 10.1109/TPEL.2007.897123.

[15] PENG, F. Z., X. YUAN, X. FANG and Z. QIAN. Z-Source Inverter for Adjustable Speed Drives. *IEEE Power Electronics Letters*. 2003, vol. 99, iss. 2, pp. 33–35. ISSN 1540-7985. DOI: 10.1109/LPEL.2003.820935.

[16] PENG, F. Z., A. JOSEPH, J. WANG, M. SHEN, L. CHEN, Z. PAN, E. O. RIVERA and Y. HUANG. Z-source inverter for motor drives. *IEEE Transactions on Power Electronics*. 2005, vol. 20, iss. 4, pp. 857–863. ISSN 1941-0107. DOI: 10.1109/TPEL.2005.850938.

[17] YAZDANI, A. and R. IREVANI. *Voltage-Sourced Converters in Power Systems: Modeling, Control, and Applications*. 1st ed. New Jersey: Wiley, 2010. ISBN 978-0-470-52156-4. DOI: 10.1002/9780470551578.

[18] LIU, J., J. HU and L. XU. Dynamic Modeling and Analysis of Z Source Converter—Derivation of AC Small Signal Model and Design-Oriented Analysis. *IEEE Transactions on Power Electronics*. 2007, vol. 22, iss. 5, pp. 1786–1796. ISSN 1941-0107. DOI: 10.1109/TPEL.2007.904219.

[19] YU, K., F. L. LUO and M. ZHU. Voltage harmonic control of Z-source inverter for UPS applications. In: *5th IEEE Conference on Industrial Electronics and Applications (ICIEA)*. Taichung: IEEE, 2010, pp. 2169–2174. ISBN 978-1-4244-5046-6. DOI: 10.1109/ICIEA.2010.5515153.

[20] GAJANAYAKE, C. J., D. M. VILATHGAMUWA and P. C. LOH. Small-signal and signal-flow-graph modeling of switched Z-source impedance network. *IEEE Power Electronics Letters*. 2005, vol. 3, iss. 3, pp. 111–116. ISSN 1540-7985. DOI: 10.1109/LPEL.2005.859771.

[21] LOH, P. C., D. M. VILATHGAMUWA, C. J. GAJANAYAKE, Y. R. LIM and C. W. TEO. Transient Modeling and Analysis of Pulse-Width Modulated Z-Source Inverter. *IEEE Transactions on Power Electronics*. 2007, vol. 22, iss. 2, pp. 498–507. ISSN 1941-0107. DOI: 10.1109/TPEL.2006.889929.

[22] SHEN, M., A. JOSEPH, J. WANG, F. Z. PENG and D. J. ADAMS. Comparison of Traditional Inverters and Z-Source Inverter for Fuel Cell Vehicles. *IEEE Transactions on Power Electronics*. 2007, vol. 22, iss. 4, pp. 1453–1463. ISSN 1941-0107. DOI: 10.1109/TPEL.2007.900505.

[23] KUNDUR, P. *Power system stability and control*. 1st ed. New York: McGraw-Hill, 1994. ISBN 0-07-035958-X.

About Authors

Masoud Jokar KOUHANJANI was born in Shiraz, Iran. He received the B.Sc. degree in electrical engineering from Shiraz University of Technology, Shiraz, Iran, in 2011 and the M.Sc. degree in electrical engineering (power) from Shiraz University, Shiraz, Iran, in 2014. His research interests include power electronic interfaces, power system programing.

Ali Reza SEIFI was born in Shiraz, Iran. He received his B.Sc. in Electrical Engineering from Shiraz University, Shiraz, Iran, in 1991, his M.Sc. in Electrical Engineering from the University of Tabriz, Tabriz, Iran, in 1993 and his Ph.D. in Electrical Engineering from Tarbiat Modarres University (T.M.U.), Tehran, Iran, in 2001. He is currently the Professor in Department of Power and Control Engineering, School of Electrical and Computer Engineering, Shiraz University, Shiraz, Iran. His research interests include power plant simulations, power systems, electrical machine simulations, power electronics and fuzzy optimization.

9

Effect of the Dielectric Inhomogeneity Factor's Range on the Electrical Tree Evolution in Solid Dielectrics

Hemza MEDOUKALI, Mossadek GUIBADJ, Boubaker ZEGNINI

Laboratoire d'Etude et Developpement des Materiaux Semi Conducteurs et Dielectriques, LEDMaScD, Amar Telidji University-Laghouat, BP 37G road of Ghardaia, Laghouat 03000, Algeria

hamzamedou@gmail.com, m.guibadj@lagh-univ.dz, b.zegnini@lagh-univ.dz

Abstract. The main contribution of the presented paper is to investigate the influence of the Dielectric Inhomogeneity Factor on the electrical tree evolution in solid dielectrics using cellular automata. We have a sample of the XLPE which is located between needle-to-plane electrodes under DC voltage. The electrical tree emanates from the end of the needle in which the electric stress attains a dielectric strength of the material. At every time step, Laplace's equation is solved to calculate the potential distribution which changes according to electrical tree development. Dynamic simulations clearly demonstrate the influence of the range of the Dielectric Inhomogeneity Factor on the electrical tree growth. Simulation results confirm the published technical literature.

Keywords

Cellular automata, cross-linked polyethylene, dielectric inhomogeneity factor, electrical tree, Laplace's equation.

1.   Introduction

The Cross-linked Polyethylene (XLPE) is widely used in medium and high voltage cables due to its excellent characteristics such as high breakdown strength, low dielectric permittivity, and low dielectric loss [1] and [2].

Electrical treeing phenomenon is one of the principal reasons for failure and deterioration of the XLPE dielectric. In this area, trees initiate from the regions where the electric stress enhances due to many factors such as protrusions in the high voltage electrode, material inhomogeneity due to manufacturing defects and presence of conducting particles or gas-filled cavities. Therefore, inception and propagation of electrical tree are accompanied by the Partial Discharge (PD) activity within developing dendrites [3], [4], [5], [6] and [7]. The phenomenon of treeing is initiated by the formation of micro-channels and completed by electrical tree which deteriorates the cable.

To remind, the growth of electrical tree is a complex mechanism that took a great deal of researches and studies. In literature, many works are published to analyze and clarify this phenomenon experimentally, theoretically and by simulation. In this field, multiple studies investigated the influence of the temperature, distance between electrodes, impurities and frequency on the behavior of the electrical tree in XLPE samples using different experimental technics [2], [3], [4], [8], [9], [10], [11], [12], [13] and [14]. Other authors have analysed both initiation and propagation mechanisms of the electrical tree inside solid insulation from experiment to theory [5], [15] and [16].

Many models are proposed to explain the formation mechanism and structures of trees in solid dielectrics. The breakdown phenomenon based on the fractal dimension is analysed in [17]. A non-lattice three-dimensional model is created to simulate both electrical tree growth and Partial Discharges (PD) activity within the growing tree channels [16]. Other researchers have presented the model which termed Deterministic Discharge-Avalanche; which is deterministic in concept and associates the branching with local field fluctuations generated by the mechanism itself [18].

The propagation of the electrical tree in solid dielectrics is simulated in various cases with the aid of Cellular Automata: the presence of voids, the existence of conducting or insulating particles and the influence of spaces charges on the development of dendrites [19], [20], [21] and [22]. Unfortunately, these previous references do not discuss the effect of the Dielectric Inhomogeneity Factor's (DIF) range.

Taking advantage of this fact, this paper is focused on the effect of the DIF's range variation on the electrical tree formation inside the XLPE, which is located between needle-to-plane electrodes (see Fig. 1). The study is based on the inherent inhomogeneity of the dielectric which gives a significant fluctuation on the electric field value. Dynamical simulations (using Cellular Automata (CA)) are presented to demonstrate practical potential of the proposed approach.

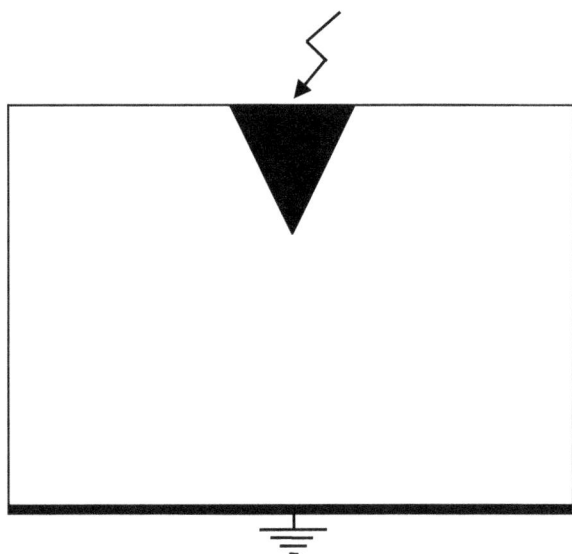

Fig. 1: The point/plane electrode arrangement, the white area corresponds to the solid dielectric and black areas correspond to the electrodes.

2.   Cellular Automata

The behavior of a physical system is not determined only by macroscopic parameters. The important aspects of the microscopic laws of physics are a great challenge for anyone who tries to simulate physical system. Cellular Automata (CA) are an idealization of dynamic systems where space, time, and variables are discrete and interactions are only local [23] and [24]. CA are first introduced by John von Neumann in the late 1940s [25] and [26]. It has been used extensively to model natural phenomena and complex systems [27]. Despite the simplicity of its structure, it is able to describe the behavior of the complex physical system. Jon Conway in 1970, and Stephen Wolfram in the beginning of the 80's have developed architecture of CA, the former proposed what is called "Game of Life" and the latter studied in much detail a family of simple one-dimensional CA rules, known as: Wolfram rules [28].

More precisely, CA consists of a regular uniform n-dimensional matrix. At each site of the matrix (cell), a physical quantity takes values. This physical quantity is the global state of the CA, and the value of this quantity at each cell is the local state of this cell. Each cell is restricted only to local neighborhood interaction, and as a result, it is incapable of immediate global communication [23]. Figure 2 shows the neighborhood of the cell which is taken to be the cell itself and some or all of the immediately adjacent cells. The state of each cell is updated simultaneously at discrete time steps, based on the states in its neighborhood at the preceding time step. The algorithm which is used to compute its successor state is referred as the CA local rule. Usually, the same local rule is applied to all cells of the CA. The state of a cell at time step $(t+1)$ is affected by the states of all eight cells in its neighborhood at time step t and by its own state at time step t:

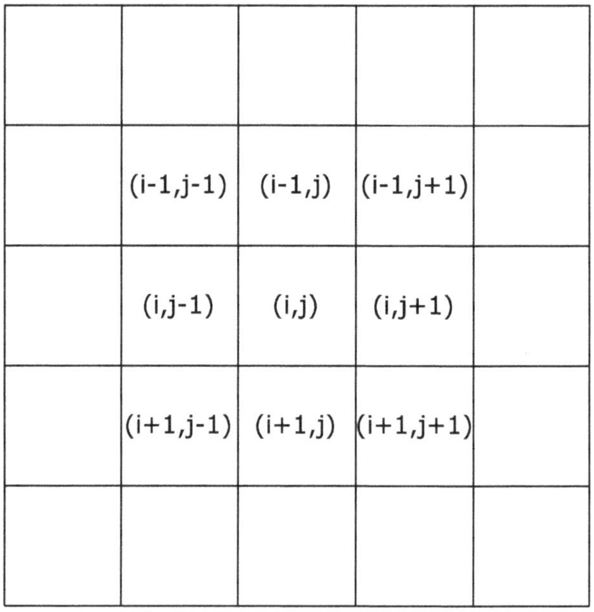

Fig. 2: The neighborhood of the (i,j) cell is formed by the (i,j) cell itself and the eight marked cells.

The CA local rule is given by:

$$S_{i,j}^{t+1} = F \begin{pmatrix} S_{i-1,j-1}^t, S_{i-1,j}^t, S_{i-1,j+1}^t, \\ S_{i,j-1}^t, S_{i,j}^t, S_{i,j+1}^t, \\ S_{i+1,j-1}^t, S_{i+1,j}^t, S_{i+1,j+1}^t \end{pmatrix}, \quad (1)$$

where $S_{i,j}^{t+1}$ and $S_{i,j}^t$ are the states of the (i,j) cell at time steps $(t+1)$ and t, respectively.

3.   Simulation

Laplace's equation Eq. (2) is solved to calculate the potential distribution inside dielectric at every time step by Partial Differential Equation (PDE) Toolbox of Matlab:

$$\nabla^2 V = 0. \tag{2}$$

For the correct calculation of the potential distribution (see Fig. 3), the definition of the appropriate boundary conditions is crucial:

- Dirichlet boundary conditions at the interface between the needle electrode and XLPE, plane electrode and XLPE are expressed as follow:

$$h \cdot V = r, \tag{3}$$

where V is the electrostatic potential value, h is the weight factor equation (normally $h = 1$) and r is the applied potential (at the needle $r = 80$ kV, at plane $r = 0$).

- Neumann boundary conditions between the dielectric sample and surrounding air are formulated as follow:

$$\vec{n} \cdot \varepsilon \cdot \vec{\nabla} V + q \cdot V = g, \tag{4}$$

where $\vec{n}$ is the outward unit normal, q is the charge ($q = 0$), V is the electrostatic potential value, g is the surface charge ($g = 0$) and ε is the relative permittivity of the medium, for XLPE $\varepsilon = 2.3$.

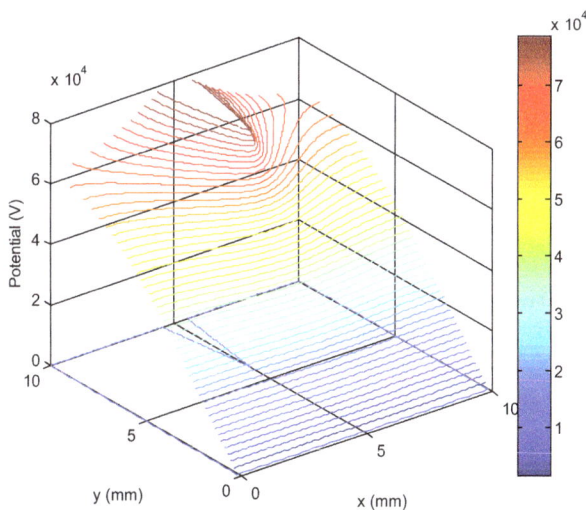

Fig. 4: CA representation of an XLPE dielectric with a needle-plane electrode arrangement.

The physical system with needle-plane electrodes is divided into a matrix of identical square cells (100×100), in which every cell has dimensions (0.1×0.1) mm. Thus, dimensions of XLPE sample are (10×10) mm (see Fig. 4). In this model, the internal state of each cell is defined by two parameters: the potential value V and the value of dielectric inhomogeneity factor g_{dif} which is generated randomly between two values.

Before simulation, the value of the electric stress $E_{\max}$ at the end of the needle tip is assumed by the formula [29]:

$$E_{\max} = \frac{2V}{r(\ln(1 + \frac{4s}{r}))}, \tag{5}$$

where V is the applied voltage, s is the electrode gap distance, and r is the radius of the needle tip.

If the Emax attains a dielectric strength $E_c = 40$ kV·mm^{-1} [29], the electrical tree initiates from the end of the needle electrode because the value of the electric stress is transferred into this latter [30] and [31], and consequently the state of the cell at the end of the needle tip at time step $t + 1$ is 1.

The common local rule (see Tab. 1) which is applied at every time step to all cells to simulate electrical tree evolution is:

Tab. 1: Local rule applied to simulate electrical tree evolution.

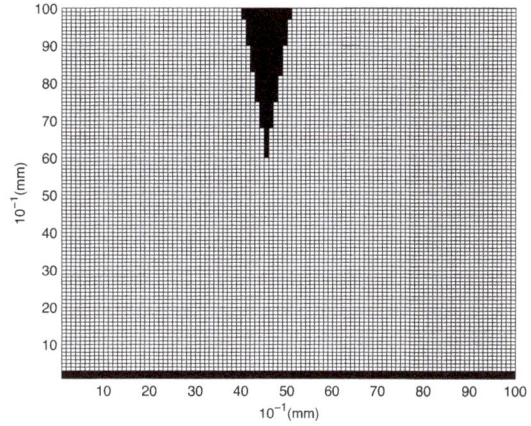

Fig. 3: Potential distribution at the needle-plane geometry; the applied voltage is $V_{ap} = 80$ kV at time step $t = 0$.

	Time step		Conditions
	t	$t + 1$	
	1	1	/
State of the (i, j) cell	0	0	None of its neighbor's state is 1.
	0	1	One or more of its neighbor's state are 1 and $E/E_c > 1$.

The potential values obtained from PDE Toolbox of Matlab are classified with the aid of a Matlab algorithm in a matrix of (100×100) cells in every time step.

The potential distribution gained from solving Laplace's equation with PDE of Matlab at every step

Tab. 2: Parameters of three simulations for the electrical tree evolution.

	Simulation 1.	Simulation 2.	Simulation 3.
Matrix dimensions	100×100	100×100	100×100
Cell dimensions (mm)	0.1×0.1	0.1×0.1	0.1×0.1
XLPE permittivity ϵ	2.3	2.3	2.3
Local dielectric strength (kV·mm^{-1})	40	40	40
Applied voltage (kV)	80	80	80
DIF's range (g_{dif})	0.95–1.04	0.95–1	0.97–1

is used to calculate electric field E between every cell of the electrical tree and the eight surrounding cells by the following equation [14] and [32]:

$$E \rightarrow g_{dif} \frac{\Delta V}{\Delta x}, \qquad (6)$$

where g_{dif} is the dielectric inhomogeneity factor of a cell (it is generated randomly), ΔV is the potential difference between the two neighboring cells (horizontal, vertical or diagonal) and Δx is the distance between centers of cells.

The algorithm checks in every step cells that can belong to an electrical tree by applying cellular automata rule (see Fig. 2). If the tree progresses, then its structure will change, which means new boundaries conditions are applied in the next step, and Laplace's equation is solved again to gain the new potential distribution.

The tree stops growing if:

- The electric field is less than local dielectric strength $E/E_c < 1$.

- The dendrites reach the plane electrode.

In this paper, three simulations are presented to display the influence of the Dielectric Inhomogeneity Factor's range on the tree propagation. In each case, the range variation of the g_{dif} is chosen.

Details of three simulations are included in the Tab. 2.

All stages of each simulation are summarized in the following flow-chart (see Fig. 5).

4.   Results and Discussion

The effect of the DIF's range on the electrical tree evolution in XLPE is simulated in three different cases as shown in Fig. 6, Fig. 7 and Fig. 8. The applied voltage at the needle was taken to be equal to 80 kV and the distance between electrodes was chosen to be 6 mm.

The electrical tree initiates from the needle tip and advances toward the plane electrode. In every time step, the Laplace's equation is solved to calculate a new distribution of the potential due to the propagation of branches which means a change in boundary conditions.

All simulations are identical regarding the sample of dielectric XLPE, the geometry, and the same applied voltage. However, the difference is only in the DIF's range variation which is randomly generated. In the case of Fig. 6 the g_{dif} is varied between 0.95–1.04, but in both cases Fig. 7 and Fig. 8; g_{dif} is varied between 0.95–1 and 0.97–1 respectively.

In these three cases, first, the tree emanates from the point electrode where the electric stress is higher

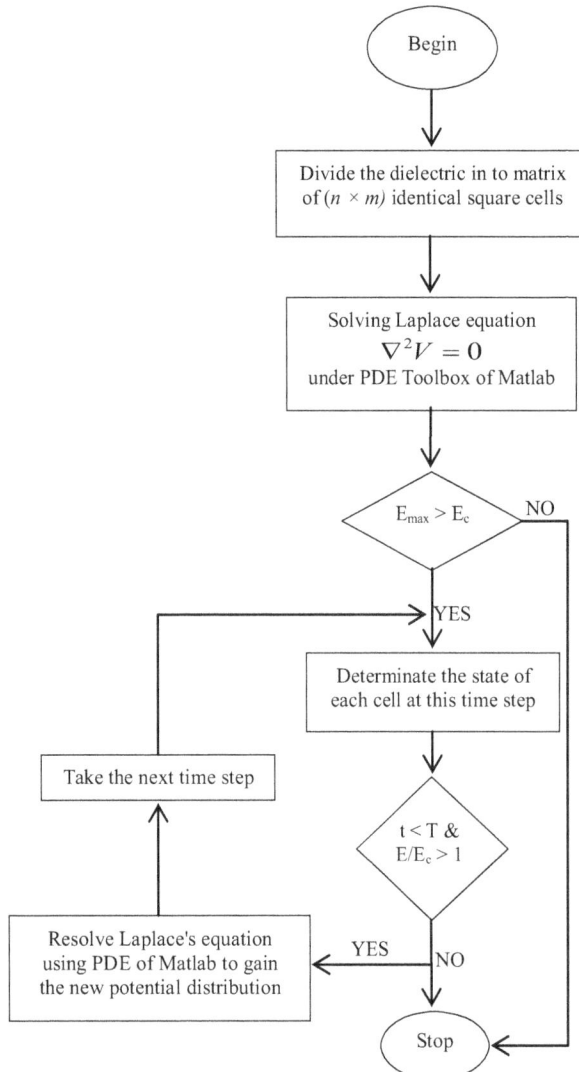

Fig. 5: Flow-chart of the dynamical simulation.

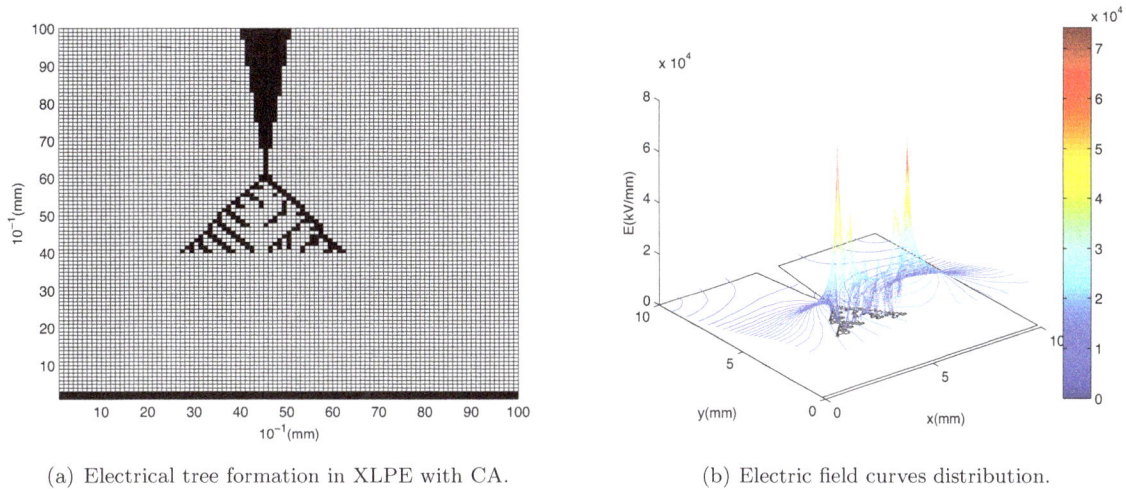

(a) Electrical tree formation in XLPE with CA.

(b) Electric field curves distribution.

Fig. 6: Simulation results at $t = 20$ steps and the range of g varied between 0.95–1.04.

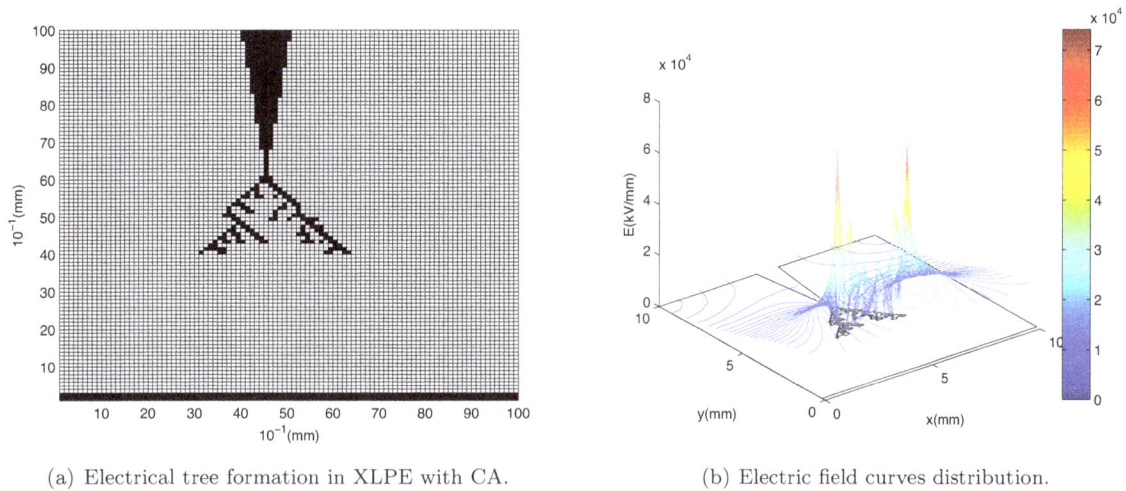

(a) Electrical tree formation in XLPE with CA.

(b) Electric field curves distribution.

Fig. 7: Simulation results at $t = 20$ steps and the range of g varied between 0.95–1.

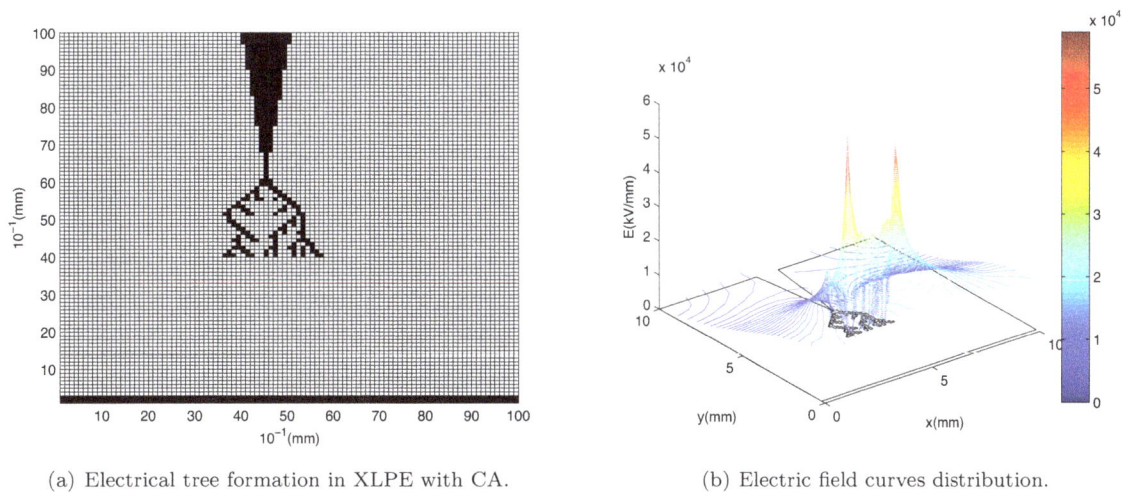

(a) Electrical tree formation in XLPE with CA.

(b) Electric field curves distribution.

Fig. 8: Simulation results at $t = 20$ steps and the range of g vary between 0.97–1.

than the dielectric strength of the XLPE ($E_{max} > E_c$). Then, dendrites start advancing toward the plane electrode by activities of discharges within the gas-filled channels following the paths where the homogeneity of XLPE is the weakest. The tree channels consist as conducting material, (i.e. they played the same role of the point electrode), so the potential value at every tree cell is considered to be equal to the value of the potential applied at the end of the needle tip.

In the first case (see Fig. 6), the electrical tree is more extensive. Also, it contains a significant number of branches, due to weak chemical bonds i.e. electric field is strong in numerous cells which means that the dielectric is less crosslinking since its smaller degree of homogeneity.

In contrast, the tree in the case of (Fig. 8) is narrow due to the strong chemical bonds since the range of the variation of the dielectric inhomogeneity factor is too small. Therefore, many structures of tree are formed, Cascade-tree is formed in both cases of (see Fig. 6) and (see Fig. 7) [8], but the tree's shape (see Fig. 8) is branch-tree [3].

Finally, it is clear that the range of the DIF, i.e. the homogeneity of the dielectric material is crucial factor for electrical tree behavior. The results of this simulation are similar to results published in literature [33].

5. Conclusion

The presented contribution consists of a confirmation that the range variation of the Dielectric Inhomogeneity Factor has a pronounced effect on the process of the electrical tree growth in XLPE dielectric. Furthermore, the current study demonstrated that:

- When the range variation of the DIF is very tight, the electrical tree becomes narrow and consists of a minimum number dendrites and vice-versa.

- For the reliable solid dielectric, a tree is produced with only one dendrite that advances perpendicularly toward the plane electrode.

- The range variation of DIF is almost constant for ideal solid dielectric.

References

[1] POLANSKY, R. Thermal Analyses of Cross-Linked Polyethylene. *Advances in Electrical and Electronic Engineering*. 2007, vol. 6, no. 3, pp. 121–126. ISSN 1336-1376.

[2] BAO, M., X. YIN and J. HE. Analysis of electrical tree propagation in XLPE power cable insulation. *Physica B: Condensed Matter*. 2011, vol. 406, no. 8, pp. 1556–1560. ISSN 0921-4526. DOI: 10.1016/j.physb.2011.01.069.

[3] CHEN, X., Y. XU, X. CAO and S. M. GUBANSKI. Electrical treeing behavior at high temperature in XLPE cable insulation samples. *IEEE Transactions on Dielectrics and Electrical Insulation*. 2015, vol. 22, iss. 5, pp. 2841–2851. ISSN 1070-9878. DOI: 10.1109/TDEI.2015.004784.

[4] NOBREGA, A. M., M. MARTINEZ and A. A. DE QUEIROZ. Investigation and analysis of electrical aging of XLPE insulation for medium voltage covered conductors manufactured in Brazil. *IEEE Transactions on Dielectrics and Electrical Insulation*. 2013, vol. 20, iss. 2, pp. 628–640. ISSN 1070-9878. DOI: 10.1109/TDEI.2013.6508767.

[5] DISSADO, L. A. Understanding electrical trees in solids: from experiment to theory. IEEE *Transactions on Dielectrics and Electrical Insulation*. 2002, vol. 9, iss. 4, pp. 483–497. ISSN 1070-9878. DOI: 10.1109/TDEI.2002.1024425.

[6] EL-ZEIN, A., M. TALAAT and M. M. EL BAHY. A numerical model of electrical tree growth in solid insulation. *IEEE Transactions on Dielectrics and Electrical Insulation*. 2009, vol. 16, iss. 6, pp. 1724–1734. ISSN 1070-9878. DOI: 10.1109/TDEI.2009.5361596.

[7] IDDRISSU, I. and S. M. ROWLAND. The impact of DC bias on electrical tree growth characteristics in epoxy resin samples. In: *IEEE Conference on Electrical Insulation and Dielectric Phenomena (CEIDP)*. Ann Arbor: IEEE, 2015, pp. 876–879. ISBN 978-1-4673-7496-5. DOI: 10.1109/CEIDP.2015.7352082.

[8] DISSADO, L. A., S. J. DODD, J. V. CHAMPION, P. I. WILLIAMS and J. M. ALISION. Propagation of electrical tree structures in solid polymeric insulation. *IEEE Transactions on Dielectrics and Electrical Insulation*. 1997, vol. 4, iss. 3, pp. 259–279. ISSN 1070-9878. DOI: 10.1109/94.689435.

[9] SARATHI, R. and P. G. RAJU. Diagnostic study of electrical treeing in underground XLPE cables using acoustic emission technique. *Polymer testing*. 2004, vol. 23, iss. 8, pp. 863–869. ISSN 0142-9418. DOI: 10.1016/j.polymertesting.2004.05.007.

[10] SARATHI, R. and P. G. RAJU. Study of electrical treeing phenomena in XLPE cable samples using acoustic techniques. *Electric power systems research*. 2005, vol. 73, iss. 2, pp. 159–168. ISSN 0378-7796. DOI: 10.1016/j.epsr.2004.07.003.

[11] CHO, Y.-S., M.-J. SHIM and S.-W. KIM. Electrical tree initiation mechanism of artificial defects filled XLPE. *Materials chemistry and physics.* 1998, vol. 56, iss. 1, pp. 87–90. ISSN 0254-0584. DOI: 10.1016/S0254-0584(98)00160-6.

[12] SARATHI, R., A. NANDINI and M. DANIKAS. Understanding Electrical Treeing Phenomena in XLPE Cable Insulation Adopting UHF Technique. *Journal of Electrical Engineering.* 2011, vol. 62, iss. 2, pp. 73–79. ISSN 1335-3632. DOI: 10.2478/v10187-011-0012-4.

[13] SHIMIZU, N. and C. LAURENT. Electrical tree initiation. *IEEE Transactions on Dielectrics and Electrical Insulation.* 1998, vol. 5, iss. 5, pp. 651–659. ISSN 1070-9878. DOI: 10.1109/94.729688.

[14] DANIKAS, M. G., I. KARAFYLLIDIS, A. THANAILAKIS and A. M. BRUNING. Simulation of electrical tree growth in solid dielectrics containing voids of arbitrary shape. *Modelling and Simulation in Materials Science and Engineering.* 1996, vol. 4, iss. 6, pp. 535–552. ISSN 0965-0393. DOI: 10.1088/0965-0393/4/6/001.

[15] SCHURCH, R., S. M. ROWLAND, R. S. BRADLEY and P. J. WITHERS. Imaging and analysis techniques for electrical trees using X-ray computed tomography. *IEEE Transactions on Dielectrics and Electrical Insulation.* 2014, vol. 21, iss. 1, pp. 53–63. ISSN 1070-9878. DOI: 10.1109/TDEI.2013.003911.

[16] NOSKOV, M. D., A. S. MALINOVSKI, M. SACK and J. SCHWAB. Self-consistent modeling of electrical tree propagation and PD activity. *IEEE Transactions on Dielectrics and Electrical Insulation.* 2000, vol. 7, iss. 6, pp. 725–733. ISSN 1070-9878. DOI: 10.1109/94.891982.

[17] NIEMEYER, L., L. PIETRONERO and H. J. WIESMANN. Fractal dimension of dielectric breakdown. *Physical Review Letters.* 1984, vol. 52, no. 12, pp. 1033–1036. ISSN 1079-7114. DOI: 10.1103/PhysRevLett.52.1033.

[18] DISSADO, L. A., J. C. FOTHERGILL, N. WISE and J. COOPER. A deterministic model for branched structures in the electrical breakdown of solid polymeric dielectrics. *Journal of Physics D: Applied Physics.* 2000, vol. 33, no. 19, pp. L109–L112. ISSN 0022-3727. DOI: 10.1088/0022-3727/33/19/103.

[19] VARDAKIS, G. and M. DANIKAS. Simulation of tree propagation in polyethylene including air void by using cellular automata: The effect of space charges. *Electrical Engineering.* 2002, vol. 84, no. 4, pp. 211–216. ISSN 1432-0487. DOI: 10.1007/s00202-002-0123-9.

[20] VARDAKIS, G. E. and M. G. DANIKAS. Simulation of electrical tree propagation using cellular automata: the case of conducting particle included in a dielectric in point-plane electrode arrangement. *Journal of electrostatics.* 2005, vol. 63, no. 2, pp. 129–142. ISSN 0304-3886. DOI: 10.1016/j.elstat.2004.06.008.

[21] VARDAKIS, G. E. and M. G. DANIKAS. Simulation of electrical tree propagation in a solid insulating material containing spherical insulating particle of a different permittivity with the aid of cellular automata. *Facta universitatis-series: Electronics and Energetics.* 2004, vol. 17, no. 3, pp. 377–389. ISSN 0353-3670. DOI: 10.2298/FUEE0403377V.

[22] VARDAKIS, G. E., M. G. DANIKAS and I. KARAFYLLIDIS. Simulation of space-charge effects in electrical tree propagation using cellular automata. *Materials Letters.* 2002, vol. 56, no. 4, pp. 404–409. ISSN 0167-577X. DOI: 10.1016/S0167-577X(02)00512-8.

[23] VICHNIAC, G. Y. Simulating physics with cellular automata. *Physica D: Nonlinear Phenomena.* 1984, vol. 10, no. 1, pp. 96–116. ISSN 0167-2789. DOI: 10.1016/0167-2789(84)90253-7.

[24] CHOPARD, B. and M. DROZ. *Cellular automata modeling of physical systems.* 1st ed. Cambridge: Cambridge University Press, 2012. ISBN 978-0-521-67345-7.

[25] VON NEUMANN, J. and A. W. BURKS. *Theory of self-reproducing automata.* 1st ed. Illinois: Illinois University Press, 1966.

[26] KARI, J. Theory of cellular automata: A survey. *Theoretical computer science.* 2005, vol. 334, iss. 1–3, pp. 3–33. ISSN 0304-3975. DOI: 10.1016/j.tcs.2004.11.021.

[27] WOLFRAM, S. *Cellular automata and complexity: collected papers.* 1st ed. Reading: Addison-Wesley, 1994. ISBN 0-201-62664-0.

[28] ATHANASSOPOULOS, S., C. KAKLAMANIS, G. KALFOUTZOS and E. PAPAIOANNOU. Cellular Automata: Simulations Using Matlab. In: *The Sixth International Conference on Digital Society.* Valencia: ICDS, 2012, pp. 63–68. ISBN 978-1-61208-176-2.

[29] MASON, J. H. Breakdown of solid dielectrics in divergent fields. *Proceedings of the IEE-Part C: Monographs.* 1955, vol. 102, no. 2, pp. 254–263. ISSN 0369-8904. DOI: 10.1049/pi-c.1955.0030.

[30] MASON, J. H. The deterioration and breakdown of dielectrics resulting from internal discharges.

Proceedings of the IEE-Part I: General. 1951, vol. 98, no. 109, pp. 44–59. ISSN 2054-0620. DOI: 10.1049/pi-1.1951.0019.

[31] ATTEN, P. and A. SAKER. Streamer propagation over a liquid/solid interface. *IEEE Transactions on Electrical Insulation.* 1993, vol. 28, no. 2, pp. 230–242. ISSN 0018-9367. DOI: 10.1109/14.212248.

[32] KREUGER, F. H. *Industrial high DC voltage: 1. fields, 2. breakdowns, 3. tests.* 1st ed. Delft: Delft University Press, 1995. ISBN 90-407-1110-0.

[33] DISSADO, L. A. and P. J. J. SWEENEY. Physical model for breakdown structures in solid dielectrics. *Physical Review B.* 1993, vol. 48, no. 22, pp. 16261–16268. ISSN 0163-1829. DOI: 10.1103/PhysRevB.48.16261.

About Authors

Hemza MEDOUKALI was born on 24.06.1988 in Algeria. He received his M.Sc. from the Universite de Constantine1 Algeria, in 2011. Since then, He prepares a Ph.D. thesis at the Dielectric materials Laboratory LeDMaScD in the Department of Electrical Engineering at the University Amar Telidji of Laghouat-Algeria. His research interests include: numerical modeling and simulation, high voltage, ageing of dielectric materials.

Mossadek GUIBADJ received his Engineer degree from Czech Technical University in Prague. In 2001, he received, the M.Sc. degree from Universite Ammar Telidji Laghouat- Algeria. He received his Ph.D. degree from Ecole Nationale Polytechnique (ENP) Algeria in 2009. Currently, he is professor and is head of research team in the Dielectric materials Laboratory LeDMaScD in the Department of Electrical Engineering at the University Amar Telidji of Laghouat, in Algeria. His research interests include numerical modeling and simulation, high voltage, partial discharges, dielectric materials.

Boubakeur ZEGNINI was born on 25.01.1968. He received the applied electrical engineering degree from the Ecole Normale d'Enseignement Technique ENSET Laghouat Algeria, in1991, the M.Sc. degree from the institute of Electrical Engineering, Center University of Laghouat in 2001. He received his Ph.D. degree in the field of Dielectric Materials from Universite des sciences et de la technologie USTO Oran Algeria in 2007, from 1991 to 2001 he was professor of technical secondary school. Since 2001 he is working as associate professor with the Department of Electrical Engineering at Amar Telidji University of Laghouat, Algeria. He joined the Laboratory of Electrical Engineering at Paul Sabatier University of Toulouse, France, "solid dielectrics and reliability" research team from 2005 to 2007. Currently, he is head of research team in the Dielectric materials Laboratory LeDMaScD in the Department of Electrical Engineering at the University Amar Telidji of Laghouat, in Algeria. Following this, he became a Full Professor at University Amar Telidji of Laghouat-Algeria, in 2012. His main research interests include high voltage, dielectric materials, outdoor insulation, numerical modeling and simulation. He is author and co-authors of many scientific publications.

10

PID, 2-DOF PID AND MIXED SENSITIVITY LOOP-SHAPING BASED ROBUST VOLTAGE CONTROL OF QUADRATIC BUCK DC-DC CONVERTER

Fateh OUNIS, Noureddine GOLEA

Electrical Engineering Department, Sciences and Applied Sciences Faculty, Larbi Ben M'hidi University, 04000 Oum El Bouaghi, Algeria

ounisfateh_01@yahoo.fr, nour_golea@yahoo.fr

Abstract. *DC-DC Quadratic Buck Converter (QBC) is largely used in applications where the high step-down conversion ratio is required. In the practical implementation, QBC is subject to uncertainties, disturbance, and sensor noise. To address the QBC control problems, a two-degree of freedom PID (2-DOF PID) is designed in the robust control framework. Further, for comparison purpose, a one-degree of freedom PID (1-DOF PID) and mixed sensitivity loop-shaping (MS-LS) controller are also proposed. Considering QBC parasitic components, the QBC small-signal transfer function is derived based on a practical approach. Sensitivity functions are used to specify the desired design requirements, and non-smooth optimization is used to tune both PID's parameters. The three control structures are implemented and tested in the Matlab/Simulink environment. As attested by simulation results, the 2-DOF PID exhibits a better regulation accuracy with enhanced robust stability and robust performance for a wide range of supply voltage/load variation and sensor noise effect.*

Keywords

DC-DC converters, mixed sensitivity, loop-shaping, PID control, quadratic buck converter, robust control.

1. Introduction

DC-DC converters are key elements in power energy modulation and conversion. Basic DC-DC converters, such as buck, boost and buck-boost, are widely used in various fields of technology [1]. Recently, new applications, such as LED lamps, microprocessors, portable devices and GPS, require very low dc voltages and they operate at very high currents. Such applications require converters with low ripples in the voltage and current and high efficiency in order to achieve precise output voltage regulation against parameter, line and load disturbances. Basic step-down converters are not suitable for high step-down voltage conversion since operating at a small duty ratio affects the converter dynamic performance and cause asymmetry in the on and off times of the switches. Moreover, very small duty ratio limits the converters switching frequency and increases peak switch current that leads to more switching losses, severe reverse-recovery problems and converter's efficiency degradation [2]. Some cascade interconnected power converters structures were developed to face this problem [3]. However, a notable disadvantage of cascaded converters is that the overall efficiency is reduced by losses in switching devices. To improve overall efficiency, Quadratic Buck Converters (QBC) were developed in [4], [5], [6] and [7]]. QBC is designed based on cascade connection of two buck converters and has only one active switching device. The DC conversion ratio is the product of the conversion ratios of the two single buck converters. QBC operates at higher switching frequencies with wide load range and achieves an improved step-down conversion ratio. The efficiency is also enhanced since only one active switch is used [8] and [9].

As QBC exhibits complex nonlinear dynamics subject to parameters uncertainties and input/load variations, control loops must be introduced to guarantee stability and operating performance. Several QBC control techniques such as linear state feedback, feedback linearization, sliding mode control and passivity based control, were presented in [10]. In [11], QBC nonlinear control scheme is proposed. QBC with LC input filter and damping control is developed in [12].

Robust QBC control based on identified Hammerstein model is proposed in [13]. Average current-mode control for the QBC is proposed in [14], where the outer voltage control loop bandwidth is limited by inner current control loop bandwidth. In [15], robust control for QBC is designed based on Kharitonov's theorem and D-stability concept. To ensure robust output regulation, robust state feedback stabilizer with saturated internal model, is proposed in [16]. In [17], inner current loop PI parameters are selected from QBC large-signal model, and outer voltage loop is controlled using a conventional PI regulator. To ensure robustness in the presence of disturbances and uncertainties, H_∞ based control is investigated in [18].

In this paper, 2-DOF PID is developed to solve the QBC robust control problem. For the comparison, a 1-DOF PID and MS-LS control are also proposed. Taking the parasitic components into account, the QBC small signal transfer function from output voltage to control signal is derived. Robust performance requirements are defined using the same weighting functions for both PID and by another set of weighting functions for MS-LS control. Contrary to MS-LS control and 1-DOF PID, the 2-DOF PID provides a wide range of the crossover frequency to specify response time-performance compromise. The non-smooth approach presented in [19] is used to tune both PID's parameters. Based on the model reduction methods, MS-LS controller is reduced from 8 to 5, and the reduced version is presented and used in simulations. Simulation results illustrate the 2-DOF PID in term of accuracy and stability robustness.

The remainder of this paper is organized as follows. Section 2. presents the QBC nominal transfer function computation. The MS-LS control is developed in Section 3. PID controller's design is provided in Section 4. The three controllers' robustness analysis is established in Section 5. Simulation results are shown in Section 6. Concluding remarks are given in Section 7.

2. QBC Nominal Model

As a first step for the control design, the QBC open loop small-signal control-to-output voltage transfer function should be established. The objective can be reached using analytical modeling and averaging techniques [20] and [21]. In this work, a practical approach is adopted. Based on QBC parameters and operating point given in App. A, a Simulink implementation of the open loop excitation is realized (Fig. 1). The PWM control signal duty cycle is adjusted to get the desired output voltage level. Taking into account the parasitic components, the QBC discrete-time transfer function $G(z) = B(z)/A(z)$ is assumed of 6th order. Hence,

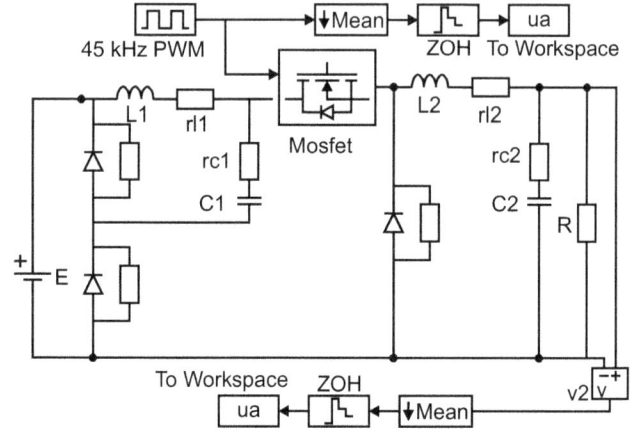

Fig. 1: QBC open loop excitation setup.

applying the Steiglitz-McBride recursive identification method [22] on converter averaged input/output signals for 5 iterations, yields the estimated transfer function:

$$
\begin{aligned}
B(z) &= 0.7108 - 1.6654z^{-1} + 1.9263z^{-2} \\
&\quad -1.6483z^{-3} + 1.2154z^{-4} - 0.5151z^{-5}, \\
A(z) &= 1 - 4.829z^{-1} + 10.1746z^{-2} - 11.9259z^{-3} \\
&\quad +8.1753z^{-4} - 3.1028z^{-5} + 0.5094z^{-6}.
\end{aligned}
\tag{1}
$$

Further, using the numerical algorithm proposed in [23], the equivalent continuous-time transfer function $G(s) = N(s)/D(s)$ is given by:

$$
\begin{aligned}
N(s) &= 544300s^5 + 1.618 \cdot 10^{11}s^4 + \\
&\quad +2.184 \cdot 10^{18}s^3 + 1.524 \cdot 10^{23}s^2 + \\
&\quad +7.774 \cdot 10^{29}s + 3.487 \cdot 10^{34}, \\
D(s) &= s^6 + 674400s^5 + 7.804 \cdot 10^{11}s^4 + \\
&\quad +3.354 \cdot 10^{17}s^3 + 1.502 \cdot 10^{23}s^2 + \\
&\quad +2.677 \cdot 10^{28}s + 2.342 \cdot 10^{33}.
\end{aligned}
\tag{2}
$$

The $G(s)$ Bode plot is shown in Fig. 2. It is clear that QBC has two second-order filters with high quality-factor Q, which depends on the selected circuit values. All poles and zeros are located on the right half of the s-plane as shown the Fig. 3. The right half plane zeros are responsible for the excessive phase lag in the ideal case. The Equivalent Series Resistances (ESR) provide some damping into the system, which is beneficent as it will ease feedback control design.

3. MS-LS Control Design

A diagram of the control design is shown in Fig. 4, where G is the quadratic buck converter transfer function. W_1, W_2 and W_3 are the performance, control and noise weighting functions, respectively. Further, w denote input signals, z output vector that includes

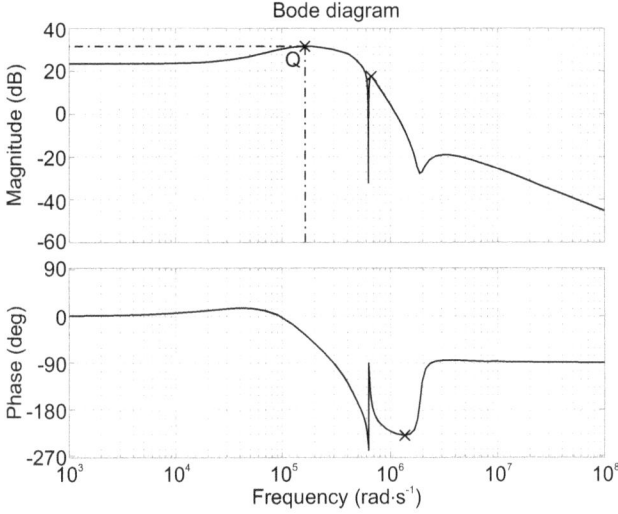

Fig. 2: Frequency open loop QBC response.

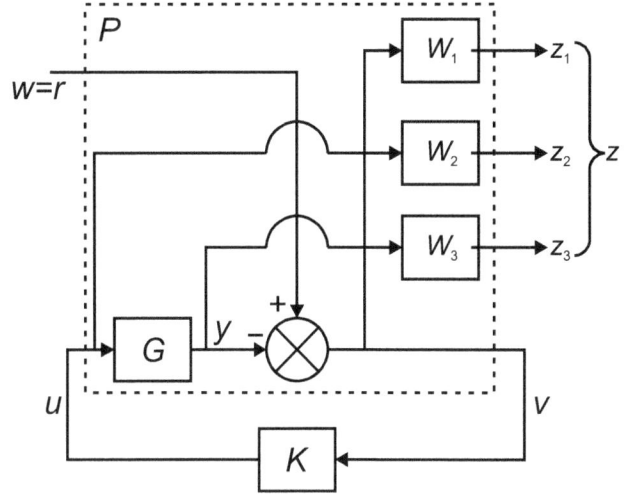

Fig. 4: MS-LS control design.

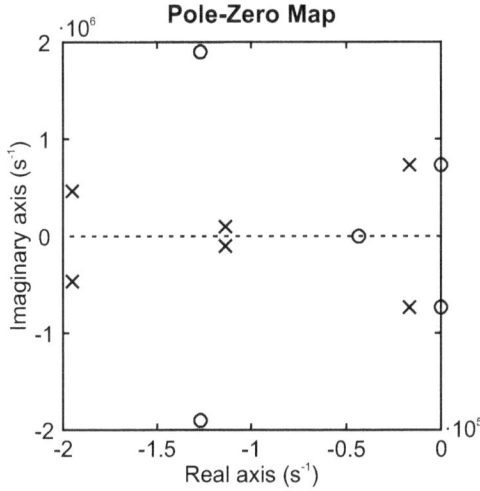

Fig. 3: Transfer function pole-zero map.

transfer function T_{zw} such that:

$$||T_{zw}||_\infty < 1. \tag{6}$$

3.1. Weighting Functions Selection

The closed loop performance of the system is largely dependent on the shape of the weighting function. The weight function W_1 specifies the control performance and W_1 is selected according to methodology suggested by Zhou [24],

$$W_1 = \frac{s/M_s + w_s}{s + w_s e_s}, \tag{7}$$

where e_s is the maximum allowed steady-state offset fixed to $e_s = 0.001$, w_s is the desired bandwidth fixed to $6 \cdot 10^3$ rad·s^{-1} and M_s is the sensitivity peak (typically $M = 1.6$). Therefore,

$$W_1 = \frac{0.625(s + 9600)}{s + 6}. \tag{8}$$

In order to avoid impulsive input effect on the converter, W_2 is chosen as:

$$W_2(s) = 0.01. \tag{9}$$

W_3 is used to shape the complementary sensitivity function T, and thus it must be large at high frequencies. Hence W_3 is chosen as:

$$W_3 = \frac{s + w_b/M_b}{e_b s + w_b}. \tag{10}$$

To keep the system stable, the complementary sensitivity T must be small for high frequencies. Thus, the

both performance and robustness measures, v is the vector of measurements available to the controller K and u the control signal. In the context, the necessary definitions are given by:

$$S(s) = \big(I + G(s)K(s)\big)^{-1}, \tag{3}$$

$$T(s) = G(s)K(s)\big(I + G(s)K(s)\big)^{-1}, \tag{4}$$

where $S(s)$ is the sensitivity function and $T(s)$ is the complementary sensitivity function. The generalized closed loop transfer function is given by:

$$T_{zw} = \begin{bmatrix} W_1 S \\ W_2 R S \\ W_3 T \end{bmatrix}, \tag{5}$$

where $R(s) = K(s)\big(I + G(s)K(s)\big)^{-1}$. In this mixed problem, the control objective is to design a stable controller that minimizes the norm of the generalized value 0.001 is selected for the parameter e_b. In order to limit the closed loop bandwidth, the parameter M_b is fixed as 1.6 and w_b is fixed to 10^4 rad·s^{-1}. Then:

$$W_3 = \frac{1000(s + 6250)}{s + 10^7}. \tag{11}$$

In order to adopt a unified solution procedure, the above matrix inequality Eq. (6) can be recast into a standard configuration as in Fig. 4. This can be obtained by using the Linear Fractional Transformation (LFT), and the generalized plant P is obtained by grouping signals into sets of external inputs, outputs, input to the controller and output from the controller, which yields:

$$\begin{bmatrix} z \\ \hline v \end{bmatrix} = \underbrace{\begin{bmatrix} W_1 & -W_1 G \\ 0 & W_2 G \\ 0 & W_3 G \\ \hline I & -G \end{bmatrix}}_{P} \begin{bmatrix} r \\ \hline u \end{bmatrix}, \tag{12}$$

where $r = w$ is the reference voltage, $z = \begin{bmatrix} z_1 & z_2 & z_3 \end{bmatrix}^T$ is the output signals vector, is the control signal and is the controlled QBC output voltage. W_1, W_2 and W_3 are the weighting functions described by Eq. (8), Eq. (9) and Eq. (11), respectively.

Based on the above configuration, the generalized plant can be built up, and consequently the controller can be calculated using Matlab robust control toolbox. Hence, the obtained 8th order controller is:

$$K(s) = \frac{N_K(s)}{D_K(s)}, \tag{13}$$

with

$$\begin{aligned} N_K(s) = &\, 5.504 \cdot 10^6 s^7 + 5.875 \cdot 10^{13} s^6 + 4.141 \cdot 10^{19} s^5 \\ &+ 4.48 \cdot 10^{25} s^4 + 1.929 \cdot 10^{31} s^3 + 8.414 \cdot 10^{36} s^2 \\ &+ 1.486 \cdot 10^{42} s + 1.289 \cdot 10^{47}, \end{aligned}$$

and

$$\begin{aligned} D_K(s) = &\, s^8 + 5.444 \cdot 10^9 s^7 + 5.525 \cdot 10^{15} s^6 \\ &+ 2.3 \cdot 10^{22} s^5 + 1.718 \cdot 10^{28} s^4 + 8.868 \cdot 10^{33} s^3 \\ &+ 5.922 \cdot 10^{39} s^2 + 2.5 \cdot 10^{44} s + 1.5 \cdot 10^{45}. \end{aligned}$$

The obtained controller Eq. (13) has high order, and can be further, reduced by examining $K(s)$ Hankel singular values σ_i. Hankel singular values based model reduction routines are grouped by the types of error bound. In Balanced Truncation (BT) and related methods, an error bound is a measure of how close the reduced order controller $K_r(s)$ is to the original system and is computed based on the infinity norm of the additive error,

$$\|K(s) - K_r(s)\|_\infty = \sum_{r+1}^{n} (\sigma_i), \tag{14}$$

with

$$K_r = \frac{N_{K_r}(s)}{D_{K_r}(s)}. \tag{15}$$

The basic idea of BT relies on balancing the two controllers' controllability Gramian and operability Gramian [25]. The Hankel singular values plotted in Fig. 5 are used to decide which states of the controller can be safely discarded. To achieve at least 1 % relative accuracy, the lowest-order controller $K_r(s)$ should be compatible with the desired level of accuracy chosen to be 5.

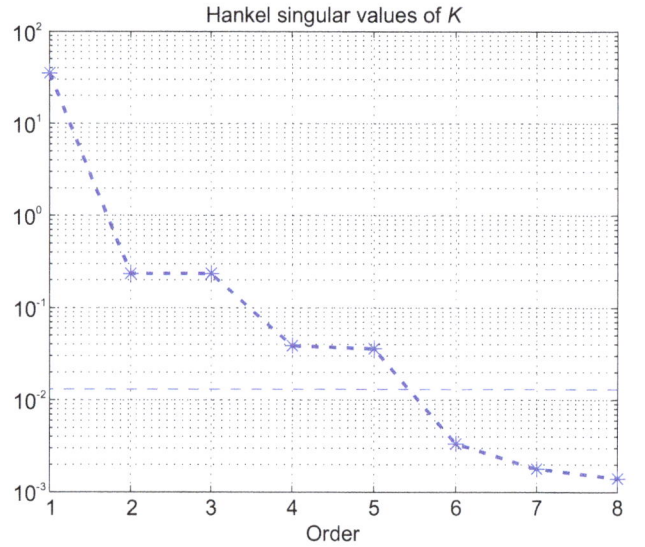

Fig. 5: Hankel singular values of $K(s)$.

The function "reduce" is the gateway to all model reduction routines available in the toolbox MATLAB. We use the default, square-root balance truncation ('balancmr') option of "reduce" as the first step. This method uses an "additive" error bound for the above described reduction method, meaning that it tries to keep the absolute approximation error uniformly small for all frequencies.

The error bound for additive-error algorithms is defined as:

$$\|K(s) - K_r(s)\|_\infty = 2(\sigma_6 + \sigma_7 + \sigma_8) = 0.0088, \tag{16}$$

which yields:

$$\begin{aligned} N_{K_r}(s) = &\, 1.045 \cdot 10^4 s^4 - 3.36 \cdot 10^9 s^3 + 5.937 \cdot 10^{15} s^2 \\ &- 1.393 \cdot 10^{21} s + 6.808 \cdot 10^{26}, \end{aligned}$$

$$\begin{aligned} D_{K_r}(s) = &\, s^5 + 2.848 \cdot 10^5 s^4 + 4.018 \cdot 10^{12} s^3 \\ &+ 1.128 \cdot 10^{17} s^2 + 1.43 \cdot 10^{24} s + 7.464 \cdot 10^{24}. \end{aligned}$$

According to condition Eq. (6), it is necessary that the magnitude response of S lies bellow the magnitude response of W_1^{-1} in the whole frequency range, and the magnitude response of T should lie bellow the response of W_3^{-1}. Figure 6 shows that these conditions are verified using the reduced controller.

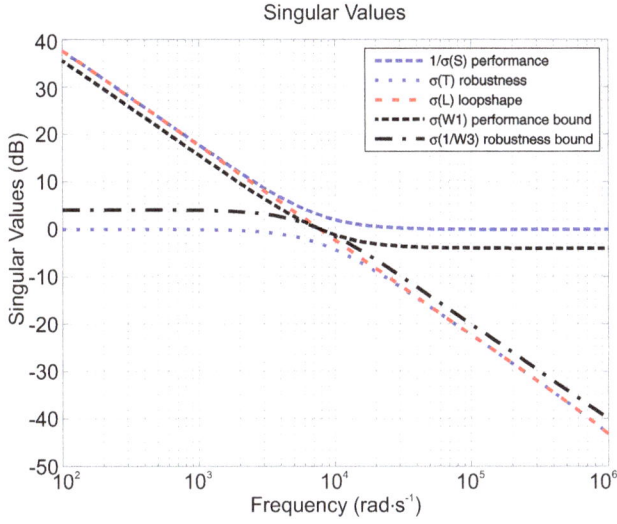

Fig. 6: MS-LS control design results.

4. 1-DOF PID and 2-DOF PID Controllers Design

In the following section, the control system structure of Fig. 7 is adopted, where $C(s)$ is the standard controller, $C_F(s)$ the input filter, and G is the converter transfer function.

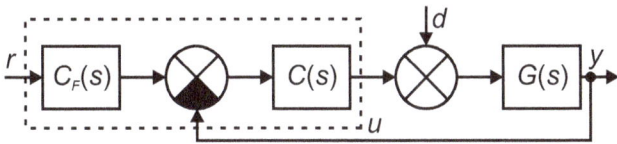

Fig. 7: 2-DOF PID structure.

The standard PID controller is used with the transfer function:

$$C(s) = K_P + \frac{K_I}{s} + \frac{K_D s}{T_f s + 1}, \qquad (17)$$

with the proportional gain K_P, the integrator gain K_I, the derivative gain K_D, and the derivative filter time constant T_f.

4.1. 2-DOF PID Controller

The output signal of a 2-DOF regulator is defined as:

$$u(s) = K_P e_p + K_I e_I + K_D e_D, \qquad (18)$$

where

$$\begin{cases} e_p = br(s) - y(s) \\[2mm] e_I(s) = \dfrac{1}{s}\big(r(s) - y(s)\big) \quad , \\[2mm] e_D(s) = \big(cr(s) - y(s)\big) \end{cases} \qquad (19)$$

where b, c are weighting parameters for proportional term and derivative term, respectively. The 2-DOF controller can be transformed into a 1-DOF controller, if b and c are selected to be equal to 1. To formulate the closed loop transfer function, the output of controller Eq. (19) is rewritten as:

$$u(s) = \big(C_F(s)r(s) - y(s)\big)C(s). \qquad (20)$$

The closed loop control system output to the perturbation is given by:

$$y(s) = \frac{C_F(s)G(s)}{1 + C(s)G(s)}r(s) + \frac{G(s)}{1 + C(s)G(s)}d(s). \qquad (21)$$

The system closed loop transfer function is defined as:

$$T_{yr}(s) = C_F(s)\frac{C(s)G(s)}{1 + C(s)G(s)}. \qquad (22)$$

The parameters $\{K_P, T_I, T_D, b, c\}$ are obtained considering the targeted specifications.

4.2. Frequency Specifications

To ensure that the output voltage tracks the reference with a desired response time and tracking error, transfer function is used to specify the maximum frequency-domain tracking error:

$$e_{\max} = \frac{A_e s + \omega_c D_e}{s + \omega_c}, \qquad (23)$$

where $\omega_c = 2/t_s$ (t_s is the settling time) is the tracking bandwidth, D_e is the maximum relative steady-state error and A_e is the peak relative error across all frequencies. For the QBC, we set $D_e = 0.001$, $A_e = 1$ and $\omega_c = 10^3$ rad·s^{-1}, since the open loop-gain should be high within the control bandwidth. To ensure good disturbance rejection, the minimum loop gain profile is chosen as:

$$W_s = \frac{0.03 w_c}{s}. \qquad (24)$$

To ensure insensitivity to measurement noise, the open loop gain should be less than 1 outside the control bandwidth, so the maximum loop gain profile is chosen as:

$$W_T = \frac{0.3 w_c}{s}. \qquad (25)$$

Using software provided by Matlab, the above requirements are converted into normalized scalars functions $f(x)$ and $g(x)$ such as:

$$g(x) = \left\| \frac{1}{e_{\max}}\big(T(s, x)\big) \right\|_{\infty}. \qquad (26)$$

$$f(x) = \left\| \begin{array}{c} W_s S_a \\ W_T^{-1} T_a \end{array} \right\|_\infty, \qquad (27)$$

where $T(s, x) = \dfrac{L}{1 + L}$ is the output complementary sensitivity function, $L(s, x)$ is the open-loop response being shaped, $T_a = D^{-1}TD$ is the scaled output complementary sensitivity function, $S_a = D^{-1} \left[(1 + L(s, x)) \right]^{-1} D$ is the scaled output sensitivity function, x is the vector of free (tunable) parameters K_P, K_I and K_D.

Then, determining the PID parameters is equivalent to solving the optimization problem:

$$\min_x \max\big(af(x), g(x)\big), \qquad (28)$$

where $a > 0$ is a parameter weighting the subproblems importance in order to get the most optimal solution for the optimization problem. Nonsmooth optimization algorithms [26] and [27] are used to solve the QBC converter control problem. According to the required desirable performance, the ultimate PIDs parameters are achieved as follows:

1-DOF: $K_p = 0.00865$, $K_I = 230$, $K_D = -8.03 \cdot 10^{-5}$,
2-DOF: $K_p = 0.00824$, $K_I = 298$, $K_D = 1.68 \cdot 10^{-5}$,
$b = 0.00015$, $c = 0.226$.

5.   Robust Analysis

We can test the robustness properties of the three controllers by executing the proper μ tests for the QBC uncertain feedback system shown in Fig. 8, where the dashed box represents the QBC real transfer function G_{unc}. The transfer functions W_{del} and Δ_G parameterize the multiplicative uncertainty at the converter input. The transfer function W_{del} is assumed known, and the transfer function Δ_G is assumed to be stable and unknown, except for the norm condition $\|\Delta_G\|_\infty < 1$. The uncertainty weight W_{del} is described as:

$$W_{del}(s) = \frac{100s + 7.035 \cdot 10^7}{s + 7.035 \cdot 10^8}. \qquad (29)$$

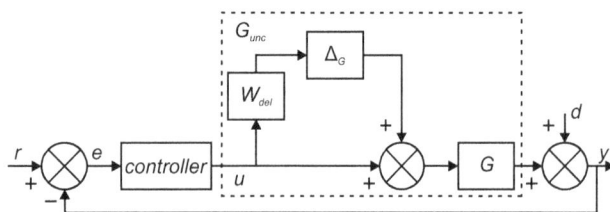

Fig. 8: Uncertain feedback system.

The uncertainty in the input is 10 % in the low frequency range, 100 % at $w = 10^6$ Hz and 1000 % in the high frequency range.

Figure 9 compares the upper bounds of the structured singular values, for the robust stability analysis of the closed-loop systems with the three controllers (1-DOF PID, 2-DOF PID and MS-LS control). To achieve robust stability, it is necessary that the μ-values are less than 1 over the frequency range [28] and [29]. It is clear that the controllers achieve a robust stability. The best robustness is obtained by the MS-LS controller.

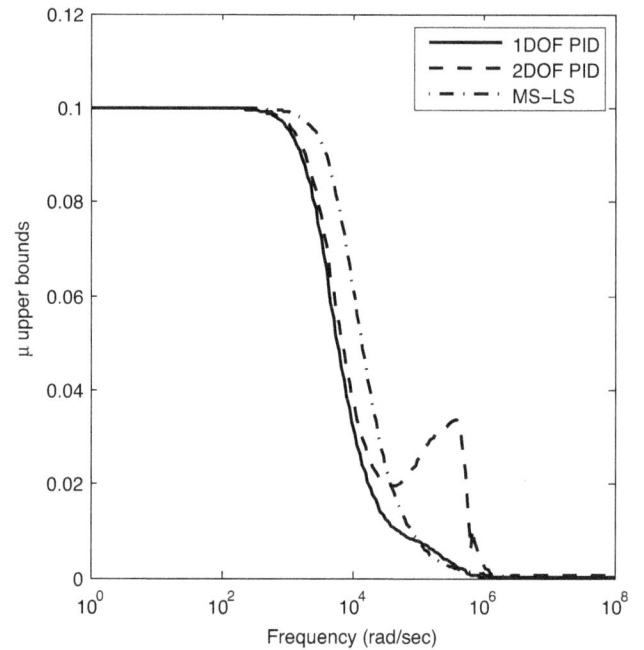

Fig. 9: QBC closed-loop robust stability.

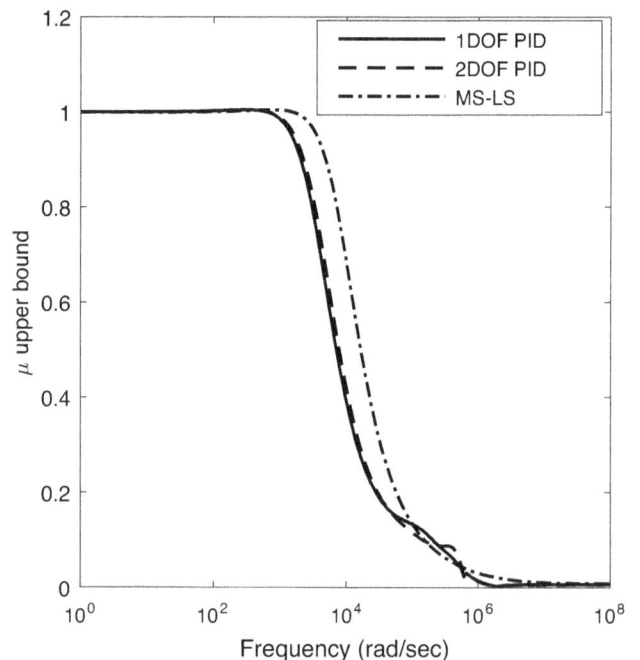

Fig. 10: QBC closed-loop robust performance.

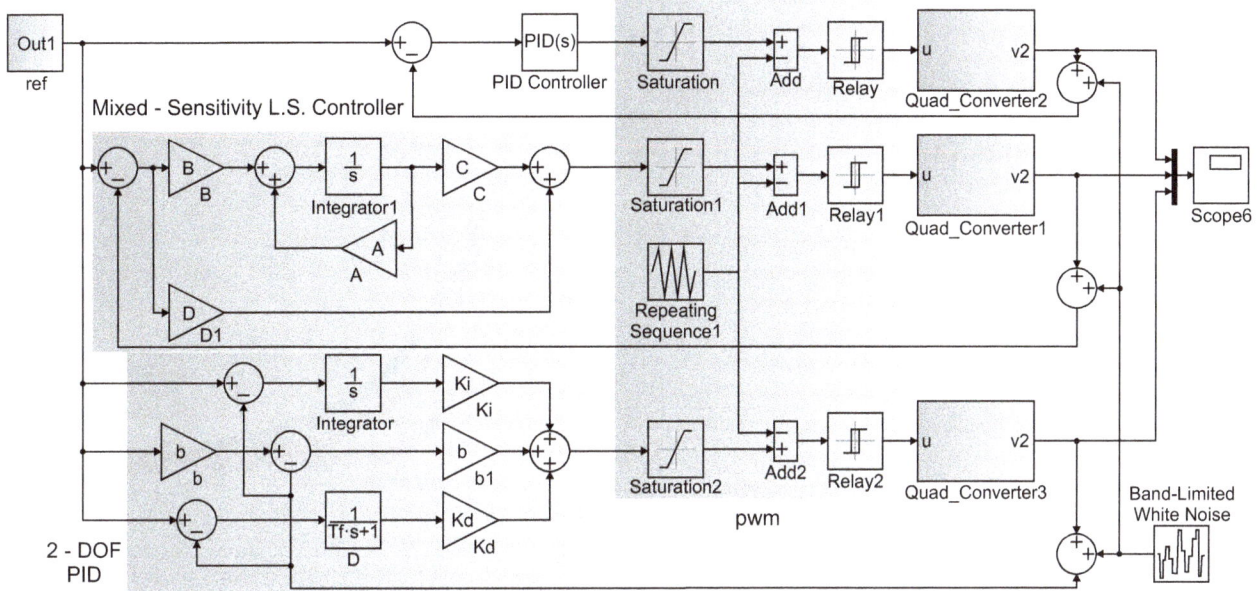

Fig. 11: Simulation block diagram.

The robust performance is achieved if and only if for each frequency computed for the closed-loop frequency response is less than 1. The robust performance tests for the three controllers are shown in Fig. 10. Again, the MS-LS controller shows large μ values over the low-frequency range.

6.   Simulation Results

As shown in Fig. 11, the three controllers designed in the above section are implemented in Matlab/Simulink. Note that MS-LS controller $K_r(s)$ is implemented using the state space realization (A, B, C, D). To compare the three controllers' performances and robustness, the following tests are performed.

6.1.   Set Point Tracking

The QBC response for a 10 V constant reference voltage is shown in Fig. 12. It can be observed that QBC settling time is 1 ms for the MS-LS control and 2-DOF PID controls, which is faster as compared to the 1.6 ms settling time for the 1-DOF PID control. Another aspect is that MS-LS control exhibits an overshoot of 16.5 %; while the overshoot for the 1-DOF PID and 2-DOF PID is of the order of 15.5 %.

In addition, when the reference input voltage changes from 10 to 12 V, as shown in Fig. 13, an oscillatory behavior is observed for MS-LS control. Comparison to that PID controllers provide a more dumped behavior, in response to the reference voltage increase

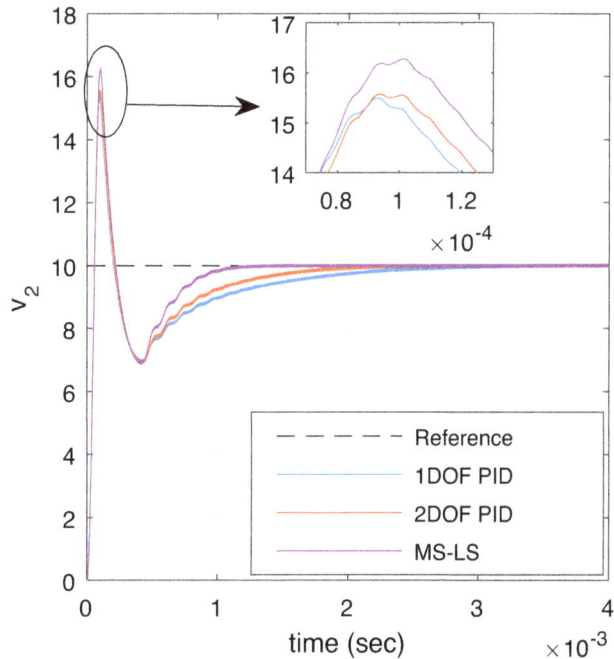

Fig. 12: QBC response to 10 V reference voltage.

or decrease, as can be noticed from Fig. 14 and Fig. 15.

6.2.   Load Variation

A load variation of 100 % (from 10 to 20 Ω) is introduced between 10 and 30 ms. QBC response shown in Fig. 16 indicates that all control methods provide almost the same performance. The output voltage exhibits an undershoot of 6 % and an overshoot of 8 %.

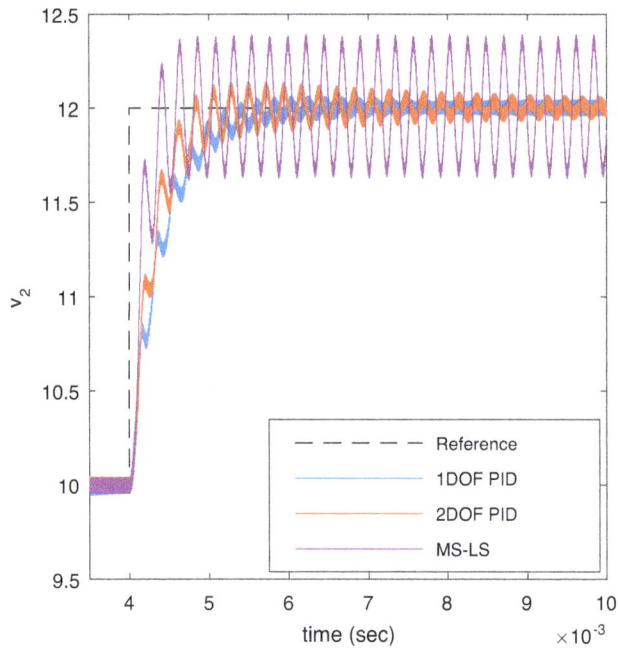

Fig. 13: QBC response to reference voltage change from 10 to 12 V.

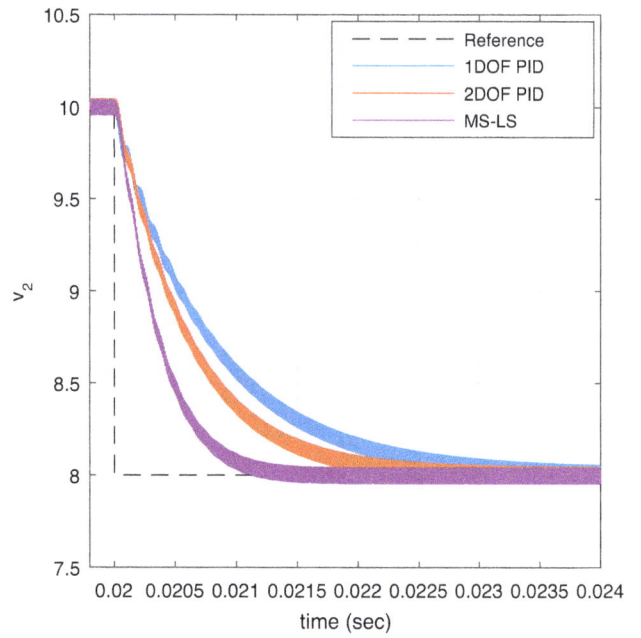

Fig. 15: QBC response to reference voltage change from 10 to 8 V.

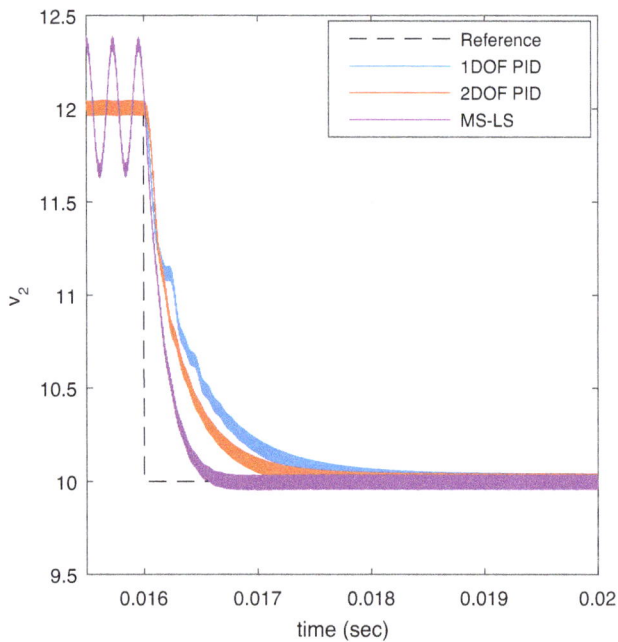

Fig. 14: QBC response to reference voltage change from 12 to 10 V.

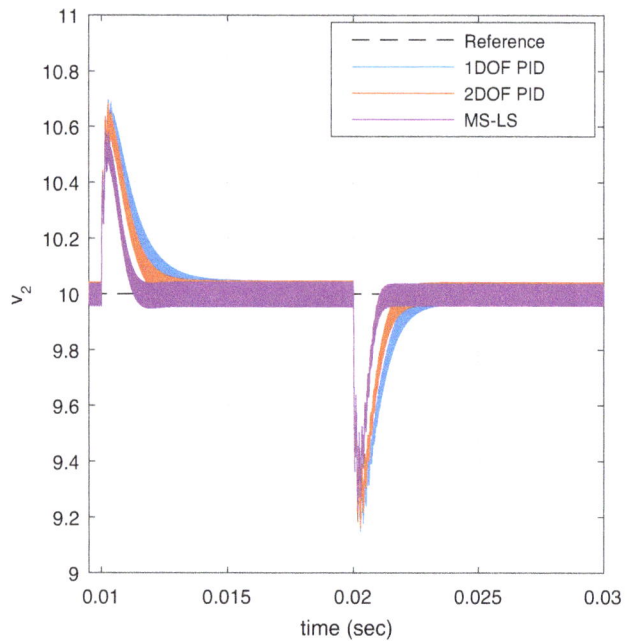

Fig. 16: QBC response to load resistance change.

methods exhibit a null undershoot compared to 3 % for MS-LS control.

6.3.   Supply Voltage Variation

A voltage drop of 3 V is introduced in the supply voltage between 5 and 10 ms. Figure 17 shows that the three control methods obtained almost the same stabilizing time with same undershoot (about 7 % at 5 ms). At 10 ms, similar overshoot (35 %) is observed for the three control methods. However, at 10 ms, PID control

6.4.   Disturbance Rejection

The supply voltage is perturbed by a sinusoidal component of 100 Hz frequency and 2 V peak-to-peak amplitude. Further, sensor white noise, of 10^8 rad·s^{-1} frequency and 10^{-5} power, is assumed to be superposed on the output voltage. Figure 18 shows that

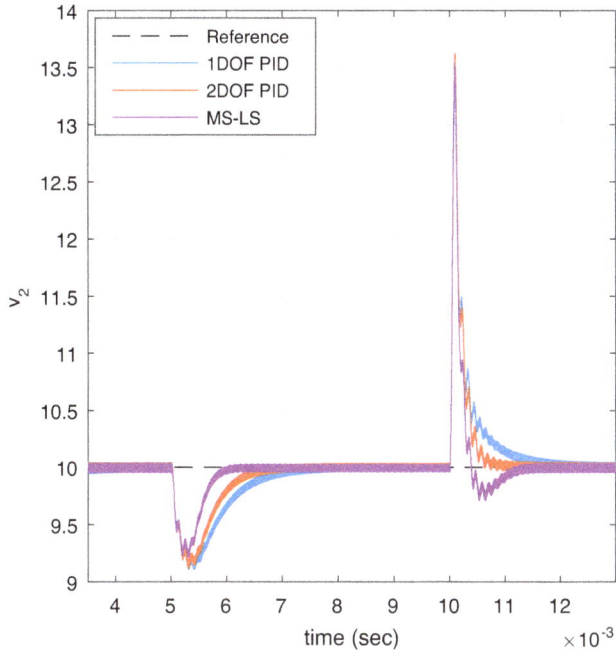

Fig. 17: QBC response to supply voltage change.

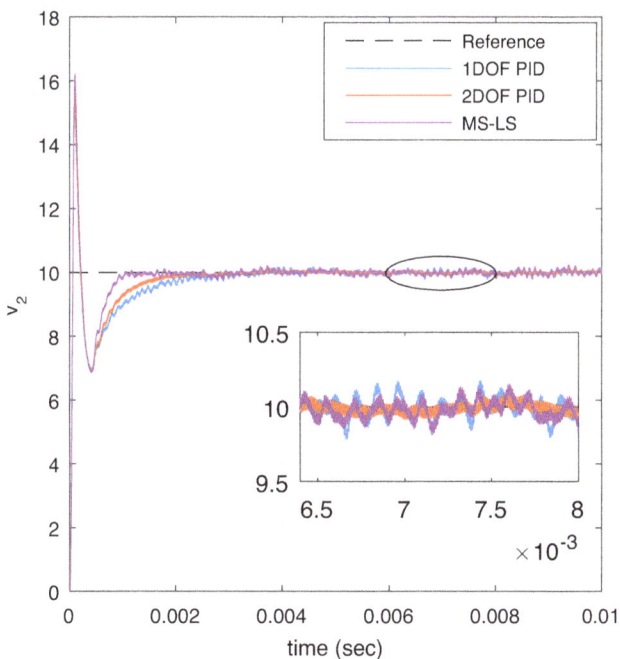

Fig. 18: QBC response to sinusoidal component in power source and white noise.

2-DOF PID controller has better disturbance rejection than 1-DOF PID and MS-LS controllers in this band of frequencies.

7. Conclusion

A 2-DOF PID controller is proposed, designed and simulated for the quadratic buck converter. For com-parison purpose, 1-DOF PID and MS-LS control con-trollers are also tested. Even if MS-LS control shows a faster response, 2-DOF PID provides more duped and accurate response. Further, under perturbations and uncertainties, 2-DOF PID control exhibits better robustness in performance and stability compared to MS-LS control. Another practically important advan-tage of the 2-DOF PID is a lower structure complexity compared to MS-LS control.

References

[1] BACHA, S., I. MUNTEANU and A. I. BRATCU. *Power electronic converters modeling and con-trol: with case studies.* London: Springer, 2014. ISBN 978-1-4471-5477-8.

[2] TOLLE, T., T. DUERBAUM and R. ELFERICH. Switching loss contributions of synchronous recti-fiers in VRM applications. In: *IEEE 34th Annual Power Electronics Specialist Conference 2003.* Pis-cataway: IEEE, 2003, pp. 144–149. ISBN 0-7803-7754-0. DOI: 10.1109/PESC.2003.1218287.

[3] MATSUO, H. and K. HARADA. The cas-cade connection of switching regulators. *IEEE Transactions on Industrial Applications.* 1976, vol. IA-12, iss. 2, pp. 192–198. ISSN 0093-9994. DOI: 10.1109/TIA.1976.349401.

[4] MAKSIMOVIC, D. and S. CUK. Switching converters with wide DC conversion range. *IEEE Transactions on Power Electronics.* 1991, vol. 6, iss. 6, pp. 151–157. ISSN 0885-8993. DOI: 10.1109/63.65013.

[5] BARBOSA, L. R., J. B. VIEIRA, L. C. FRE-ITAS, M. S. VILELA and V. J. FARIAS. A buck quadratic PWM soft-switching converter us-ing a single active switch. *IEEE Transactions on Power Electronics.* 1999, vol. 14, iss. 3, pp. 445–453. ISSN 0885-8993. DOI: 10.1109/63.761688.

[6] PACHECO, V. M., A. J. DO NASCIMENTO, V. J. FARIAS, J. B. VIEIRA and L. C. DE FRE-ITAS. A quadratic buck converter with lossless commutation. *IEEE Transactions on Industrial Electronics.* 2000, vol. 47, no. 2, pp. 264–272. ISSN 0278-0046. DOI: 10.1109/41.836341.

[7] REYES-MALANCHE, J. A., N. VAZQUEZ and J. LEYVA-RAMOS. Switched-capacitor quadratic buck converter for wider conversion ratios. *IET Power Electronics.* 2015, vol. 8, iss. 12, pp. 2370–2376. ISSN 1755-4535. DOI: 10.1049/iet-pel.2014.0755.

[8] CARBAJAL-GUTIERREZ, E. E., J. A. MORALES-SALDANA and J. LEYVA-RAMOS. Modeling of a single-switch quadratic buck converter. *IEEE Transactions on Aerospace and Electronic Systems.* 2005, vol. 41, iss. 4, pp. 1450–1456. ISSN 0018-9251. DOI: 10.1109/TAES.2005.1561895.

[9] AYACHIT, A. and M. K. KAZIMIERCZUK. Steady-state analysis of PWM quadratic buck converter in CCM. In: *IEEE 56th International Midwest Symposium on Circuits and Systems (MWSCAS).* Columbus: IEEE, 2013, pp. 49–52. ISBN 978-1-4799-0065-7. DOI: 10.1109/MWS-CAS.2013.6674582.

[10] SIRA-RAMIREZ, H. and R. SILVA-ORTIGOZA. *Control Design Techniques in Power Electronics Devices.* London: Springer-Verlag, 2006. ISBN 978-1-84628-458-8.

[11] WEI, X. L., K. M. TSANG and W. L. CHAN. Non-linear PWM control of single-switch quadratic buck converters using internal model. *IET Power Electronics.* 2009, vol. 2, iss. 5, pp. 475–483. ISSN 1755-4535. DOI: 10.1049/iet-pel.2008.0170.

[12] SOSA, J. M., E. D. SILVA-VERA, G. ES-COBAR, P. R. MARTINEZ-RODRIGUEZ and A. A. VALDEZ-FERNANDEZ. Control design for a quadratic buck converter with LC input filter. In: *IEEE 13th International Conference on Power Electronics (CIEP).* Guanajuato: IEEE, 2016, pp. 149–154. ISBN 978-1-5090-1775-1. DOI: 10.1109/CIEP.2016.7530747.

[13] ALONGE, F., R. RABBENI, M. PUCCI and G. VITALE. Identification and robust control of a quadratic DC/DC boost converter by Hammerstein model. *IEEE Transactions on Industry Applications.* 2015, vol. 51, iss. 5, pp. 3975–3985. ISSN 0093-9994. DOI: 10.1109/TIA.2015.2416154.

[14] MORALES, J. A. S., J. LEYVA-RAMOS, E. E. G. CARBAJAL and M. G. ORITZ-LOPEZ. Average current-mode control scheme for a quadratic buck converter with a Single switch. *IEEE Transactions on Power Electronics.* 2008, vol. 23, iss. 1, pp. 485–490. ISSN 0885-8993. DOI: 10.1109/TPEL.2007.910907.

[15] BEVRANI, H., P. BABAHAJYANI, F. HABIBI and T. HIYAMA. Robust control design and implementation for a quadratic buck converter. In: *International Power Electronics Conference IPEC-2010.* Sapporo: IEEE, 2010, pp. 99–103. ISBN 978-1-4244-5394-8. DOI: 10.1109/IPEC.2010.5543644.

[16] JIN, Q. B. and Q. LIU. Analytical IMC-PID design in terms of performance/robustness tradeoff for integrating processes: From 2-Dof to 1-Dof. *Journal of Process Control.* 2014, vol. 24, iss. 3, pp. 22–32. ISSN 0959-1524. DOI: 10.1016/j.jprocont.2013.12.011.

[17] MORALES, J. A. S., L. P. RODRIGO and P. H. ELVIA. Parameters selection criteria of proportional-integral controller for a quadratic buck converter. *IET Power Electronics.* 2014, vol. 7, iss. 6, pp. 1527–1535. ISSN 1755-4535. DOI: 10.1049/iet-pel.2013.0343.

[18] XU, G., D. SHA and X. LIAO. Decentralized inverse-droop control for input-series-output-parallel DC-DC converters. *IEEE Transactions on Power Electronics.* 2015, vol. 30, iss. 9, pp. 4621–4625. ISSN 0885-8993. DOI: 10.1109/TPEL.2015.2396898.

[19] APKARIAN, P. and D. NOLL. Nonsmooth H_∞ Synthesis. *IEEE Transactions on Automatic Control.* 2006, vol. 51, iss. 1, pp. 71–86. ISSN 0018-9286. DOI: 10.1109/TAC.2005.860290.

[20] AYACHIT, A. and M. K. KAZIMIERCZUK. Open-loop small-signal transfer functions of the quadratic buck PWM DC-DC converter in CCM. In: *IEEE 40th Annual Conference of the Industrial Electronics Society.* Dallas: IEEE, 2014, pp. 1643–1649. ISBN 978-1-4799-4032-5. DOI: 10.1109/IECON.2014.7048723.

[21] WLODZIMIERZ, J. Small-signal transmittances of DC-DC step-down PWM converter in various operation modes. *Archives of Electrical Engineering.* 2015, vol. 64, iss. 3, pp. 505–529. ISSN 2300-2506. DOI: 10.2478/aee-2015-0038.

[22] STEIGLITZ, K. and L. E. MCBRIDE. A technique for the identification of linear systems. *IEEE Transactions on Automatic Control.* 1965, vol. 10, iss. 4, pp. 461–464. ISSN 0018-9286. DOI: 10.1109/TAC.1965.1098181.

[23] KOLLAR, I., G. FRANKLIN and R. PINTELON. On the equivalence of z-domain and s-domain models in system identification. In: *IEEE Instrumentation and Measurement Technology Conference.* Brussels: IEEE, 1996, pp. 14–19. ISBN 0-7803-3312-8. DOI: 10.1109/IMTC.1996.507191.

[24] ZHOU, K. and J. C. DOYLE. *Essentials of Robust Control.* Upper Saddle River: Prentice-Hall, 1999. ISBN 978-0135258330.

[25] SAFONOV, M. G. and R. Y. CHIANG. A Schur method for balanced-truncation model reduction. *IEEE Transactions on Automatic Control.* 1989,

vol. 34, iss. 7, pp. 729–733. ISSN 0018-9286. DOI: 10.1109/9.29399.

[26] APKARIAN, P. and P. GAHINET. Decentralized and fixed-structure H∞ control in Matlab. In: *50th IEEE Conference on Decision and Control and European Control Conference*. Piscataway: IEEE, 2011, pp. 8205–8210. ISBN 978-1-61284-800-6. DOI: 10.1109/CDC.2011.6160298.

[27] APKARIAN, P., M. N. DAO and D. NOLL. Parametric robust structured control design. *IEEE Transactions on Automatic Control*. 2015, vol. 60, iss. 7, pp. 1857–1869. ISSN 0018-9286. DOI: 10.1109/TAC.2015.2396644.

[28] FAN, M. K. H., A. L. TITS and J. DOYLE. Robustness in the presence of mixed parametric uncertainty and unmodeled dynamics. *IEEE Transactions on Automatic Control*. 1991, vol. 36, iss. 1, pp. 25–38. ISSN 0018-9286. DOI: 10.1109/9.62265.

[29] YEDAVALLI, R. K. *Robust control of uncertain dynamic systems. A linear state space approach*. New York: Springer, 2014. ISBN 978-1-4614-9131-6.

About Authors

Fateh OUNIS was born in Oum El Bouaghi, Algeria. He received the Master degree in electrical engineering from Oum El Bouaghi University, in 2009. Actually, he is finalising Ph.D. thesis. His research interests are power converters design and control.

Noureddine GOLEA received his Doctorate degree in 2001 from Batna University, Algeria. Currently, he is a full Professor in Electrical Engineering Department at Oum El Bouaghi University, Algeria. His research interests are non-linear control, intelligent control and power systems control.

Appendix A
Quadratic Buck Converter Parameters

Circuit parametrs	Values
Input voltage E	24 V
Reference voltage v_2	10 V
Switching frequency f	110 kHz
L_1	39 µH
L_2	27 µH
C_1	16 µF
C_2	18 µF
$r_{L_1}, r_{L_2}, r_{C_1}, r_{C_2}$	0.25 Ω
Load R	10 Ω

11

Design and Control of Parallel Three Phase Voltage Source Inverters in Low Voltage AC Microgrid

El Hassane MARGOUM[1], Nissrine KRAMI[1], Luis SECA[2],
Carlos MOREIRA[2], Hassan MHARZI[1]

[1]Department of Electrical Engineering, National School of Applied Science,
Ibn Tofail University, Kenitra, Morocco
[2]Institute for Systems and Computer Engineering, Technology and Science (INESC TEC),
Faculty of Engineering, University of Porto, R. Dr. Roberto Frias, 4200 Porto, Portugal

margoum.elhassane@gmail.com, nissrine.krami@gmail.com, lseca@inescporto.pt,
carlos.moreira@inesctec.pt, h_mharzi@yahoo.fr

Abstract. *Design and hierarchical control of three phase parallel Voltage Source Inverters are developed in this paper. The control scheme is based on synchronous reference frame and consists of primary and secondary control levels. The primary control consists of the droop control and the virtual output impedance loops. This control level is designed to share the active and reactive power correctly between the connected VSIs in order to avoid the undesired circulating current and overload of the connected VSIs. The secondary control is designed to clear the magnitude and the frequency deviations caused by the primary control. The control structure is validated through dynamics simulations. The obtained results demonstrate the effectiveness of the control structure.*

Keywords

Droop control, energy storage systems, hierarchical control, MicroGrid, Smart grid.

1. Introduction

The MicroGrid (MG) concept has been proposed in order to increase controllability and observability of the distribution grid [1]. Hence, MG can integrate different types of Distributed Generation (DG) such as Renewable Energy Sources (RES) (PhotoVoltaic (PV) panels, micro windturbines), low carbon technologies, Energy Storage Systems (ESS) and loads. All these components are connected together and smartly managed in order to improve reliability and security of supply at the distribution level [1] and [2].

MG is a flexible system [3]. It can operate in two modes: connected to the main network (grid connected mode) at the Low Voltage (LV) level or autonomously (islanded mode) [3]. In the grid connected mode, the MG can exchange the power with the main grid and can also provide ancillary services to the upstream distribution network, such as reactive power, grid frequency support and so forth [4] and [5]. The amount of power exchanged with the external grid is calculated by the tertiary control functions implemented generally in a MicroGrid Central Controller (MGCC) [3], [5], [6] and [7]. In the case of contingencies or faults occurrence in the main grid, the MG changes the mode of operation from grid connected to islanded mode [4], where the power demanded by the local loads is supplied by local distributed generations, since the MG is operating disconnected from the main grid. The Energy Management System (EMS) takes into account the State of Charge (SoC) of ESSs, controllable sources and controllable loads in order to balance generation and consumption, thus improving reliability, efficiency and security of supply [1], [5], [8] and [9].

The MG requirements and the nature of power generated by DGs make the power converters indispensable components. In the case of AC MG, the final stage is an inverter (DC-AC converter). The inverters are very flexible devices that allow implementation of advanced control solutions in order to increase controllability of the MG, thus enabling further integration of intermittent RES [3], [4], [7] and [10]. The power inverters in AC MG can be classified into grid forming and grid feeding units [4]. The grid forming units operate as Voltage Source Inverters (VSI). They are responsible for regulation of frequency and magnitude of the AC bus voltage. Moreover, they are mandatory when the MG operates in islanded mode, since they are the only ones responsible for the voltage regulation. In contrast, the grid feeding units are designed to inject power into the grid and they are often connected to the DGs based on RES such as PV systems and microwind turbines. However, unlike grid forming units, the grid feeding units cannot constitute a MG [4] and [5].

When the MG is operating in the islanded mode, it is necessary to use fast energy storage technologies in order to balance generation and consumption and keep the amplitude and frequency of the common load bus voltage in acceptable limits [10]. The ESS are interfaced by VSIs that are connected in parallel through the MG [1] and [10]. Adequate power sharing is required in order to avoid the undesired circulating current as well as theoverload power converters [3]. Droop control technique has been widely used as an autonomous power sharing solution. This technique is based on local information [3], [4], [5], [7] and [11]. However, it suffers from several drawbacks, especially when it is applied to low voltage grids that have a high R/X ratio due to resistive distribution feeders [3] and [12]. Hence, the active and reactive power are affected by the coupling impedance [13]. An additional control loop called Virtual Output Impedance (VOI) has been proposed [14] and [15], this control loop makes the distribution grid behaving like an inductive grid, thus ensuring the active and reactive power decoupling and improving the system stability. By adding this control loop, the active Power-Frequency (P-ω) and the reactive Power-Voltage (Q-V) can be applied to share the active and reactive power, respectively [4] and [12]. Althoughthe frequency and the voltage are directly involved (frequency and amplitude deviations). For this reason, a centralized controller based on low bandwidth communication links has been proposed in order to restore the frequency and the magnitude to their nominal values [3], [5] and [7].

Hierarchical control structure of MG has been proposed in several works [3], [5], [7] and [16]. This control structure is a compromise between fully centralized and fully decentralized approaches [5]. It includes three main control levels: primary, secondary and tertiary control [3] and [7]. The primary control is responsible for power sharing between the connected inverters in order to avoid the undesired circulating current and converters overload. The secondary control restores the amplitude and the frequency deviations of the load bus voltage caused by the power sharing controller. The

tertiary control manages the power between the MG and the main grid [3].

Primary and secondary control for parallel connected VSIs forming an autonomous AC MG are developed and discussed in this paper. The control scheme consists of a local controller and a centralized controller. The local controller includes voltage and current Proportional Integral (PI) controllers based on synchronous reference frame and power sharing controller that consists of droop control and VOI. The secondary control implemented in a centralized controller restores the deviations produced by the power sharing controller.

The remainder of this paper is structured as follows: Section 2. presents a design of the MG local controller that consists of inner loops, droop control and virtual output impedance. Section 3. presents a design of the centralized secondary control. Simulation results for two parallel connected VSIs forming an islanded AC MG are presented and discussed in Sec. 4.

2. Local Controller Design: Inner Loops and Primary Control

The VSIs local controller is based on local measurements [3], it generally includes droop control, output virtual impedance and inner control loops (Voltage and current loops) [3], [7], [13], [17] and [18]. The block diagram of the VSI local controller is depicted in Fig. 1. The power stage of each VSI consists of a three-leg three phase inverter connected to a Battery Energy Storage System (BESS) through the DC link Capacitor (C), loaded by an LC filter and connected to the MG AC bus through a line impedance.

The VSI control system consists of the power sharing controller (droop control and VOI), voltage loop, current loop and the Pulse Width Modulator (PWM). The park transformation is used to transform variables from the natural frame to the synchronous reference frame. More details are presented in the following subsections.

2.1. Inner Loops

The VSIs inner control loops are based on synchronous reference frame including an outer voltage loop and an inner current loop as shown in Fig. 2. Proportional-Integral (PI) controllers are used in two control loops [19]. Voltage and current feed-forward have been added in order to improve performances of the regulators [5].

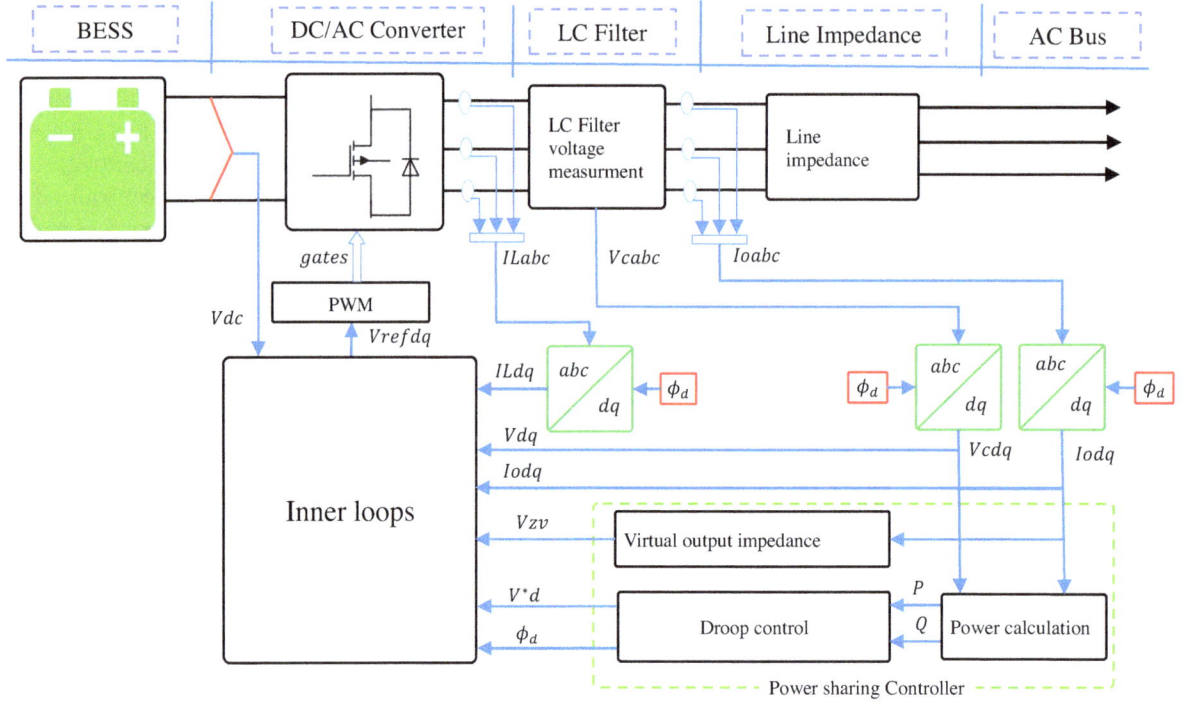

Fig. 1: VSI local controller.

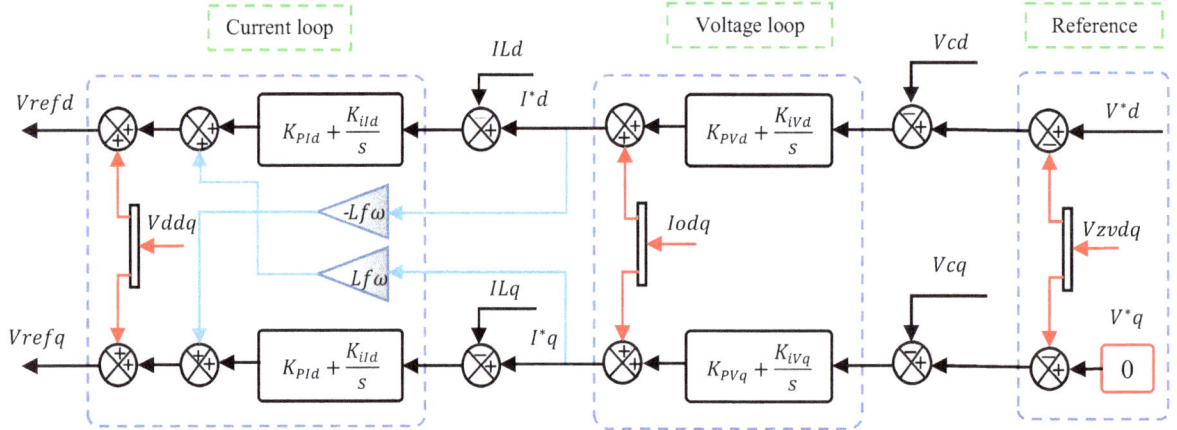

Fig. 2: VSI inner control loops.

Transfer functions of the voltages (V_d and V_q) and currents (I_d and I_q) controllers are given as follows [19]:

$$R_{Vd}(s) = K_{PVd} + \frac{K_{iVd}}{s}, \qquad (1)$$

$$R_{Vq}(s) = K_{PVq} + \frac{K_{iVq}}{s}, \qquad (2)$$

$$R_{Id}(s) = K_{PId} + \frac{K_{iId}}{s}, \qquad (3)$$

$$R_{Iq}(s) = K_{PIq} + \frac{K_{iIq}}{s}, \qquad (4)$$

where K_{PVd} (K_{PVq}) are proportional term coefficients of the direct (quadrature) voltage controller, and K_{iVd} (K_{iVq}) are integral term coefficients of the direct (quadrature) voltage controller.

K_{PId} (K_{PIq}) are proportional term coefficients of the direct (quadrature) current controller, and K_{iId} (K_{iIq}) are integral term coefficients of the direct (quadrature) current controller. A block diagram of the local controller is depicted in Fig. 2.

2.2. Primary Control

If two or more power converters are parallel connected to a common Load Bus (LB), the undesired circulating current can appear and the power will not be shared properly between the parallel connected power converters [3]. In order to solve this power sharing issue, the primary control has been proposed [3]. This control level adapts the frequency and the amplitude of the

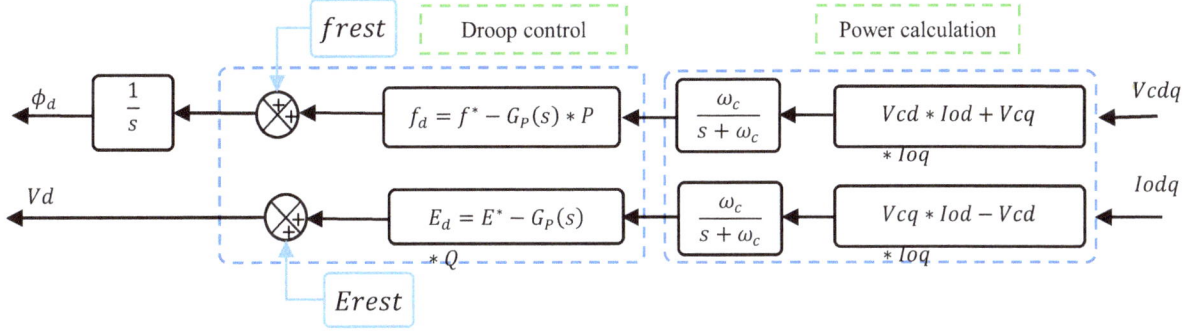

Fig. 3: Block diagram of power calculation and droop control.

voltage reference in order to share the active and reactive power properly between the connected VSIs. Thus avoiding the undesired circulating current and the converters overload [3], [4] and [7]. This control level consists of droop control and the VOI [3].

1) Power Calculation and Droop Control

The equations of active and reactive power generated by a VSI are given in dq-coordinates variables by Eq. (5) and Eq. (6), respectively [4].

$$P = \frac{3}{2}\left(V_{cd} \cdot I_{od} + V_{cq} \cdot I_{oq}\right), \tag{5}$$

$$Q = \frac{3}{2}\left(V_{cq} \cdot I_{od} - V_{cd} \cdot I_{oq}\right), \tag{6}$$

where $Vcdq$ and $Iodq$ are the capacitor voltage and the current after the LC filter.

A block diagram of the power calculation and droop control is shown in Fig. 3.

The Park transformation is used to obtain the dq-coordinates variables.

$$\begin{bmatrix} Xd \\ Xq \\ X0 \end{bmatrix} = \\ = \frac{2}{3}\begin{bmatrix} \cos(\theta) & \cos(\theta - \frac{2\pi}{3}) & \cos(\theta + \frac{2\pi}{3}) \\ \sin(\theta) & \sin(\theta - \frac{2\pi}{3}) & \sin(\theta + \frac{2\pi}{3}) \\ \frac{1}{2} & \frac{1}{2} & \frac{1}{2} \end{bmatrix}\begin{bmatrix} Xa \\ Xb \\ Xc \end{bmatrix}, \tag{7}$$

where $(Xd, Xq, X0)$ is the variable vector in synchronous reference frame and (Xa, Xb, Xc) is the variable vector in the natural reference frame.

In order to filter the active and reactive power ripples a first order Low Pass Filter (LPF) is used. The transfer function of the LPF is given by Eq. (8):

$$LPF(s) = \frac{\omega_c}{s + \omega_c}, \tag{8}$$

where ω_c is the LPF cut-off frequency.

A MG consists of a number of parallel connected converters [11]. Figure 4 shows the equivalent circuit of

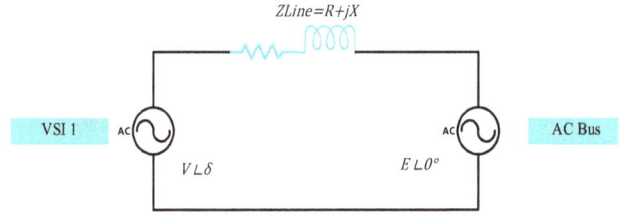

Fig. 4: Equivalent circuit of two parallel connected VSIs.

a VSI connected in parallel with an AC bus. The active and reactive power delivered by the VSI are expressed as follows [20]:

$$P = \left(\frac{EV}{Z}\cos\delta - \frac{V^2}{Z}\right)\cos\theta + \frac{EV}{Z}\sin\delta \cdot \sin\theta, \tag{9}$$

$$Q = \left(\frac{EV}{Z}\cos\delta - \frac{V^2}{Z}\right)\sin\theta - \frac{EV}{Z}\sin\delta \cdot \cos\theta, \tag{10}$$

where V and E are the VSI voltage amplitude and AC bus voltage amplitude, respectively, Z and θ are the amplitude and the angle of the line impedance, respectively, δ is the load angle.

When the coupling impedance is mainly inductive, the active and reactive power equations become as follows:

$$P \approx \frac{EV}{X}\sin\delta, \tag{11}$$

$$Q \approx \frac{EV}{X}\cos\delta - \frac{V^2}{X}. \tag{12}$$

Low voltage grids have a high $\frac{R}{X}$. In this work we have used the VOI to ensure the inductive behavior of the line impedance. We assumed that the equivalent line impedance is mainly inductive.

The frequency and voltage droop expressions are given by Eq. (13) and Eq. (14), respectively:

$$f_d = f_{ref} - G_P(S) \cdot P, \tag{13}$$

$$E_d = E_{ref} - G_Q(S) \cdot Q. \tag{14}$$

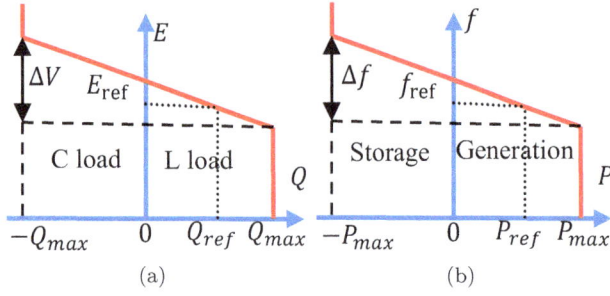

Fig. 5: Frequency (a) and voltage (b) droop characteristics for inductive line impedance.

These equations are graphically presented as depicted in Fig. 5.

Where E_{ref} and f_{ref} are the voltage magnitude and frequency reference, respectively, E_d, f_d and ϕ_d are the drooped voltage magnitude, frequency and the phase reference, respectively, P and Q are the measured active and reactive power.

$G_P(s)$ and $G_P(s)$ can be designed as follows:

$$G_P(s) = Kp = \frac{\Delta f}{P_{\max}}, \qquad (15)$$

$$G_q(s) = Kq = \frac{\Delta E}{2Q_{\max}}, \qquad (16)$$

where Kp and Kq are the proportional coefficients of the frequency and the voltage droop, respectively, Δf and ΔE are the maximum acceptable deviation of the frequency and the voltage, respectively, $P_{\max}$ and $Q_{\max}$ are the maximum active and reactive power delivered by the VSI, respectively.

2) Virtual Output Impedance (VOI)

When the coupling impedance is not purely inductive, the traditional droop control scheme cannot be applied directly. However, the droop control is not enough to ensure the system stability and reactive power sharing between the connected VSIs. For this reason, a virtual impedance loop was proposed in the technical literature [3], [11], [12], [13], [15], [21], [22] and [23]. The objective of this control loop is to decouple active and reactive power control [4] and [24], thus ensuring the system stability and improving the performances of the droop control without causing any power losses.

The VOI loop consists of dropping the voltage reference calculated by the droop control loop proportionally to the output current. The dropped voltage reference is expressed as follows:

$$V_{ref} = V_d - Z_v(s) \cdot I_o, \qquad (17)$$

where $Z_v(s)$ is the VOI transfer function, V_d voltage reference calculated by the droop control loop and I_o is the output current after the LC filter.

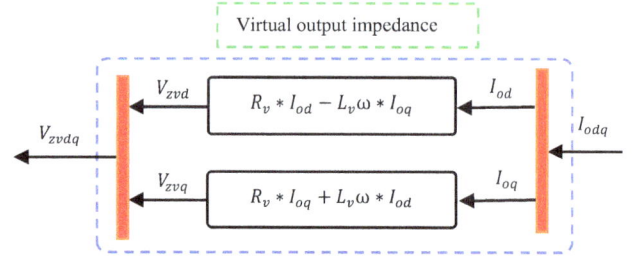

Fig. 6: Synchronous reference frame virtual output impedance.

The synchronous reference frame VOI depicted in Fig. 6 is adopted in this work. The VOI equations are given in dq-coordinates as follows [12] and [24]:

$$V_{zvd} = R_v \cdot I_{od} - L_v \omega \cdot I_{oq}, \qquad (18)$$

$$V_{zvq} = R_v \cdot I_{oq} + L_v \omega \cdot I_{od}, \qquad (19)$$

where R_v and L_v being the virtual resistance and the virtual inductance and $I_o dq$ is the dq-coordinates output current of the VSI after the LC filter (as shown in Fig. 1).

The virtual resistance has been added in order to damp the oscillations [22]. As the virtual inductance is high as good power sharing performances are achieved. However, its size is limited by the maximum voltage deviations in the LB. Since, the VOI drops the voltage reference generated by the droop control loop proportionally to the output current.

3. Centralized Controller Design: Secondary Control

Droop control and virtual output impedance control loop produce voltage and frequency deviations. The secondary control is designed to restore these deviations and keep the frequency and the amplitude at the LB in the allowable limits [3] and [6]. The secondary controller is based on two PI controllers, one restores the voltage deviations and the other one restores the frequency deviations. The secondary control block diagram is depicted in Fig. 7.

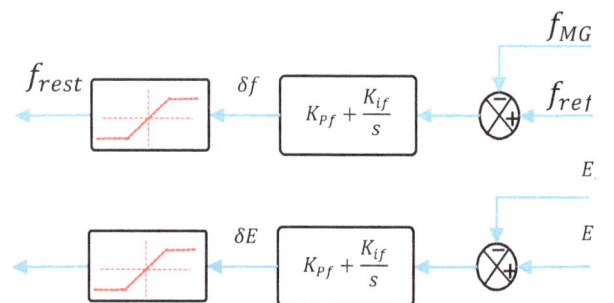

Fig. 7: Block diagram of Secondary control level.

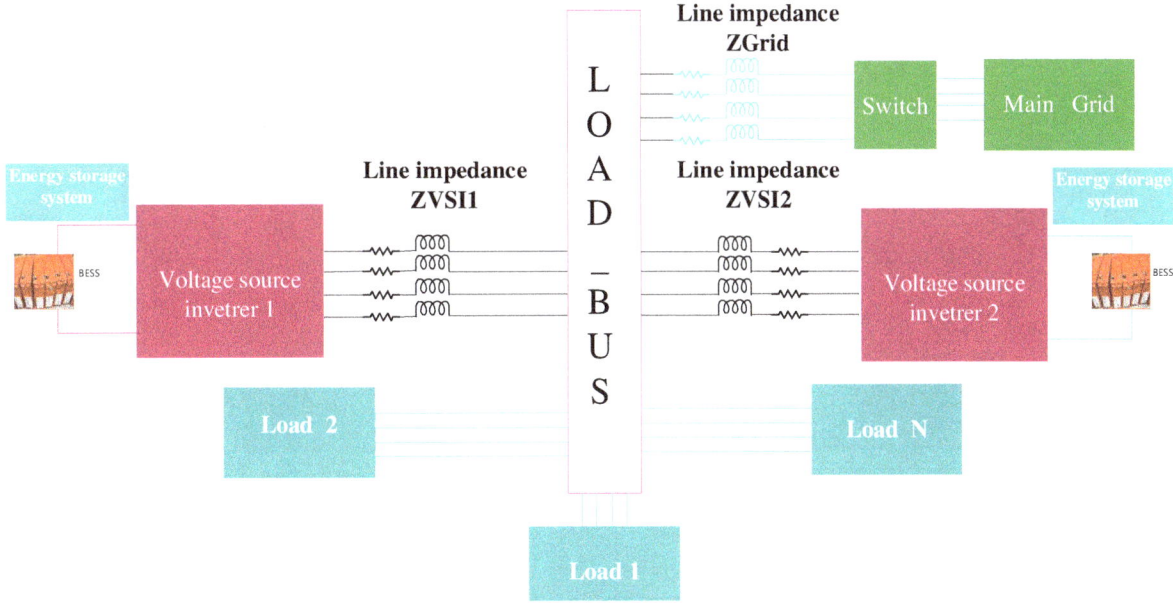

Fig. 8: Block diagram of two parallel connected VSIs with distributed loads forming an AC MicroGrid.

The voltage and frequency restoration signals can be calculated as follows [3]:

$$\delta f = k_{pf}(f^*_{MG} - f_{MG}) + k_{if} \int (f^*_{MG} - f_{MG})\mathrm{dt}, \quad (20)$$

$$\delta E = \\ = k_{pE}(E^*_{MG} - E_{MG}) + k_{iE} \int (E^*_{MG} - E_{MG})\mathrm{dt}, \quad (21)$$

where δf (δE) are the frequency (amplitude) restoration signals, f^*_{MG} (E^*_{MG}) and f_{MG} (E_{MG}) are the reference of the frequency (amplitude) and the measured frequency (amplitude) at the LB, respectively, K_{Pf} (K_{PE}) are the proportional term coefficients of the frequency (voltage) PI controller, and K_{if} (K_{iE}) are the integral term coefficients of the frequency (amplitude) PI controller, respectively.

4.   Simulation Results

In order to evaluate the control system performances, an islanded MG formed by two parallel connected VSIs, as shown in Fig. 8, are simulated using Matlab/Simulink environment. Each inverter is connected to the LB through a line impedance, the output impedance of two VSIs are not equal, ZVSI2= 2·VSI1 as seen in Fig. 8. The system parameters are listed in Tab. 1 The switching frequency of the VSIs is set to 20 kHz. In order to show clearly the active and reactive power sharing between the connected converters at different load conditions, three loads are connected to the load bus at different times (unbalanced loads are not considered in this study). Figure 9 shows the power sharing between the VSIs. After the connection of load 1 to the LB from $t = 0$ s to $t = 5$ s, the active power is shared properly between the VSIs and

the reactive power is shared with a small error. Then, load 2 is connected at $t = 5$ s, and disconnected at $t = 10$ s. Meanwhile, the active power at the LB increases from 5 kW to 10 kW. As a consequence, the reactive power error between the two VSIs increases. This error is due to the strong coupling between the active and reactive power. At $t = 15$ s, the second VSI is disconnected from the LB and thenthe connected loads are supplied by one VSI. At $t = 17$, another active load of 1 kW is connected to the LB. Figure 10 shows the output voltage at the LB. Figure 11 and Fig. 12 show the output current of VSI1 and VSI2, respectively.

The main disadvantage of the droop control is the steady state voltage and frequency deviations, the power sharing is achieved through dropping the amplitude and the frequency of the voltage. Figure 13 shows the nominal and measured values of the frequency of the load bus voltagethat drops proportionally to the active power. Figure 14 shows the nominal voltage (220 V) and the load bus voltage, the load bus voltage drops proportionally to the reactive power.

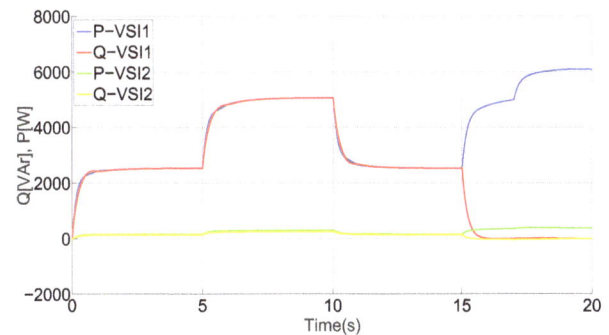

Fig. 9: Active and reactive power sharing.

Tab. 1: System parameters.

Power stage	
Parameter	**Value**
Nominal RMS voltage: V_{rms} (V)	220
Nominal frequency: f (Hz)	50
DC voltage: V_{dc} (V)	650
Inverter filter inductance: L_f (H)	2.0E-3
Inverter filter capacitance: C_f (F)	11.0E-6
VSI 1 line impedance: ZVSI1=R1+jX1	0.2+j0.0126
VSI 2 line impedance: ZVSI2=R2+jX2	0.4+j0.0252
load 1: P (kW), Q (KVAR)	P=5, Q=250
load 2: P (kW), Q (KVAR)	P=5
load 3: P (kW), Q (KVAR)	P=1

Inner loops	
Voltage controller PI	**Current controller PI**
KpVd = 0.5, KiVd = 1000, KpVq = 0.15, KiVq = 300.	KpId = 40, KiId = 85, KpIq = 40, KiIq = 85.

Primary control			
Droop control		**Virtual output impedance**	
Kp=3.33E-5	Kq=22E-3	Rv=1Ω	Lzv=7 mH

Secondary control			
Frequency restoration		**Voltage restoration**	
Kpf=15E-4	Kif=500	KpE=15E-4	KiE=1000

Power calculation filters: T = 0.2

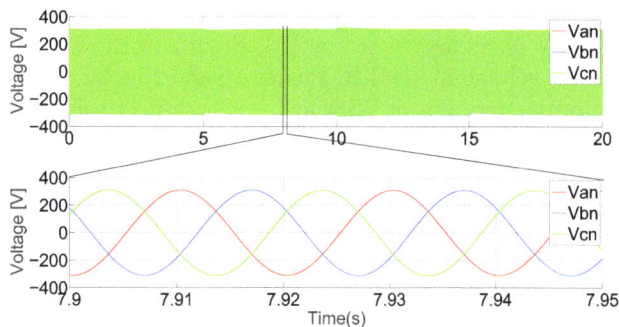

Fig. 10: Output voltage of a VSI.

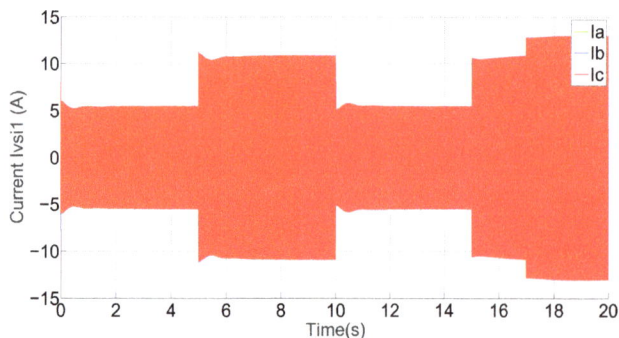

Fig. 11: Output current of a VSI1.

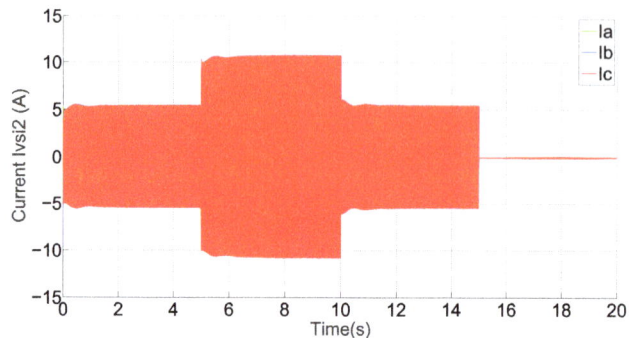

Fig. 12: Output current of a VSI2.

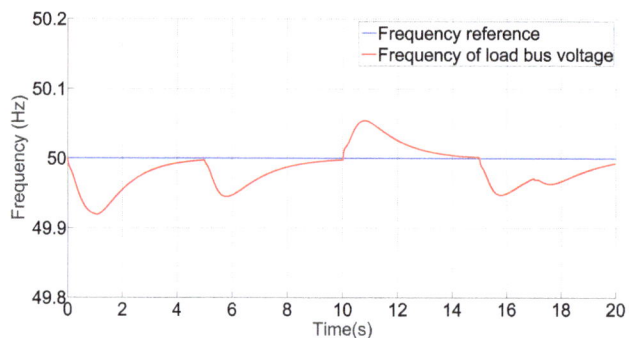

Fig. 13: Frequency reference and LB frequency without secondary control.

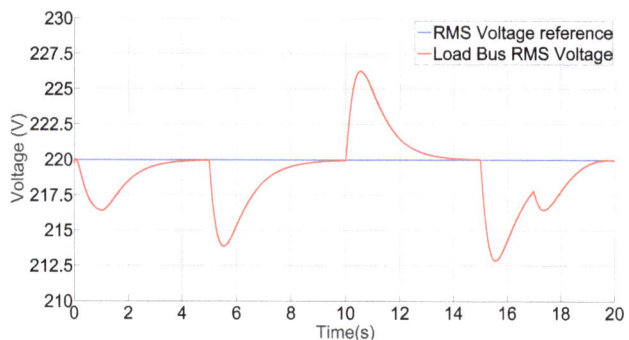

Fig. 14: RMS voltage reference and LB RMS voltage without secondary control.

In order to keep the frequency and the amplitude of the load bus voltage in acceptable limits, the secondary control is used. Figure 15 and Fig. 16 show the reference and the measured frequency and voltage in the LB respectively. The secondary control restores correctly the deviations produced by the primary control. The frequency and RMS voltage of the LB are kept in the allowable limits.

The load bus frequency and voltage overshoots in Fig. 15 and Fig. 16 are causedby the disconnection of load 2 at $t = 10$ s, the secondary control regulated the voltage and the frequencyto their nominal values.

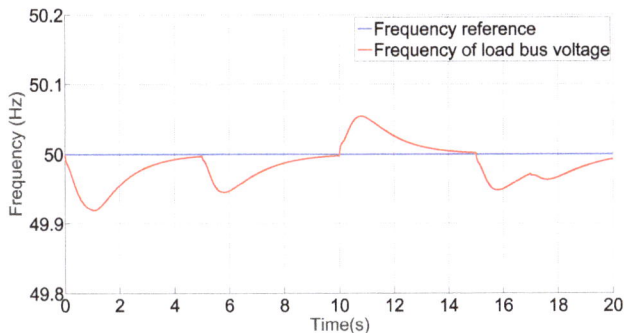

Fig. 15: Frequency reference and LB frequency with secondary control.

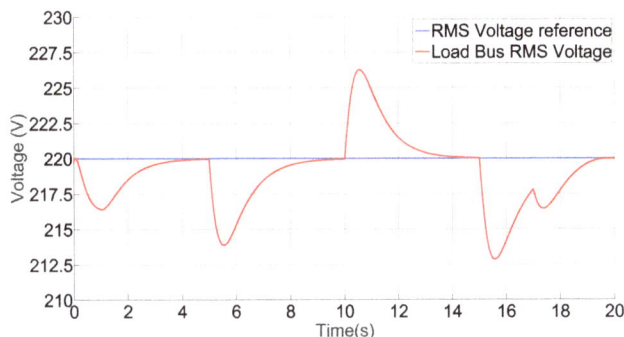

Fig. 16: RMS voltage reference and LB RMS voltage with secondary control.

5. Conclusion

Design and analysis of inner control loops, Primary and secondary control for parallel connected VSIs forming a low voltage MG are presented in this paper. The inner control loops are based on synchronous reference frame. PI controllers are used in voltage and current loops. The primary control includes droop control and a VOI loop. The centralized secondary control includes voltage and frequency restoration controllers.

The simulation results are also presented and show that the active and reactive power are shared properly. The amplitude and the frequency of the LB voltage are restored by the centralized secondary control and kept in the allowable limits.

Acknowledgment

This work was supported by IRESEN (Institut de Recherche en Energie Solaire et Energies Nouvelles) of Morocco in the framework of the Inno-PV research project "SECRETS - Sustainable Energy Clusters REalized Through Smart Grids".

References

[1] GOUVEIA, C., J. MOREIRA, C. L. MOREIRA and J. A. PECAS LOPES. Coordinating storage and demand response for microgrid emergency operation. *IEEE Transactions on Smart Grid.* 2013, vol. 4, iss. 4, pp. 1898–1908. ISSN 1949-3053. DOI: 10.1109/TSG.2013.2257895.

[2] BLAABJERG, F. and J. M. GUERRERO. Smart Grid and Renewable Energy Systems. In: *International Conference on Electrical Machines and Systems.* Beijing: IEEE, 2011, pp. 1–10. ISBN 978-1-4577-1044-5. DOI: 10.1109/ICEMS.2011.6073290.

[3] GUERRERO, J. M., J. C. VASQUEZ, J. MATAS, L. G. DE VICUNA and M. CASTILLA. Hierarchical control of droop-controlled AC and DC microgrids - A general approach toward standardization. *IEEE Transactions on Industrial Electronics.* 2011, vol. 58, iss. 1, pp. 158–172. ISSN 0278-0046. DOI: 10.1109/TIE.2010.2066534.

[4] ROCABERT, J., A. LUNA, F. BLAABJERG and P. RODRIGUEZ. Control of Power Converters in AC Microgrids. *IEEE Transactions on Power Electronics.* 2012, vol. 27, iss. 11, pp. 4734–4749. ISSN 0885-8993. DOI: 10.1109/TPEL.2012.2199334.

[5] DANIEL, E. O., A. MEHRIZI-SANI, A. H. ETEMADI, C. A. CANIZARES, R. IRAVANI, M. KAZERANI, A. H. HAJIMIRAGHA, O. GOMIS-BELLMUNT, M. SAEEDIFARD, R. PALMA-BEHNKE, G. A. JIMENEZ-ESTEVEZ and N. D. HATZIARGYRIOU. Trends in Microgrid Control. *IEEE Transactions on Smart Grid.* 2014, vol. 5, iss. 4, pp. 1905–1919. ISSN 1949-3053. DOI: 10.1109/TSG.2013.2295514.

[6] MENG, L., M. SAVAGHEBI, F. ANDRADE, J. C. VASQUEZ, J. M. GUERRERO and M. GRAELLS. Microgrid central controller development and hierarchical control implementation in the intelligent microgrid lab of Aalborg University. In: *Applied Power Electronics Conference and Exposition.* Charlotte: IEEE, 2015, pp. 2585–2592. ISBN 978-1-4799-6736-0. DOI: 10.1109/APEC.2015.7104716.

[7] VASQUEZ, J. C., J. M. GUERRERO, J. MIRET, M. CASTILLA and L. GARCIA DE VICUNA. Hierarchical Control of Intelligent Microgrids. *IEEE Industrial Electronics Magazine.* 2010, vol. 4, no. 4, pp. 23–29. ISSN 1932-4529. DOI: 10.1109/MIE.2010.938720.

[8] LOPES, J. A. P., C. L. MOREIRA and A. G. MADUREIRA. Defining control strategies for

microgrids islanded operation. *IEEE Transactions on Power Systems*. 2006, vol. 21, iss. 2, pp. 916–924. ISSN 0885-8950. DOI: 10.1109/TP-WRS.2006.873018.

[9] GOUVEIA, C., C. L. MOREIRA, J. ABEL, P. LOPES and D. VARAJAO. Microgrid Service Restoration: the role of plugged-in electric vehicles. *IEEE Industrial Electronics Magazine*. 2013, vol. 7, iss. 4, pp. 26–41. ISSN 1932-4529. DOI: 10.1109/MIE.2013.2272337.

[10] WU, D., F. TANG, T. DRAGICEVIC, J. C. VASQUEZ and J. M. GUERRERO. Autonomous Active Power Control for Islanded AC Microgrids With Photovoltaic Generation and Energy Storage System. *IEEE Transactions on Energy Conversion*. 2014, vol. 29, iss. 4, pp. 882–892. ISSN 0885-8969. DOI: 10.1109/TEC.2014.2358612.

[11] VASQUEZ, J. C., J. M. GUERRERO, M. SAVAGHEBI and R. TEODORESCU. Modeling, analysis, and design of stationary reference frame droop controlled parallel three-phase voltage source inverters. *IEEE Transactions on Industrial Electronics*. 2011, vol. 60, iss. 4, pp. 272–279. ISSN 0278-0046. DOI: 10.1109/TIE.2012.2194951.

[12] HE, J. and Y. W. LI. Analysis and design of interfacing inverter output virtual impedance in a low voltage microgrid. In: *IEEE Energy Conversion Congress and Exposition*. Atlanta: IEEE, 2010, pp. 2857–2864. ISBN 978-1-4244-5286-6. DOI: 10.1109/ECCE.2010.5618181.

[13] GUAN, Y., J. M. GUERRERO, X. ZHAO, J. C. VASQUEZ and X. GUO. A New Way of Controlling Parallel-Connected Inverters by Using Synchronous Reference Frame Virtual Impedance Loop - Part I: Control Principle. *Power Electron.* 2016, vol. 31, iss. 6, pp. 4576–4593. ISSN 0885-8993. DOI: 0.1109/TPEL.2015.2472279.

[14] WANG, X., F. BLAABJERG and Z. CHEN. An improved design of virtual output impedance loop for droop-controlled parallel three-phase voltage source inverters. In: *IEEE Energy Conversion Congress and Exposition (ECCE)*. Raleigh: IEEE, 2012, pp. 2466–2473. ISBN 978-1-4673-0802-1. DOI: 10.1109/ECCE.2012.6342404.

[15] GUERRERO, J. M., L. GARCIADEVICUNA, J. MATAS, M. CASTILLA and J. MIRET. Output Impedance Design of Parallel-Connected UPS Inverters With Wireless Load-Sharing Control. *IEEE Transactions on Industrial Electronics*. 2005, vol. 52, iss. 4, pp. 1126–1135. ISSN 0278-0046. DOI: 10.1109/TIE.2005.851634.

[16] PALIZBAN, O. and K. KAUHANIEMI. Hierarchical control structure in microgrids with distributed generation: Island and grid-connected mode. *Renewable and Sustainable Energy Reviews*. 2015, vol. 44, iss. 1, pp. 797–813. ISSN 1364-0321. DOI: 10.1016/j.rser.2015.01.008.

[17] BIDRAM, A. and A. DAVOUDI. Hierarchical structure of microgrids control system. *Smart Grid*. 2012, vol. 3, iss. 4, pp. 1963–1976. ISSN 1364-0321. DOI: 10.1109/TSG.2012.2197425.

[18] SAVAGHEBI, M., J. M. GUERRERO, A. JALILIAN and J. C. VASQUEZ. Hierarchical control scheme for voltage unbalance compensation in islanded microgrids. In: *37th Annual Conference of the IEEE Industrial Electronics Society*. Melbourne: IEEE, 2011, pp. 3158–3163. ISBN 978-1-61284-972-0. DOI: 10.1109/IECON.2011.6119815.

[19] BLAABJERG, F., R. TEODORESCU, M. LISERRE and A. V. TIMBUS. Overview of control and grid synchronization for distributed power generation systems. *IEEE Transactions on Industrial Electronics*. 2006, vol. 53, iss. 5, pp. 1398–1409. ISSN 0278-0046. DOI: 10.1109/TIE.2006.881997.

[20] GUERRERO, J. M., J. MATAS, L. G. DE VICUNA, M. CASTILLA and J. MIRET. Wireless-control strategy for parallel operation of distributed-generation inverters. In: *Proceedings of the IEEE International Symposium on Industrial Electronics*. Dubrovnik: IEEE, 2005, pp. 845–850. ISBN 0-7803-8738-4. DOI: 10.1109/ISIE.2005.1529025.

[21] GUERRERO, J. M., M. CHANDORKAR, T.-L. LEE and P. C. LOH. Advanced Control Architectures for Intelligent Microgrids Part I: Decentralized and Hierarchical Control. *IEEE Transactions on Industrial Electronics*. 2013, vol. 60, no. 4, pp. 1254–1262. ISBN 0278-0046. DOI: 10.1109/TIE.2012.2196889.

[22] GUERRERO, J. M., J. MATAS, L. GARCIA DE VICUNA, M. CASTILLA and J. MIRET. Decentralized Control for Parallel Operation of Distributed Generation Inverters Using Resistive Output Impedance. *IEEE Transactions on Industrial Electronics*. 2007, vol. 54, iss. 2, pp. 994–1004. ISSN 0278-0046. DOI: 10.1109/TIE.2007.892621.

[23] WANG, X., Y. W. LI, F. BLAABJERG and P. C. LOH. Virtual-impedance-based control for voltage-source and current-source converters. *IEEE Transactions on Power Electronics*. 2015,

vol. 30, iss. 12, pp. 7019–7037. ISSN 0885-8993. DOI: 10.1109/TPEL.2014.2382565.

[24] SUMMERS, C. D., T. J. BETZ, R. E. MOORE and T. G. TOWNSEND. Implementing the virtual output impedance concept in a three phase system utilising cascaded PI controllers in the dq rotating reference frame for microgrid inverter. In: *Power Electronics and Applications*. Lille: IEEE, 2013, pp. 1–10. ISBN 978-1-4799-0116-6. DOI: 10.1109/EPE.2013.6634691.

About Authors

El Hassane MARGOUM received the M.Sc. degree in electrical engineering from the Higher Normal School of Technical Education, Mohamed V University, Rabat, Morocco in 2014. He is currently pursuing his Ph.D. degree at National school of applied sciences /University of Ibn-Tofail, Kenitra, Morroco. He is working on the SECRETS project (Sustainable Energy Clusters Realized Through Smart Grids) in the development of smart power electronic converters for microgeneration systems, and he is a guest Ph.D. student at the Smart Grids with the Electric Vehicles Laboratory of the Institute for Systems and Computer Engineering (INESC), Porto.

Nissrine KRAMI has been an assistant professor in the Department of Electrical Engineering of the national school of applied sciences of the University of Ibn Tofail since 2010. His research interests are focused on microgrid dynamics and control, smart grids.

Carlos Leal MOREIRA received the B.Sc. and Ph.D. degrees in electrical engineering from the University of Porto, Portugal, in 2003 and 2008, respectively. Since 2010, he has been the smart grid area leader in the Power Systems Unit of the Institute for Systems and Computer Engineering of Porto (INESC Porto). He has also been an assistant professor in the Department of Electrical Engineering of the Faculty of Engineering of the University of Porto since 2008. His research interests are focused on MicroGrid dynamics and control, smart grids, and smart metering.

Luis SECA received the Electrical Engineering degree (five-year course) and the M.Sc. degree (two-year course) in electrical engineering, both from FEUP, in 2002 and 2006, respectively. He is currently a senior Researcher/Consultant in the Power Systems Unit of INESC Porto. His current research interests include steady-state and dynamic studies on integration of distributed energy resources in transmission and distribution networks, voltage and frequency control, and ancillary services provision.

Hassan MHARZI he is professor research at the National School of Applied Sciences, Kenitra, Morocco. Member of the Laboratory for Systems Engineering and Head of Team: Intelligent Energy, Power and Industrial Systems (Eisei). Nantes in France since 1996 and has the ability to supervise research since 2004. His research interests are focused on development and characterization of semiconductor materials for the production of electronic components (diodes, transistors, ptohovoltaic, cells, etc.).

12

PMSM MODEL WITH PHASE-TO-PHASE SHORT-CIRCUIT AND DIAGNOSIS BY ESA AND EPVA

Chourouk BOUCHAREB, Mohamed Said NAIT SAID

Electrical Engineering Department, Laboratory LSPIE Batna 2000, Batna University, Route de Biskra, 05078, Algeria

c.bouchareb@live.fr, medsnaitsaid@yahoo.fr

Abstract. *One of the most frequent faults in PMSM stator is the insulation failure due to the degradation of the main isolation in the motor winding. This paper is aimed at suggesting a dynamic model of PMSM with phase-to-phase fault based on an equivalent electric circuit model including the real form of back EMF. The faulty model is used for studying the machine behavior and extracting the fault signatures for diagnosis. Two diagnostic techniques the Spectral Analysis (ESA) and Extend Park's Vectors Approach (EPVA) based on frequency analysis are applied to detect this kind of fault.*

Keywords

EPVA, ESA, Inter-turn fault, phase-to-phase fault, PMSM model.

1. Introduction

In recent years, Permanent Magnet Synchronous Motor (PMSM) has become one of most important electric machines because of the inherent advantages of high power density, high efficiency, small weight, high reliability and easy control of external torque of stator's current control. Consequently, it is widely used in industry, e.g. in traction, automobiles, robotics and aerospace technology, as well as electric vehicles and ship propulsion systems [1], [2] and [3].

The fault diagnosis of electrical machines had been the target of an intense amount of interesting researches during the last 30 years. Reducing maintenance costs and preventing unscheduled down-times, which result in losses of production and financial incomes and benefitting from their utility in safety-sensitive applications, are the priorities of electrical drives for manufacturers and operators [4], [5] and [6].

In fact, correct diagnosis and early detection of incipient faults require the development of an accurate model for electrical machine, able to simulate electrical faults and to apply an effective diagnostic technique.

However, model accuracy and computation time represents two opposite criteria. Conventional model (equivalent electric circuit or equivalent magnetic circuit) obtained with Park transformation for instance is based on restrictive assumptions and does not require long computation time [7] and [8]. On the other hand, model obtained with the finite elements method is based on minimal assumption and requires long computation time [9] and [10]. There is a real need to establish an alternative model, which offers a good balance between accuracy and computation time.

One of the most common faults, called insulation failure, is the inter-turn short circuit in one of the stator coils. Since the coil insulation material is under the high voltage and temperature stress, it degrades gradually and finally loses the insulating characteristic [6]. The inter-turn fault is mostly caused by mechanical stress, moisture and partial discharge, which is accelerated for inverter supplied electrical machines [11].

In this paper, a dynamic model of a stator surface mounted PMSM with inter-turn fault is presented. We focus on phase-to-phase fault of the stator winding. This model based on equivalent electric circuit exhibits a trade-off between simplicity and precision, and it is used for studying a machine behavior under fault conditions for different levels of fault severity using MATLAB Simulink software.

Exploiting this faulty model to extract fault signatures in order to diagnose and to predict the insulation failure breakdown when the fault is not very severe in order to avoid the machine winding damages. To detect this fault, we chose two simple and useful techniques based on frequency analysis. These techniques are Electric Spectral Analysis (ESA) and Extend Park's Vectors Approach (EPVA). The contribution of this work is the addition of the real waveform of back Electro Motive Force (EMF) of healthy machine which contains a harmonic at $3 \cdot f_s$ of supply frequencies, because if the model does not take the uncertainties, like real back-EMF, the indicator will give a wrong diagnostic.

2. PMSM Fault Dynamic Model

2.1. Phase-to-Phase Fault Dynamic Model

The phase-to-phase fault denotes insulation failures between two windings of two phases at the stator. The insulation failure is modeled by a resistance, where its value depends on the fault severity. The stator winding of a PMSM machine with phase-to-phase fault is represented by Fig. 1. In this figure, the fault occurs between 'a' and 'b' phases, r_f denotes the fault insulation resistance. The sub-windings (a_{s1}) and (a_{s2}) represent respectively, the healthy and faulty part of the phase winding a, and sub-windings (b_{s1}) and (b_{s2}) represent, the healthy and faulty part of the phase winding b respectively. When the fault resistance r_f decreases towards zero, the insulation fault evaluates towards an inter-turn full short-circuit.

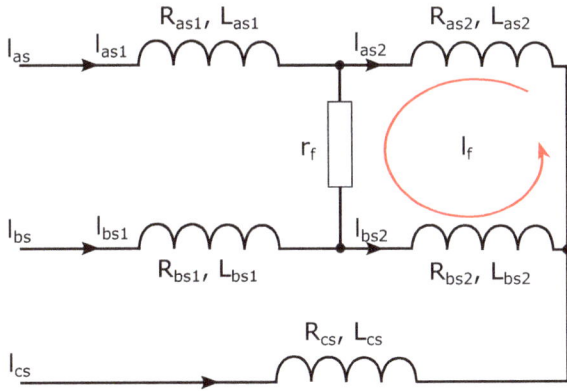

Fig. 1: Three-phases winding with phase-to-phase fault.

2.2. PMSM Healthy Model in abc-Coordinates

The voltages equations from the circuit in Fig. 1 without fault(healthy machine), given by r_f infinite value, as in [2], [12] and [13] are:

$$[V_s] = [R_s] \cdot [I_s] + [L_s] \cdot \frac{d}{dt} \cdot [I_s] + [E_s], \quad (1)$$

$$
\begin{bmatrix} v_{as} \\ v_{bs} \\ v_{cs} \end{bmatrix} = \begin{bmatrix} R_s & 0 & 0 \\ 0 & R_s & 0 \\ 0 & 0 & R_s \end{bmatrix} \begin{bmatrix} I_{as} \\ I_{bs} \\ I_{cs} \end{bmatrix} + \\
+ \begin{bmatrix} L & M & M \\ M & L & M \\ M & M & L \end{bmatrix} \cdot \frac{d}{dt} \cdot \begin{bmatrix} I_{as} \\ I_{bs} \\ I_{cs} \end{bmatrix} + \begin{bmatrix} e_{as} \\ e_{bs} \\ e_{cs} \end{bmatrix}, \quad (2)
$$

where the healthy machine variable and parameters are:

- $v_{as,bs,cs}$ - three phase stator voltages,
- $I_{as,bs,cs}$ - three phase stator currents,
- $e_{as,bs,cs}$ - three phase back EMF,
- R_s - stator resistance,
- L - self inductance of the stator,
- M - mutual inductance of the stator.

2.3. PMSM Faulty Model in abc-Coordinate

Voltage equations, which describe the faulty circuit presented in Fig. 1, can be expressed as:

$$[V_s] = \begin{bmatrix} v_{as1} & v_{as2} & v_{bs1} & v_{bs2} & v_{cs} \end{bmatrix}^T, \quad (3)$$

where:

- v_{as1} - the voltage of the healthy part phase a,
- v_{as2} - the voltage of faulty part of phase a,
- v_{bs1} - the voltage of the healthy part phase b,
- v_{bs2} - the voltage of faulty part of phase b.

The new resistances of healthy and faulty parts of phase 'a' and 'b' are calculated as follows:

$$R_{as1} = (1 - \sigma) \cdot R_{as}, \quad (4)$$

$$R_{as2} = \sigma \cdot R_{as}, \quad (5)$$

$$R_{bs1} = (1 - \sigma) \cdot R_{bs}, \quad (6)$$

$$R_{bs2} = \sigma \cdot R_{bs}, \quad (7)$$

$$\sigma = \frac{N_f}{N_s}. \quad (8)$$

The study of the elementary circuits of the phases has given the following relations:

$$v_{as} = v_{as2} + v_{as1}, \quad (9)$$

$$v_{bs} = v_{bs2} + v_{bs1}, \quad (10)$$

$$I_{as1} = I_{as}, \qquad (11)$$

$$I_{bs1} = I_{bs}, \qquad (12)$$

where R_{as1} is the stator phase resistance of healthy parts of 'a' phase while R_{as2} is the faulty stator phase resistance. R_{bs1} it is the stator phase resistance of healthy parts of 'b' phase while R_{bs2} is its faulty stator phase resistance, σ is the ratioof number of the turns (N_f) over the phase winding number of the turns (N_s).

The self-inductances of the faulty and healthy parts of winding (a_{as1}, a_{as2}), and winding (b_{bs1}, b_{bs2}) are proportional to the square of the fraction of shorted turns σ, and also the mutual inductance is proportional to this number of both parts. Therefore, we assume:

$$L_{as1} = (1 - \sigma)^2 L_{as}, \qquad (13)$$

$$L_{as2} = \sigma^2 L_{as}, \qquad (14)$$

$$M_{as2b} = \sigma M, \qquad (15)$$

$$M_{as1as2} = \sigma(1 - \sigma)L, \qquad (16)$$

where L_{as1} is the stator phase inductance of healthy parts of a phase while L_{as2} is the stator phase inductance of faulty parts of a phase, I_f is the additional current engendered by the short circuit, r_f is the insulation faulty resistance and v_f is the corresponded faulty voltage.

The stator currents become:

$$[I_s] = [I_{as} \ (I_{as} - I_f) \ I_{bs} \ (I_{bs} + I_f) \ I_{cs}]^T . \qquad (17)$$

The equation which describes the short circuit loop is in Eq. (18).

From previous analysis, we obtain the global equations governing the behavior of the machine with the presence of this short-circuit fault as the Eq. (19).

In the Eq. (19):

$$R' = R_{as} + R_{bs} + R_{cs}, \qquad (20)$$

$$L_f = -(-L_{a2} + M_{a2b2} - L_{b2} + M_{b2a2}), \qquad (21)$$

$$M_{bf} = -M_{a1a2} + M_{a1b2} - L_{a2} + M_{a2b2}, \qquad (22)$$

$$M_{cf} = -M_{ca2} + M_{cb2}. \qquad (23)$$

The expression of the electromagnetic torque can be written as follows:

$$T_e = \frac{e_{as} \cdot I_{as} + e_{bs} \cdot I_{bs} + e_{cs} \cdot I_{cs} - e_f \cdot I_f}{\Omega}, \qquad (24)$$

where Ω is the mechanical angular speed.

2.4. PMSM Faulty Model in α, β-Coordinates

The machine equations with inter-turn fault in stationary α and β axis reference frame are in Eq. (25), where:

$$R_r = \sqrt{\frac{2}{3}} \left(-R_{a2} - \frac{R_{b2}}{2} \right), \qquad (26)$$

$$r_f = R_{a2} + R_{b2} + R_f, \qquad (27)$$

$$r_{b2} = \frac{1}{2}\sqrt{2}R_{b2}, \qquad (28)$$

$$M_{f\alpha} = \sqrt{\frac{2}{3}} \left(M_{af} - \frac{M_{bf}}{2} - \frac{M_{cf}}{2} \right), \qquad (29)$$

$$M_{f\beta} = \frac{1}{2}\sqrt{2} \left(M_{bf} - M_{cf} \right), \qquad (30)$$

$$L_s = L - M, \qquad (31)$$

with,

- $I_{\alpha,\beta}$ - α and β axis components of stator currents,

- $e_{\alpha,\beta}$ - α and β components of stator back EMF.

Then the electromagnetic torque expression for the phase-to-phase fault model becomes:

$$T_e = \frac{e_\alpha \cdot I_\alpha + e_\beta \cdot I_\beta - e_f \cdot I_f}{\Omega}. \qquad (32)$$

We consider for all the studies that the electromotive force of the healthy motor has a sinusoidal form as shown in Fig. 2(a) and contains a 3rd harmonic at $3 \cdot f_s$ of supply frequencies as seen in Fig. 2(b).

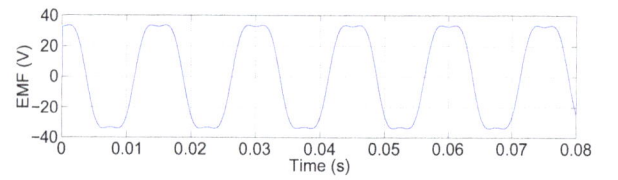

(a) Electromotive force.

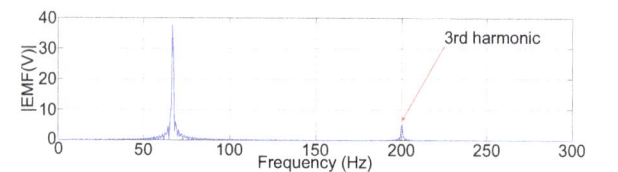

(b) Spectrum analysis.

Fig. 2: Electromotive force and its spectrum analysis.

$$0 = -R_{a2} \cdot I_{as} + R_{b2} \cdot I_{bs} - (L_{a2} + M_{a1a2} - M_{b2a1} - M_{b2a2}) \cdot \frac{dI_{as}}{dt} - (M_{a2b1} + M_{a2b2} - L_{b2} - M_{b2b1}) \cdot \frac{dI_{bs}}{dt} +$$
$$- (M_{a2c} - M_{b2c}) \cdot \frac{dI_{cs}}{dt} - e_f + (R_{a2} + R_{b2} + r_f) \cdot I_f - (-L_{a2} + M_{a2b2} - L_{b2} + M_{b2a2}) \cdot \frac{dI_f}{dt} \qquad (18)$$

$$\begin{bmatrix} v_{as} \\ v_{bs} \\ v_{cs} \\ 0 \end{bmatrix} = \begin{bmatrix} R_s & 0 & 0 & R_{a2} \\ 0 & R_s & 0 & R_{b2} \\ 0 & 0 & R_s & 0 \\ -R_{a2} & R_{b2} & 0 & R' \end{bmatrix} \begin{bmatrix} I_{as} \\ I_{bs} \\ I_{cs} \\ I_f \end{bmatrix} + \begin{bmatrix} L & M & M & M_{af} \\ M & L & M & M_{bf} \\ M & M & L & M_{cf} \\ M_{af} & M_{bf} & M_{cf} & L_f \end{bmatrix} \frac{d}{dt} \begin{bmatrix} I_{as} \\ I_{bs} \\ I_{cs} \\ I_f \end{bmatrix} + \begin{bmatrix} e_{as} \\ e_{bs} \\ e_{cs} \\ -e_f \end{bmatrix} \qquad (19)$$

3. Dynamic Fault Model Simulation Results

The study of the behavior of PMSM under fault conditions using the proposed fault dynamic model requires an accurate knowledge of circuit parameters. The PMSM parameters are given as shown in AppA [2].

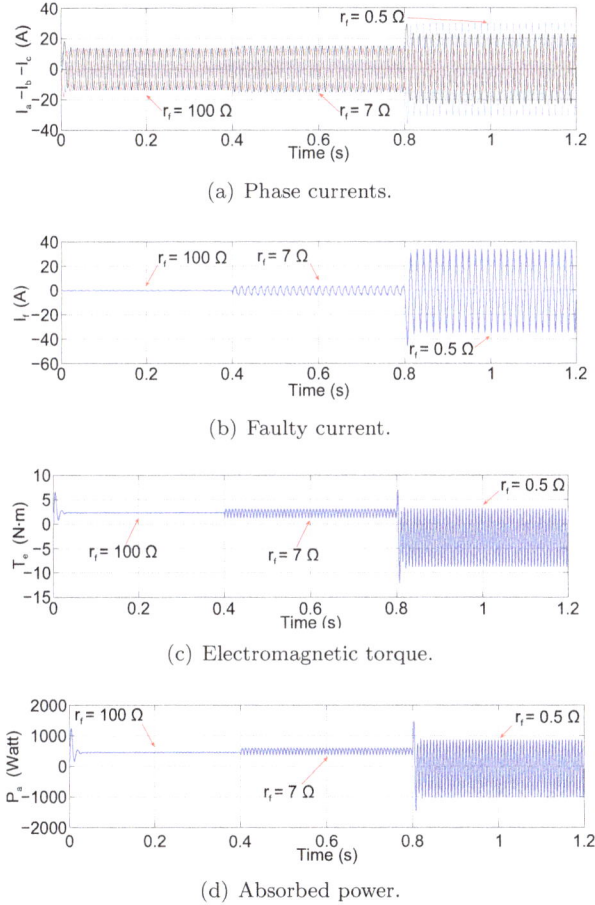

(a) Phase currents.

(b) Faulty current.

(c) Electromagnetic torque.

(d) Absorbed power.

Fig. 3: Phase currents, faulty current, electromagnetic torque and absorbed power versus time for three values of fault resistances: $r_f = 100$ Ω, $r_f = 7$ Ω and $r_f = 0.5$ Ω.

The machine is supposed to be supplied by 3-phases sinusoidal balanced voltage source with star connection and without neutral connection and operates at synchronous speed (speed and supply frequency are 1000 rpm and 66.67 Hz respectively). Simulation of the proposed model is realized using MATLAB environment.

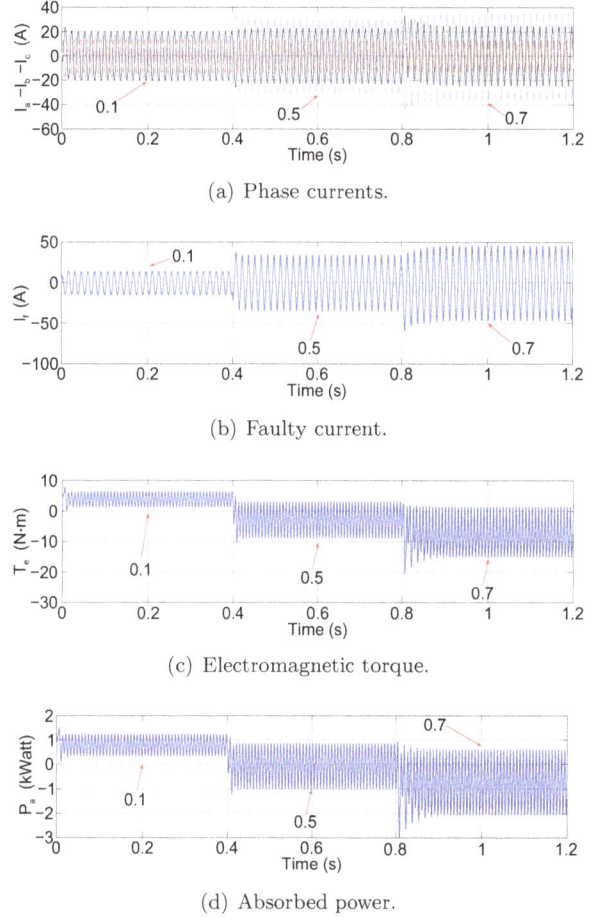

(a) Phase currents.

(b) Faulty current.

(c) Electromagnetic torque.

(d) Absorbed power.

Fig. 4: Phase currents, faulty current, electromagnetic torque and absorbed power versus time at three values of the fraction of shorted turns: ($\sigma = 0.1$, $\sigma = 0.5$ and $\sigma = 0.7$) and $r_f = 0.5$ Ω.

For this model, Fig. 3 shows the characteristics phase currents (a, b, c), faulty current (I_f), electromagnetic torque and absorbed power for different values of fault insulation resistance such as $r_f = 100$ Ω, 0.5 Ω and 7 Ω. The fraction of shorted turns is fixed at 50 %.

Figure 4 shows the characteristics (phase currents (a, b, c), faulty current (I_f), electromagnetic torque and absorbed power for different values of the fraction of shorted turns ($\sigma = 10$ %, $\sigma = 50$ % and $\sigma = 70$ %), where the fault insulation resistance is fixed to $r_f = 0.5$ Ω.

As it can be seen from Fig. 3, for three different values of fault resistances (healthy case: $r_f = 100$ Ω and

$$
\begin{bmatrix} v_\alpha \\ v_\beta \\ 0 \end{bmatrix} = \begin{bmatrix} R_s & 0 & R_r \\ 0 & R_s & r_{b2} \\ R_r & r_{b2} & r_f \end{bmatrix} \cdot \begin{bmatrix} I_\alpha \\ I_\beta \\ I_f \end{bmatrix} + \begin{bmatrix} L_s & 0 & M_{f\alpha} \\ 0 & L_s & M_{f\beta} \\ M_{f\alpha} & M_{f\beta} & L_f \end{bmatrix} \cdot \frac{d}{dt} \cdot \begin{bmatrix} I_{as} \\ I_{bs} \\ I_{cs} \\ I_f \end{bmatrix} + \begin{bmatrix} e_\alpha \\ e_\beta \\ -e_f \end{bmatrix} . \tag{25}
$$

faulty case: $r_f = 7\ \Omega$ and $r_f = 0.5\ \Omega$) when the fault resistance decreases, the three phases currents increase to compensate the negative effects of the short-circuit fault. It can cause a current unbalance in the power supply, and the increase of the absorbed power. We can observe a torque ripple when the faulty case is applied.

Changing the fraction of short-turns means changing the severity of applying fault. From Fig. 4, it is clear that the magnitude of the torque ripple is mainly determined by the severity of the fault. The magnitude of the phase currents and absorbed powerchange proportionally with the severity of the fault and become unbalanced.

It would be very helpful to predict the insulationfailure, breakdown when the fault is not high developed inorder to avoid the machine winding damages [14].

4. Diagnostic of Stator Fault by ESA and EPVA Techniques

Two techniques based on frequency analysis are applied to detect faults in stator, consecutively defined in [15], [16] and [17]. First is ESA, based on the Fast Fourier decomposition of the phase currents winding, the electromagnetic torque and the absorbed power. The second is EPVA, which is based on the frequency analysis of the module of the Park's Vector's of currents as shown below.

4.1. Electric Spectral Analysis (ESA)

We applied this technique on the phase stator currents, the instantaneously absorbed power and the electromagnetic torque. The instantaneous absorbed power is illustrated by the following equation [18]:

$$
p(t) = v_{as}(t)i_{as}(t) + v_{bs}(t)i_{bs}(t) + v_{cs}(t)i_{cs}(t). \tag{33}
$$

The phase stator currents, the instantaneous absorbed power and electromagnetic torque spectrum analysis results of both healthy and faulty conditions with different values of faulty resistance ($r_f = 100\ \Omega$, $r_f = 7\ \Omega$ and $r_f = 0.5\ \Omega$) of simulation machine are presented in Fig. 5, Fig. 6, and Fig. 7 respectively.

1) Currents Spectral Analysis

The ESA signatures reveal the existence of a spectral component in phase 'a' and 'b', with a small amplitude at the frequency with value three times higher than the supply due the existence of an inter-turn short circuit in the stator winding and its amplitude increase with the increase of severity of faultas seen in Fig. 5(b), Fig. 5(c) and Fig. 5(d), where $r_f = 0.5\ \Omega$ and in Fig. 5(e), where $r_f = 7\ \Omega$, the existence of this harmonic is due to the presence of the third harmonic of the electromotive force presented in Fig. 2(b). We can observe no existence of this harmonic in phase 'c' because the short circuit occurs between phase 'a' and 'b'. Note that at healthy conditions the current does not have this component (third harmonic), as seen in Fig. 5(a).

2) Electromagnetic Torque Spectral Analysis

It is noticeable from Fig. 7, that in case of fault, we notice the appearance of high harmonic at double value of supply frequency, especially if $r_f = 0.5\ \Omega$. The increase of the harmonic amplitude is inversely proportional to the values of fault resistance.

3) Absorbed Power Spectral Analysis

Figure 6 shows the absorbed power spectrum with and without fault. We can observe only a zero frequency component at healthy conditions. In faulty conditions the same analysis as that of the electromagnetic torque is noted. From the comparative analysis of results under healthy and faulty conditions, it is clear that the fault appears in the ESA signaturedue to the presence of harmonic of even rows on the spectrum analysis of electromagnetic torque and absorbed power and by the appearance of the harmonic of odd rows on the spectrum analysis of phase currents. The appearances of these harmonics are directly related to the existence of asymmetries caused by the short-circuit in the stator winding. With the consumption that we have a balanced voltage source, the appearance of harmonics in phase 'a' and 'b' indicates the short-circuit between these two phases.

(a) Healthy case for $r_f = 100 \; \Omega$.

(b) Faulty case for $r_f = 0.5 \; \Omega$ (Phase a).

(c) Faulty case for $r_f = 0.5 \; \Omega$ (Phase b).

(d) Faulty case for $r_f = 0.5 \; \Omega$ (Phase c).

(e) Faulty case for $r_f = 7 \; \Omega$

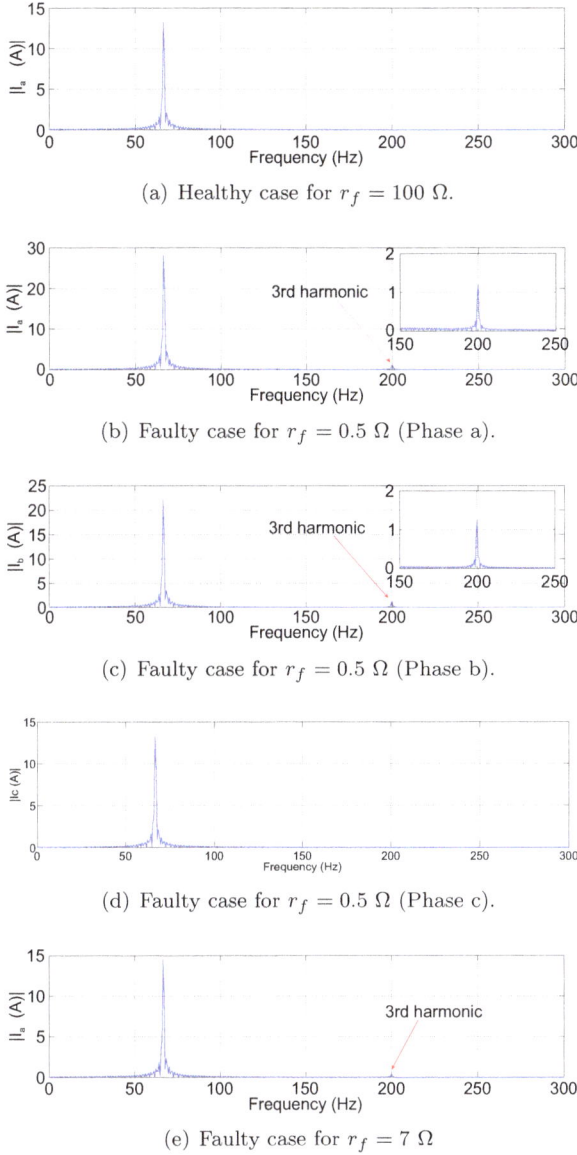

Fig. 5: Spectrum of phase currents.

4.2.   Extend Park's Vector Approach (EPVA)

This technique is based on the two equivalent currents in reference frame obtained by Park's transformation [18]:

$$I_d = \sqrt{\frac{2}{3}} \cdot I_{as} - \frac{1}{\sqrt{6}} \cdot I_{bs} - \frac{1}{\sqrt{6}} \cdot I_{cs}, \qquad (34)$$

$$I_q = \frac{1}{\sqrt{2}} \cdot I_{bs} - \frac{1}{\sqrt{2}} \cdot I_{cs}, \qquad (35)$$

where I_d and I_q are the instantaneous values of electric currents in direct and quadrature axis. Idis always a sine wave and I_q has a cosine wave in healthy conditions. These two components have the same values and their locus is a circle as seen in Fig. 8(a). In case of the inter-turn short circuit, the current becomes unbal-

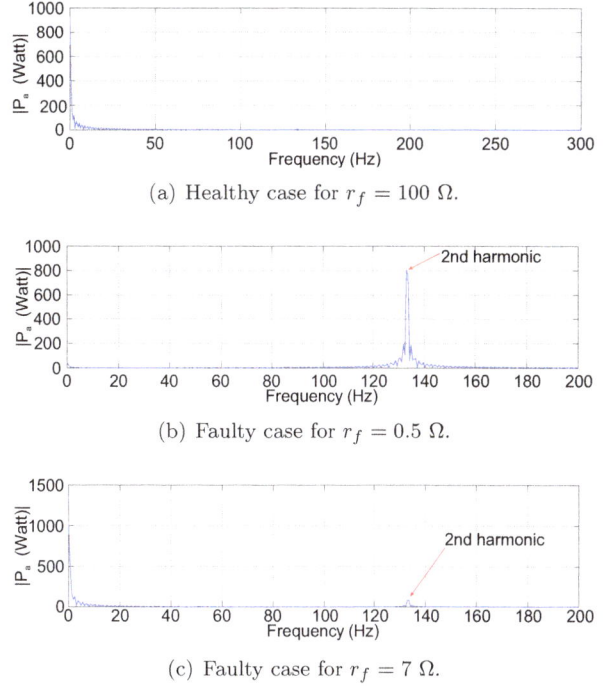

(a) Healthy case for $r_f = 100 \; \Omega$.

(b) Faulty case for $r_f = 0.5 \; \Omega$.

(c) Faulty case for $r_f = 7 \; \Omega$.

Fig. 6: Spectrum of absorbed power.

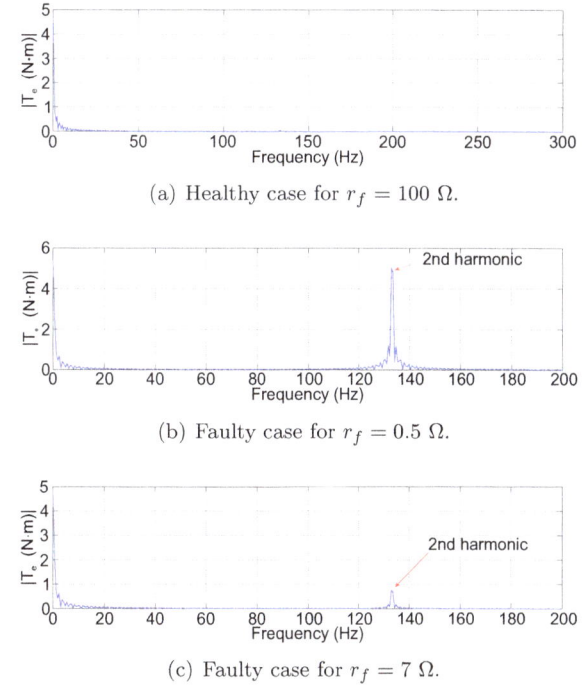

(a) Healthy case for $r_f = 100 \; \Omega$.

(b) Faulty case for $r_f = 0.5 \; \Omega$.

(c) Faulty case for $r_f = 7 \; \Omega$.

Fig. 7: Spectrum of electromagnetic torque.

anced and it can be expressed as the sum of a positive sequence and a negative sequence component. As a result of this fault, the Concordia's vector locus shape deviates and becomes elliptic as shown in Fig. 8(b).

If the motor operates under healthy conditions (i.e. under symmetrical conditions), the three currents form a balanced system and constitute a positive sequence system. Hence, i_d and i_q can be written as below [18]:

$$i_p = \sqrt{i_d^2 + i_q^2},\qquad(36)$$

$$i_d = \frac{\sqrt{6}}{2}\cdot i_{max}\cdot \sin(\omega t),\qquad(37)$$

$$i_q = \frac{\sqrt{6}}{2}\cdot i_{max}\cdot \sin\left(\omega t - \frac{\pi}{2}\right),\qquad(38)$$

where i_{max} is a maximum value of the current positive sequence, ω is the angular supply frequency, and i_p is the Park's equivalent current module. When the system is balanced, the current Park's vector modulus is constant as illustrated in Fig. 9(a). Under faulty condition the currents will contain other components besides the positive sequence component and in this case the Park's Vector modulus will contain a dominant DC and AC level of the motor current supply [15] and their existence is directly related to the asymmetries, as we can see in Fig. 9(b).

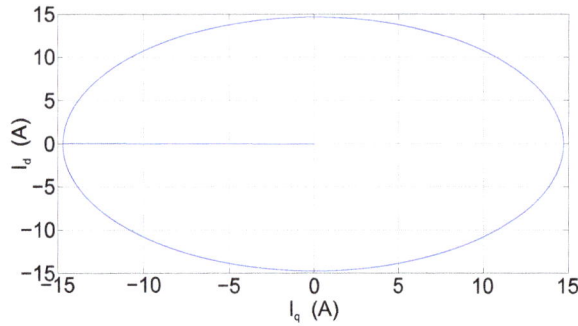

(a) Healthy case.

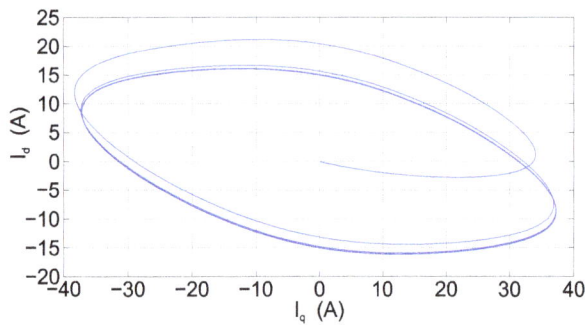

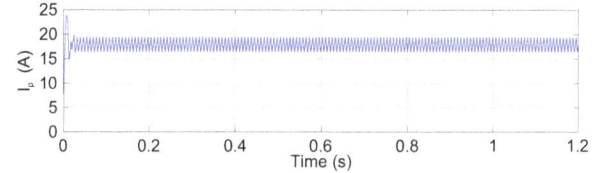

(b) Faulty case.

Fig. 8: Concordia's currents vector locus.

The aim of EPVA technique is to apply the frequency analysis to the Park's vector modulus in order to obtain the EPVA signature when the system is unbalanced. After simulation and analysis, we obtain the results for healthy condition ($r_f = 100\ \Omega$) and faulty conditions ($r_f = 7\ \Omega$ and $r_f = 0.5\ \Omega$) as shown in Fig. 10.

From these results, the EPVA signature reveals the existence of a aspectral component at a frequency of

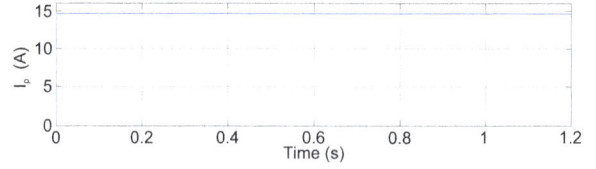

(a) Healthy case.

(b) Faulty case.

Fig. 9: Park's vector modulus.

66.67 Hz-twice the fundamental supply frequency and it is so clear from results when the fault resistance decreases (the severity of fault increases) the amplitude of the spectral component makes it a good indicator of the occurred fault.

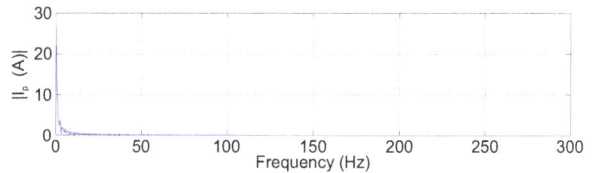

(a) Healthy case for $r_f = 100\ \Omega$.

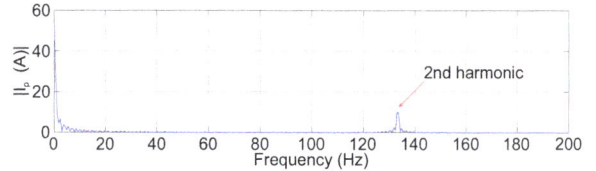

(b) Faulty case for $r_f = 0.5\ \Omega$.

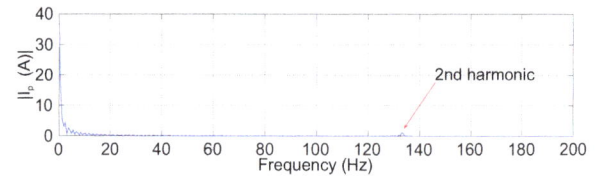

(c) Faulty case for $r_f = 7\ \Omega$.

Fig. 10: Spectrum of Park's vector modulus.

5.   Conclusion

This paperproposed a dynamic model for surface mounted PMSM machine under phase-to-phase short-circuit in the stator winding. The real form of back EMF is presented and included in the model. This faulty model is used to study the behavior of the machine under various fault conditions and severity. From the analysis of the simulation results, phase-to-phase short-circuit fault causes high torque ripples and currentunbalance in the system. Higher circulating currents could be generated by the motor winding short-circuit. More importantly, the detection of these kinds of faultsis crucial in the design and development procedure of the motor drive and its diagnosis. Two simple and effective diagnosis method as ESA and EPVA based on frequency analysis are used to analyze and to indicate the presence of the short-circuit fault between two phases in the stator. The appearance of the 2nd and 3rd harmonic indicates the presence of this fault andthe amplitude of the harmonics is proportional to the severity of this fault. The shape of Concordia's currents vector locus is a good indicator of the presence of the fault when its form changes from the circle trajectory to an elliptical one.

References

[1] HADEF, M., M. R. MEKIDECHE and A. O. N'DIAYE. Diagnosis of stator winding short circuit faults in a direct torque controlled interior permanent magnet synchronous motor. In: *IEEE Vehicle Power and Propulsion*. Chicago: IEEE, 2011, pp. 1–8. ISBN 978-61284-247-9. DOI: 10.1109/VPPC.2011.6043166.

[2] VASEGHI, B., B. NAHID-MOBAREKEH, N. TAKORABET and F. MEIBODY-TABAR. Modeling of non-salient PM synchronous machines under stator winding inter-turn fault condition: dynamic model-emfmode. In: *IEEE Vehicle Power and Propulsion Conference*. Arlington: IEEE, 2007, pp. 635–640. ISBN 0-7803-9761-4. DOI: 10.1109/VPPC.2007.4544200.

[3] CAPOLINO, G. A., C. BRUZZESE, R. PUSCA and J. ESTIMA. Trends in fault diagnosis for electrical machines: areview of diagnostic techniques. *IEEE Industrial Electronics Magazine*. 2014, vol. 8, iss. 2, pp. 31–42. ISSN 1932-4529. DOI: 10.1109/MIE.2013.2287651.

[4] PROGOVAC, D., L. Y. WANG and G. YIN. System identification of permanent magnet machines and its applications to inter-turn fault detection. In: *IEEE Transportation Electrification Conference and Expo (ITEC)*. Detrit MI: IEEE, 2013, pp. 1–5. ISBN 978-1-4799-0148-7. DOI: 10.1109/ITEC.2013.6573486.

[5] VASEGHI, B., B. NAHID-MOBAREKEH, N. TAKORABET and F. MEIBODY-TABAR. Modeling of IM with stator winding inter-turn fault validated by fem. In: *Electrical Machines*

Conrerence. Hammamat: IEEE, 2008, pp. 1–5. ISBN 978-1-4244-1736-0. DOI: 10.1109/ICELMACH.2008.4800130.

[6] GWAN GU, B., J. HYUK CHOI and I. SOUNG JUNG. Inter turn short fault model of PMSMs with series and parallel winding connection. In: *Energy Conversion Congress Conference*. Athlanta: IEEE, 2013, pp. 4388–4395. ISBN 978-1-4799-0336-8. DOI: 10.1109/ECCE.2013.6647287.

[7] TALLAM, R. M., T. G. HABTLER and R. G. HARLEY. Transient model for induction machines with stator winding turn faults. *IEEE Transactions on Industry Applicatin*. 2002, vol. 38, iss. 3, pp. 632–637. ISSN 0093-9994. DOI: 10.1109/TIA.2002.1003411.

[8] ARKAN, M., D. KASTIC-PEROVIC and P. J. NSWORTH. Modeling and simulation of induction motors with inter turn fault for diagnostics. *ELSEVIER Journal of Electric Power System Research (EPSR)*. 2005, vol. 75, iss. 1, pp. 57–66. ISSN 0378-7796. DOI: 10.1016/j.epsr.2004.08.015.

[9] DAI, M. and A. SEBASTIAN. Fault analysis of a PM brishless DC motor using finite element method. *IEEE Transaction On Energy Conversion*. 2005, vol. 20, iss. 1, pp. 1–4. ISSN 0885-8969. DOI: 10.1109/TEC.2004.841516.

[10] MOHAMMED, O. A., Z. LIU, S. LIU and N. Y. ABED. Inter turn short circuit fault diagnosis for PM machines using FE based of phase variable model and wavelet analysis. *IEEE Transaction On Magnetic*. 2007, vol. 43, iss. 4, pp. 1729–1732. ISSN 0018-9464. DOI: 10.1109/TMAG.2006.892301.

[11] JOENG, I. I. S. U., B. J. HYON and K. NAM. Dynamic modeling and control for SPMSMs with internal turn short faults. *IEEE Transactionon Power Electronic*. 2013, vol. 28, iss. 7, pp. 3495–3508. ISSN 0885-8993. DOI: 10.1106/TPEL.2012.2222049.

[12] KIM, K.-T., J. HUR, B.-W. KIM and G.-H. KANG. Circulating current calculation using fault modeling of IPM type BLCD motor of inter-turn fault. In: *IEEE Electric machines and Systemsconferenc*. Tokyo: IEEE, 2011, pp. 1–5. ISBN 9781-4577-1043-8. DOI: 10.1109/ICEMS.2011.6073686.

[13] VASEGHI, B., B. NAHID-MOBAREKEH, N. TAKORABET and F. MEIBODY-TABAR. Experimentally validation dynamic fault model for PMSM with stator winding inter-turn fault. In: *Industry Application Society Annual Meeting*. Edmonton: IEEE, 2008, pp. 1–5. ISBN 978-1-4244-2279-1. DOI: 10.1109/08IAS.2008.24.

[14] LAI, C., A. BALAMURALI, V. BOUSABA, K. L. V. IYER and N. KAR. Analysis of stator winding inter-turn short circuit fault in interior and surface mounted permanent magnet traction machines. In: *Transportation Electrification Conference and Expo*. Dearborn: IEEE, 2014, pp. 1–6. ISBN 978-1-4799-2262-8. DOI: 10.1109/ITEC.2014.6861775.

[15] CRUZ, S. M. A. and A. J. M. CARDOSO. Stator winding fault diagnosis in three phasessyncrhronous and asynchronous motor, by the extended Park's vector approach. *IEEE Transaction On Industry Applications*. 2001, vol. 37, iss. 5, pp. 1227–1233. ISSN 1939-9367. DOI: 10.1109/28.952496.

[16] CRUZ, S. M. A. and A. J. M. CARDOSO. Multiple reference frames theory: a new methode for the diagnosis of stator fault inthree phases induction motors, by the extended Park's vector approach. *IEEE Transaction On Energy Conversion*. 2005, vol. 20, iss. 3, pp. 611–619. ISSN 1558-0059. DOI: 10.1109/TEC.2005.847975.

[17] DYONOSIOS, V. S. and D. M. EPAMINONDAS. Induction Motor Stator Fault Diagnosis Technique Using Park Vector Approach and Complex Wavelets. In: *IEEE International Conference on Electric Machinery (ICEM)*. Marseille: IEEE, 2012, pp. 1730–1734. ISBN 978-1-4673-0142-8. DOI: 10.1109/ICEIMach.2012.6350114.

[18] PARRA, A. P., M. C. A. ENCICO, J. O. OCHOA and J. A. P. PENAR. Stator fault diagnosis on squirrel cage induction motor by ESA and EPVA. In: *Power Electronics and Power Quality Applications*. Bogota: IEEE, 2014, pp. 1–6. ISBN 978-1-4799-1007-6. DOI: 10.1109/PEPQA.2013.6614937.

About Authors

Chourouk BOUCHAREB was born in 1975, in Algiers, Algeria. She received an Engineer Diploma in Electrical Engineering in 1999 and an M.Sc. degree in Control engineeringin 2005, both from Electrical Engineering Department of Batna University. Her research interests include the electric machines and their control drives and diagnosis. She is a member at the Laboratory University, named Electromagnetic Induction and Propulsion Systems (LSPIE) of Batna University.

Mohamed-Said NAIT-SAID was bornin 1958, in Batna, Algeria, He received an Engineer Diploma in Electrical Engineering from the National Polytechnic High School of Algiers, Algeria (February 1983), and the M.Sc. degree in Electronics and Control Engineering from Electronics Department at Constantine University in 1992. He received the Ph.D. degree in Electrical Engineering from University of Batna after he accomplished his free scientific research accomplished in Automatic Laboratory of Amiens University in French from 1996 to 1999. Currently he is a full professor at the Electrical Engineering Department of Batna University II and is responsible for the Master course of Control and Diagnosis of the Electrical Systems. From 2000–2005, Dr. Nait-Said was the head of the first created research laboratory in Batna University, named Electromagnetic Induction and Propulsion Systems (LSPIE) of Batna and also in 2006 he has been appointed the head of the scientific committee of the same department. LSPIE has been evaluated by the Algerian ministry of the universities as the best laboratory in Batna University (100 percent satisfied. Dr. Nait-Said has supervised twenty five Masters and ten Ph.D. theses. His research interests include the electric machines and their control drives and diagnosis.

Appendix A - AC Driver Parameters

- $P_N = 5$ kW,
- $P = 4$,
- EMF at 1000 rpm $= 34$ V,
- $I_N = 19$ A.

13

The Sensitivity of the Input Impedance Parameters of Track Circuits to Changes in the Parameters of the Track

Lubomir IVANEK[1], Vladimir MOSTYN[2], Karel SCHEE[3], Jan GRUN[3]

[1]Department of Electrical Engineering, Faculty of Electrical Engineering and Computer Science,
VSB–Technical University of Ostrava, 17. listopadu 15, 708 00 Ostrava-Poruba, Czech Republic
[2]Department of Robotics, Faculty of Mechanical Engineering, VSB–Technical University of Ostrava,
17. listopadu 15, 708 00 Ostrava-Poruba, Czech Republic
[3]Prvni signalni a.s., Bohuminska 368/172, 712 00 Ostrava-Muglinov, Czech Republic

lubomir.ivanek@vsb.cz, vladimir.mostyn@vsb.cz, schee@1sig.cz, grun@1sig.cz

Abstract. *This paper deals with the sensitivity of the input impedance of an open track circuit in the event that the parameters of the track are changed. Weather conditions and the state of pollution are the most common reasons for parameter changes. The results were obtained from the measured values of the parameters R (resistance), G (conductance), L (inductance), and C (capacitance) of a rail superstructure depending on the frequency. Measurements were performed on a railway siding in Orlova. The results are used to design a predictor of occupancy of a track section. In particular, we were interested in the frequencies of 75 and 275 Hz for this purpose. Many parameter values of track substructures have already been solved in different works in literature. At first, we had planned to use the parameter values from these sources when we designed the predictor. Deviations between them, however, are large and often differ by three orders of magnitude (see Tab. 8). From this perspective, this article presents data that have been updated using modern measurement devices and computer technology. And above all, it shows a transmission (cascade) matrix used to determine the parameters.*

Keywords

Impedance, parameter estimation, railway safety, sensitivity analysis.

1. Introduction

In this article, the results of measuring the parameters of rail traction - the resistance of the track circuit, the conductance between the rails, and the inductance and the capacitance of the track circuit - are described. The sensitivity of the input impedance of the track circuit is calculated after changing the parameters of the track. These parameters were investigated to design the predictor, which detects the occupancy of the rail traction. This article does not provide a new method of measurement. The modified Ohm impedance measurement method was used in the measurement. However, elaboration of the results must provide parameter values in a form which can be directly used in a laboratory simulation of track circuits. This simulator was assembled from type two-ports (parameters G, R, L, C) and it was used to verify the device characteristics to predict the occupancy of the track circuit. Thus, cascade coefficients were used for evaluating traction parameters, as well as for elaboration of the measured values in the simulator. Therefore, it is possible to immediately correct potential deficiencies of the simulator.

All these parameters were calculated as specific parameters, which means they were recalculated to the unit length (km) of the tracks.

The following approach was chosen to determine the primary parameters:

- Measurement of the input impedance of an open circuit and input impedance of a section of a railway line terminated with a short circuit.

- Calculation of cascade coefficients.

- Calculation of impedance and admittance of two-ports of type Π, T and Γ.

- Calculation of the primary two-port parameters and their conversion to the unit length.

Measurements were performed in the Orlova locality in the rail yard of a private firm, AWT. The section of the measured track did not have any track crossings or track branches and had only a slight radius of curvature of the track. The railway yard was very dirty from coal dust, loam and clay, dry fallen leaves, twigs, and so on. The weather was rainy. It was raining slightly when the measurements were taken, and had been raining for a long time beforehand. The length of the track section was 540 m.

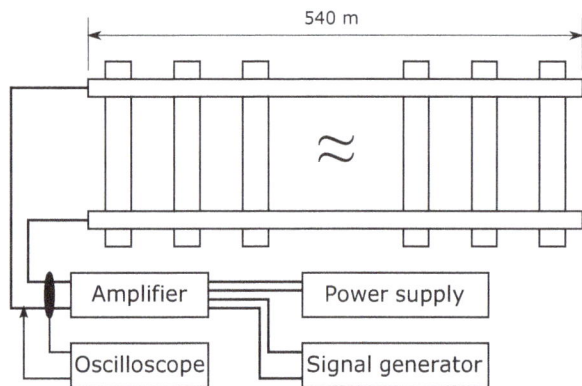

Fig. 1: Connection diagram.

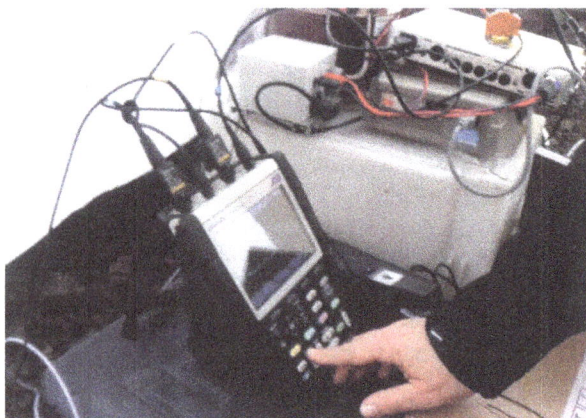

Fig. 2: View of the workplace arrangement.

Examples of display of the measured waveforms on the oscilloscope:

At lower frequencies, the current leads the voltage on empty track (see Fig. 3).

The voltage leads the current at higher frequencies and therefore the track circuit has an inductive character. Figure 4 shows an example of waveforms for a track

that was shunted to 524 m away from the measuring site.

The track circuit in Fig. 4 certainly has an inductive character.

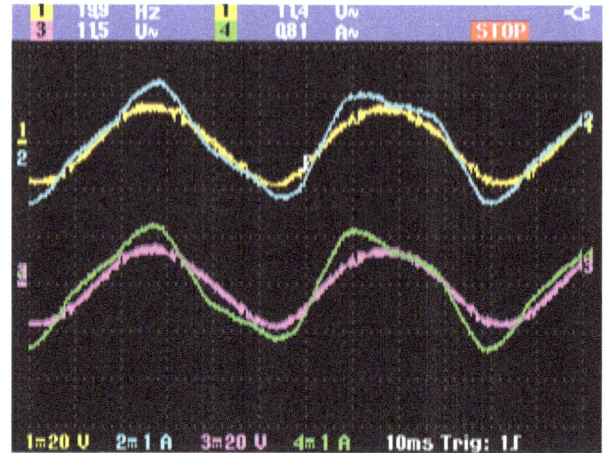

Fig. 3: Voltage and current courses for the frequency of 19.9 Hz.

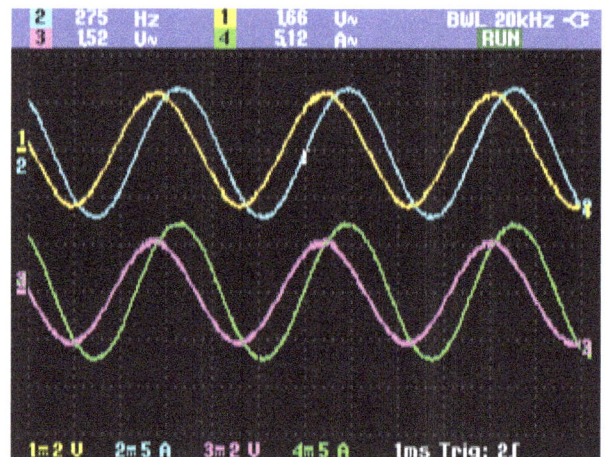

Fig. 4: Voltage and current waveforms for the frequency of 275 Hz.

A device for prediction of occupancy of the track circuit must be regularly calibrated during operation when weather conditions are changing. Calibration will be done on unoccupied track circuits. In an open circuit of the traction, mostly capacity and conductivity of the substructures may be varied. Therefore, we are

Tab. 1: Assignment of oscilloscope channels to the measured quantities.

No.	Quantity	Function
CH1	U1	Voltage at the amplifier output
CH2	'I1	The output current from an amplifier
CH3	U2	Voltage between the rails at the measuring location
CH4	'I2	Current to rails

Tab. 2: Module and phase of impedance of open traction circuit and short traction circuit.

Impedance open circuit			Impedance short circuit		
f_524m (Hz)	Zop_524m (Ω)	Fi0_524m (rad)	f_524m (Hz)	Zsh_524m (Ω)	Fik_524m (rad)
19.89	14.47	-0.26	19.43	0.44	0.36
50.26	13.44	-0.14	49.74	0.67	0.76
74.76	13.25	-0.1	75.86	0.85	0.83
100.51	13.8	-0.09	100.23	1.1	0.89
123.57	12.99	-0.06	125.32	1.18	0.93
149.86	12.94	-0.09	150.59	1.34	0.99
173.21	12.84	-0.06	174.85	1.49	0.99
202.7	12.72	-0.07	199.82	1.64	1.1
250.73	12.66	-0.08	248.55	1.9	1.4
274.19	12.62	-0.05	275.29	2.4	1.8
302.48	12.27	-0.01	303.74	2.2	1.6
491.28	11.93	0.05	500.87	3.24	1.11
1009.77	11.93	0.13	1032.62	5.81	1.17

interested in finding out how the parameters change as the weather changes, and which parameter of the device will be the most responsive one. This parameter will be determining during the calibration of the instrument.

2. Open and Short Circuit Impedance

Measurements were performed for the frequency range from 20 to 5000 Hz. The length of the measured section of rail traction was 540 m. The courses were obtained in the form of graphs (Fig. 3 and Fig. 4) and in table form. The obtained curves expressed in tabular form contained a large amount of noise and in order to use them, they had to be approximated by a sine wave function - (Fig. 5).

The measured values were the magnitude Z and the phase φ of the input impedance at the beginning of the track circuits. Measurements were performed under the open circuit Z_{op} and short circuit Z_{sh}. It is clear from Tab. 2 that the frequency at which the impedances are measured in the open and short circuits are not completely identical. Therefore, three kinds of interpolation by polynomials were performed for both curves: cubic spline interpolation, B-spline interpolation, and polynomial regression.

$$
\begin{aligned}
vsc &:= cspline(vx, vy), \\
vsb &:= bspline(vx, vy, u, 2), \\
vsr &:= regress(vx, vy, 3).
\end{aligned}
\qquad (1)
$$

From these polynomials, impedance values have been assigned to the same integer frequencies (see Tab. 3).

Impedance components were converted to the form of complex numbers using a simple Eq. (2):

$$
\underline{Z} = Z e^{j\varphi} = Z(\cos\varphi + j\sin\varphi). \qquad (2)
$$

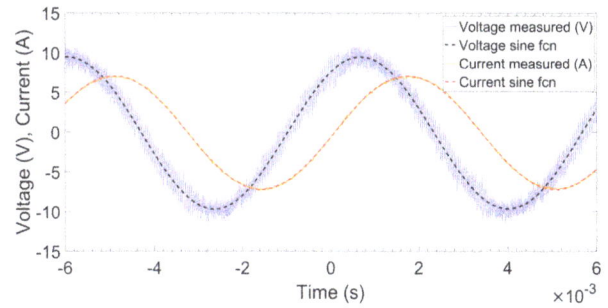

Fig. 5: Example of interleaving of the actual course of state variables by the sine function.

Tab. 3: Impedance of the open circuit and the short circuit.

Freq.	$\underline{Z}_{op}$ (Ω)	$\underline{Z}_{sh}$ (Ω)
20	13.99-3.73j	0.42+0.16j
50	13.3-1.95j	0.48+0.46j
75	13.18-1.4j	0.57+0.64j
100	13.02-1.28j	0.63+0.79j
125	12.98-0.8j	0.7+0.96j
150	12.89-1.2j	0.73+1.13j
175	12.82-0.87j	0.82+1.25j
200	12.7-0.9j	0.87+1.4j
250	12.61-1.09j	0.95+1.65j
275	12.54-0.39j	0.96+1.81j
300	12.27-0.15j	1.07+1.93j
500	11.91+0.7j	1.43+2.91j
1000	11.83+1.61j	2.25+5.36j
3000	12.01+0.46j	6.8+11.54j
5000	11.99+7.79j	11.31+14.38j

3. Calculation of the Cascade Coefficients

Matrix of cascade coefficients:

$$
\underline{A} = \begin{vmatrix} \sqrt{\dfrac{\underline{Z}_{op}}{\underline{Z}_{op}-\underline{Z}_{sh}}} & \underline{Z}_{sh}\sqrt{\dfrac{\underline{Z}_{op}}{\underline{Z}_{op}-\underline{Z}_{1k}}} \\ \dfrac{1}{\sqrt{\underline{Z}_{op}(\underline{Z}_{op}-\underline{Z}_{sh})}} & \sqrt{\dfrac{\underline{Z}_{op}}{\underline{Z}_{op}-\underline{Z}_{sh}}} \end{vmatrix}. \qquad (3)
$$

Primarily we are interested in the frequency of 75 Hz with the wavelength of 1154701 m, and secondarily, the frequency of 275 Hz with the wavelength of 314918.3 m. Consequently, the 524 m segment of the railway can be considered as an elementary line section - for example as a two-port of type T or Π with a matrix of cascade coefficients according to Tab. 4.

Coefficient A12 was negative for frequencies above 1000 Hz. Also, the determinant of the matrix coefficients from this frequency was not equal to zero. Therefore, these values were not used again. The frequencies of 75 and 275 Hz interest us most.

Cascade coefficients were also calculated from the secondary circuit parameters:

$$\underline{Z}_0 = \sqrt{\underline{Z}_{10}\ \underline{Z}_{1k}}, \tag{4}$$

$$\tanh(\underline{g}_0) = \sqrt{\frac{\underline{Z}_{1k}}{\underline{Z}_{10}}} \Rightarrow \underline{g}_0 = \frac{1}{2}\ln\left(\frac{1+\sqrt{\frac{\underline{Z}_{1k}}{\underline{Z}_{10}}}}{1-\sqrt{\frac{\underline{Z}_{1k}}{\underline{Z}_{10}}}}\right). \tag{5}$$

From these secondary parameters, the following equation is valid:

$$\underline{A}_{22} = \underline{A}_{11} = \frac{e^{g_0}+e^{-g_0}}{2} = \cosh(\underline{g}_0), \tag{6}$$

$$\underline{A}_{12} = \underline{Z}_0\sinh(\underline{g}_0). \tag{7}$$

Tab. 4: Cascade coefficients.

Freq.	$\mathbf{A_{11}}$	$\mathbf{A_{12}}$	$\mathbf{A_{21}}$
20	1.013+0.009j	0.424+0.166j	0.067+0.019j
50	1.014+0.020j	0.478+0.477j	0.074+0.012j
75	1.018+0.028j	0.563+0.667j	0.076+0.010j
100	1.020+0.034j	0.615+0.827j	0.077+0.010j
125	1.020+0.041j	0.676+1.011j	0.078+0.008j
150	1.021+0.049j	0.690+1.190j	0.078+0.011j
175	1.025+0.055j	0.772+1.327j	0.079+0.010j
200	1.026+0.062j	0.805+1.490j	0.080+0.011j
250	1.025+0.074j	0.851+1.762j	0.080+0.013j
275	1.028+0.081j	0.841+1.939j	0.082+0.009j
300	1.034+0.089j	0.935+2.091j	0.084+0.008j
500	1.046+0.140j	1.088+3.245j	0.088+0.007j
1000	1.043+0.271j	0.894+6.198j	0.089+0.011j
3000	0.826+0.547j	-0.694+13.252j	0.070+0.043j
5000	0.766+1.253j	-9.357+25.195j	0.093+0.044j

4.   Calculation of Parameters of Two-port

The measured part of traction was replaced by a two-port of type Π. This two-port can be described as a matrix Eq. (8):

$$\underline{A} = \begin{vmatrix} 1+\underline{Y}\underline{Z} & \underline{Z} \\ \underline{Y}\underline{Z}+\underline{Y}^2\underline{Z} & 1+\underline{Y}\underline{Z} \end{vmatrix}. \tag{8}$$

Thus

$$\underline{Z} = \underline{A}_{12}, \tag{9}$$

$$\underline{Y} = \frac{\underline{A}_{11}-1}{\underline{Z}}. \tag{10}$$

From the matrix Eq. (8), we have calculated the parameters of the two-port (see Tab. 5 and Tab. 6). The parameter values R, G, L, and C were converted to a unit length of 1 km.

Tab. 5: Parameters of two-port.

Freq.	$\underline{\mathbf{Z}}(\mathbf{\Omega})$	$\underline{\mathbf{Y}}(\mathbf{S})$
20	0.453+0.149j	0.032+0.016j
50	0.575+0.493j	0.037+0.004j
75	0.565+0.665j	0.038+0.003j
100	0.653+0.812j	0.038+0.005j
125	0.660+1.014j	0.039+0.001j
150	0.721+1.188j	0.039+0.001j
175	0.841+1.286j	0.038+0.010j
200	0.798+1.519j	0.040+0.001j
250	0.897+1.763j	0.039+0.007j
275	0.860+1.941j	0.040+0.001j
300	0.926+2.114j	0.040+0.002j
500	1.079+3.233j	0.041+0.003j
1000	0.839+6.020j	0.043+0.004j

Tab. 6: The resulting values for the parameters R, G, L, and C of the type Π two-port.

Freq.	$\mathbf{R\cdot km^{-1}}$	$\mathbf{G\cdot km^{-1}}$	$\mathbf{L\cdot km^{-1}}$	$\mathbf{C\cdot km^{-1}}$
20	0.839	0.118	0.0022	0.000458
50	1.6	0.139	0.0029	0.0000468
75	1.5	0.141	0.00261	0.0000258
100	1.21	0.142	0.00239	0.0000304
125	1.22	0.144	0.00239	0.00000352
150	1.34	0.145	0.00234	0.00000348
175	1.56	0.14	0.00217	0.0000361
200	1.48	0.147	0.00224	0.00000342
250	1.66	0.146	0.00209	0.0000158
275	1.59	0.149	0.00206	0.00000324
300	1.72	0.15	0.00208	0.00000328
500	2,00	0.154	0.0019	0.00000315
1000	1.55	0.158	0.00176	0.00000212

Tab. 7: The resulting values for the parameters R, G, L, and C converted to a unit length of 1 km.

Freq.	$\mathbf{R\cdot km^{-1}}$	$\mathbf{G\cdot km^{-1}}$	$\mathbf{L\cdot km^{-1}}$	$\mathbf{C\cdot km^{-1}}$
20	0.839	0.117	0.0022	$3.08\cdot10^{-4}$
50	1.6	0.12	0.0029	$9.03\cdot10^{-5}$
75	1.5	0.14	0.00261	$3.68\cdot10^{-5}$
100	1.21	0.14	0.00239	$3.43\cdot10^{-5}$
125	1.22	0.144	0.00239	$1.27\cdot10^{-5}$
150	1.34	0.142	0.00234	$2.07\cdot10^{-5}$
175	1.56	0.144	0.00217	$1.94\cdot10^{-5}$
200	1.48	0.145	0.00224	$1.03\cdot10^{-5}$
250	1.69	0.144	0.00209	$1.35\cdot10^{-5}$
275	1.56	0.151	0.00206	$6.46\cdot10^{-6}$
300	1.72	0.153	0.00208	$3.83\cdot10^{-6}$
500	2.2	0.16	0.0019	$3.58\cdot10^{-7}$
1000	1.62	0.169	0.00176	$-2.63\cdot10^{-7}$

Graphs of the dependence of these parameters on the frequency are shown in Fig. 6, Fig. 7, Fig. 8 and Fig. 9.

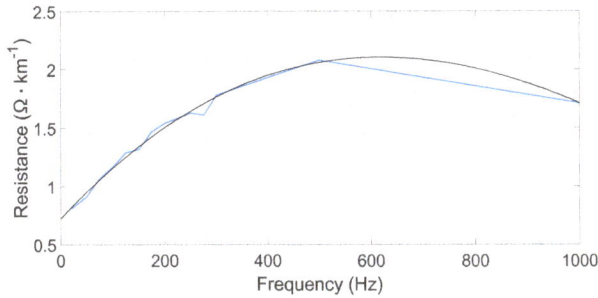

Fig. 6: Specific resistance as a function of frequency.

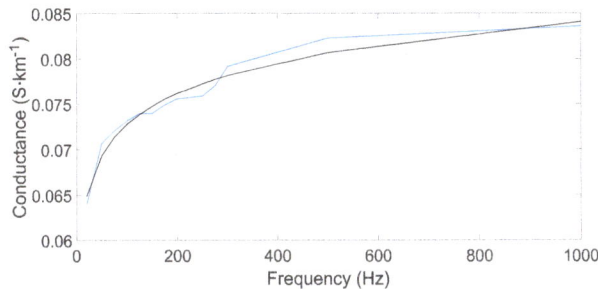

Fig. 7: Specific conductance as a function of frequency.

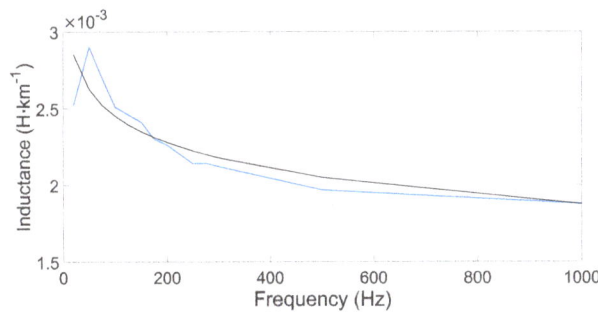

Fig. 8: Specific inductance as a function of frequency.

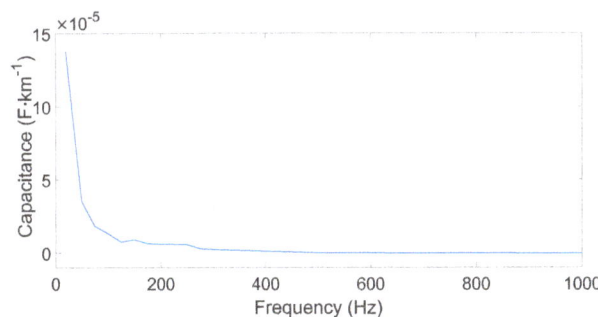

Fig. 9: Specific capacitance as a function of frequency.

5. Calculation of the Sensitivity of the Input Impedance to Changes of Parameters

The parameters of traction circuits change with the weather. First, we determined the sensitivity of the input impedance of the open circuit when changing the individual parameters. Thus we found which parameter affected the function of the predictor predominantly, and vice versa, which one can possibly be neglected. The assessed frequency was 275 Hz. Thus, we used these values:

Tab. 8: Used values of R, G, L and C.

$R \cdot km^{-1}$	$G \cdot km^{-1}$	$L \cdot km^{-1}$	$C \cdot km^{-1}$
1.56	$1.51 \cdot 10^{-1}$	$2.06 \cdot 10^{-3}$	$6.46 \cdot 10^{-6}$

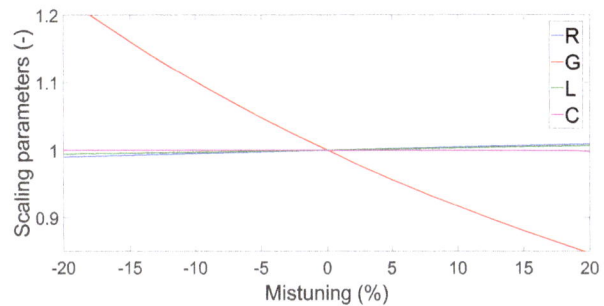

Fig. 10: Dependence of changes of normed real component of impedance in the open circuit on the percentage change of each parameter.

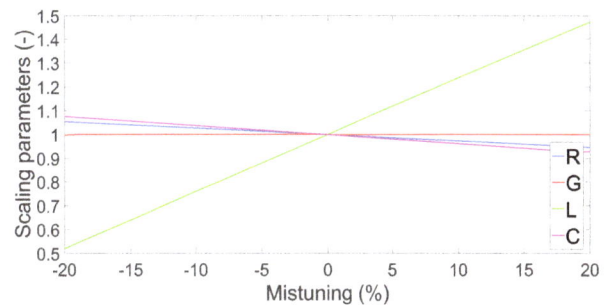

Fig. 11: Dependence of changes of the normed imaginary component of impedance in the open circuit on the percentage change of each parameter.

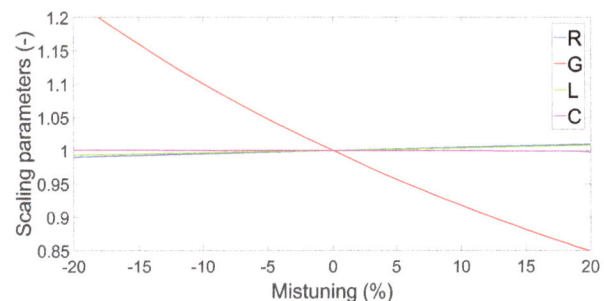

Fig. 12: Dependence of changes of normed magnitude of impedance in the open circuit on the percentage change of each parameter.

These parameters were gradually mistuning about 20 % in both the positive and the negative values while the values of the other parameters remained constant.

6. Conclusion

The measurements that are described herein are intended for the design of a predictor that searches the occupancy of the track circuit. This predictor needs to react to weather changes and changes of other conditions of traction. Therefore, repeated measurements must be carried out, especially in the open circuit condition. When measuring under open circuit conditions, it is necessary to evaluate the particular ratio of effective values of voltages and currents and thus to evaluate the magnitude of the impedance. It is most sensitive to changes of conductance G. Conductance G and capacitance C change the most with the weather conditions. Thus is enough, if only the change of the conductivity G is respected during calibration. Other parameters remain constant.

Finally, we compared the calculated values with those from various works in literature for the frequency of 275 Hz.

Tab. 9: Comparison of our results with those of other papers.

$R \cdot km^{-1}$ (Ω)	$G \cdot km^{-1}$ (S)	$L \cdot km^{-1}$ (H)	$C \cdot km^{-1}$ (F)	Ref.
0.55	2.25	$1.55 \cdot 10^{-3}$	$0.25 \cdot 10^{-6}$	[2]
1.1	$1.04 \cdot 10^{-3}$	-	$0.9 \cdot 10^{-6}$	[4]
0.18	$18 \cdot 10^{-3}$	$12.7 \cdot 10^{-3}$	$7 \cdot 10^{-6}$	[5]
0.004	-	$1.45 \cdot 10^{-3}$	-	[8]
1.56	$1.51 \cdot 10^{-1}$	$2.06 \cdot 10^{-3}$	$6.46 \cdot 10^{-6}$	[*]

[*] this paper

However, it is too difficult to compare the values of measurements under different conditions at different shunt resistances conducted in series to the rails when measuring the short circuit and so on. The measuring conditions are not comparable at all.

Acknowledgment

This research is supported by the project SP 2016/143 "Research of antenna systems; effectiveness and diagnostics of electric drives with harmonic power; reliability of the supply of electric traction; issue data anomalies." and by the project TA04031780 "AXIO - a system for measuring distance and speed of rail vehicles.

References

[1] MARISCOTTI, A. and P. POZZOBON. Determination of the electrical parameters of railway traction lines: calculation, measurement, and reference data. *IEEE Transactions on Power Delivery.* 2004, vol. 19, no. 4, pp. 1538–1546. ISSN 0885-8977. DOI: 10.1109/TPWRD.2004.835285.

[2] COLAK, K. and M. H. HOCAOGLU. Calculation of rail potentials in a DC electrified railway system. In: *38th International Universities Power Engineering Conference.* Thessaloniki: UPEC, 2003, pp. 5–8.

[3] HILL, R. J., S. BRILLANTE and P. J. LEONARD. Railway track transmission line parameters from finite element field modelling: Series impedance. *IEEE Proceeding Electric Power Applications.* 1999, vol. 146, iss. 6, pp. 647–660. ISSN 1350-2352. DOI: 10.1049/ip-epa:19990649.

[4] HILL, R. J. and D. C. CARPENTER. Rail track distributed transmission line impedance and admittance: theoretical modeling and experimental results. *IEEE Transaction Vehicular Technology.* 1993, vol. 42, iss. 2, pp. 225–241. ISSN 0018-9545. DOI: 10.1109/25.211460.

[5] HILL, R. J., D. C. CARPENTER and T. TASAR. Railway track admittance, earth-leakage effects and track circuit operation. In: *Technical Papers Presented at the 1989 IEEE/ASME Joint Railroad Conference.* Philadelphia: IEEE, 1989, pp. 55–62. ISBN 0-7803-3854-5. DOI: 10.1109/RRCON.1989.77281.

[6] GARG, R., P. MAHAJAN and P. KUMAR. Sensitivity Analysis of Characteristic Parameters of Railway Electric Traction System. *International Journal of Electronics and Electrical Engineering.* 2014, vol. 2, no. 1, pp. 8–14. ISSN 2301-380X. DOI: 10.12720/ijeee.2.1.8-14.

[7] SZELAG, A. Rail track as a lossy transmission line Part I: Parameters and new measurement methods. *Archives of Electrical Engineering.* 2000, vol. XLIX, no. 3–4, pp. 407–423. ISSN 1427-4221.

[8] HOLMSTROM, F. R. The model of conductive interference in rapid transit signaling systems. *IEEE Transactions on Industry Applications.* 1986, vol. IA–22, iss. 4, pp. 756–762. ISSN 0093-9994. DOI: 10.1109/TIA.1986.4504788.

[9] SZELAG, A. Rail track as a lossy transmission line. Part II: New method of measurements-simulation and in situ measurements. *Archives of Electrical Engineering.* 2000, vol. XLIX, no. 3–4, pp. 425–453. ISSN 1427-4221.

[10] JOURNEY, M. P., O. J. STEEL and B. K. FORTE. Electromagnetic modeling of electric railway systems to study its compatibility. *Revista Digital Lampsakos.* 2010, vol. 2010, no. 3, pp. 42–47, ISSN 2145-4086. DOI: 10.21501/issn.2145-4086.

[11] GUGLIELMINO, E. Determinacion de parametros electromagneticos de vias ferreas. *Ingenierias*. 2003, vol. VI, no. 19, pp. 39–46. ISSN 1405-7743.

[12] NOLAN, D., P. MCGREGOR and J. PAFF. TMG E1631 - Westinghouse FS2600 Track Circuits Manual. *Transport for NSW* [online]. 2015. Available at: http://www.asa.transport.nsw.gov.au/sites/default/files/asa/railcorp-legacy/disciplines/signals/tmg-e1631.pdf.

[13] HUAN, Q., Y. ZHANG. and B. ZHAO. Study on shunt state of track circuit based on transient current. *WSEAS Transactions on Circuits and Systems*. 2015, vol. 14, iss. 1, pp. 1–7. ISSN 2224-266X.

[14] GARCIA, J. C., J. A. JIMENEZ, F. ESPINOSA, A. HERNANDEZ, I. FERNANDEZ, M. C. PEREZ, J. URENA, M. MAZO and J. J. GARCIA. Characterization of railway line impedance based only on short-circuit measurements. *International Journal of Circuit Theory and Applications*. 2015, vol. 43, no. 8, pp. 984–994. ISSN 1097-007X. DOI: 10.1002/cta.1987.

About Authors

Lubomir IVANEK was born in Frydek Mistek in the Czech Republic. He graduated from the VSB–Technical University Ostrava, Faculty of Mechanical and Electrical Engineering, earned his Ph.D. degree from the Czech Technical University in Prague, Department of Theoretical Electrical Engineering. He received his Associate Professor degree at the VSB–Technical University of Ostrava in the field of Theoretical Electrical Engineering. His research involves the waves propagation and antennas, mathematical modelling of the electromagnetic field, stray current under electric tractions.

Vladimir MOSTYN was born in Prilepy in the Czech Republic. He received the M.Sc. degree in Electrical Engineering in 1979 from the VSB–Technical University of Ostrava, Czech Republic. Since 1990, he has been an Assistant Professor with the Department of Robotics and he received the Ph.D. degree in 1996 in Control Engineering at the same university. In 2006, after successful accomplishment of the professorship at the Technical University of Kosice, he was appointed Professor in the branch of Production Systems with Industrial Robots and Manipulators, by the president of the Slovak Republic. Now he is working as a professor at the Department of Robotics at the Faculty of Mechanical Engineering at the VSB–Technical University of Ostrava, Czech Republic. He is the author of two books and more than 80 articles. Prof. V. Mostyn is an Associate Editor of the journal MM Science Journal and the member of Czech Association of Robotic Surgery.

Karel SCHEE was born in Opava, Czech Republic. He received his M.Sc. from VSB–Technical University Ostrava in 1999. His research interests include automation of rail transport.

Jan GRUN was born in Ostrava, Czech Republic. He received his M.Sc. from Kiew Technical University (KPI) in 1985. He is engaged in design automation HW rail transport.

14

Design of Sensor Systems for Long Time Electrodermal Activity Monitoring

Erik VAVRINSKY[1,2], Viera STOPJAKOVA[1], Martin DONOVAL[1], Martin DARICEK[1], Helena SVOBODOVA[2], Jozef MIHALOV[1], Michal HANIC[1], Vladimir TVAROZEK[1]

[1]Institute of Electronics and Photonics, Faculty of Electrical Engineering and Information Technology, Slovak University of Technology, Ilkovicova 3, 812 19 Bratislava, Slovakia
[2]Institute of Medical Physics, Biophysics, Informatics and Telemedicine, Faculty of Medicine, Comenius University, Sasinkova 2, 813 72 Bratislava, Slovakia

erik.vavrinsky@stuba.sk, viera.stopjakova@stuba.sk, jozef.mihalov@stuba.sk, martin.daricek@stuba.sk, martin.donoval@stuba.sk, svobodova11@uniba.sk, michal.hanic@stuba.sk, vladimir.tvarozek@stuba.sk

Abstract. *This article describes successive development of electrodermal activity monitoring sensor system. Our aim is to improve existing systems to be more practical and suitable for long-term monitoring. Therefore, compared to conventional devices, our system must be easily wearable, without limiting the examined person in ordinary life, with low power consumption, battery operated and reducing the impact of negative artefacts. Specifically, we describe here three devices. The first is serving mainly to familiarize with the methodology, extensive testing and optimization of measurement parameters. Based on the obtained result, we constructed second system in form of small ring - "EDA ring". Last sensor system is developed with the effort to integrate the monitoring of electrodermal activity in e-health and smart clothes.*

Keywords

EDA-ring, e-Health, electrodermal activity, IDAE electrodes, smart clothes.

1. Introduction

Interconnection of microelectronics and medicine has been particularly in last decade very interesting technical field. The increase of computing power of new microprocessors, production of novel integrated chips and continual development of measurable human health parameters sensing techniques positively affect the expansion of new integrated medical devices. These health monitoring devices [1], [2] and [3] present not only a powerful healthcare assistance for end consumer, but also useful instrument for many biomedical research groups.

2. Theory

Electrodermal Activity (EDA), also known as galvanic skin response, electrodermal response or psychogalvanic reflex is used widely in physiological research due to its low cost and high utility. Typical EDA measuring systems [4] measure skin conductance in macroscopic level. That means that electric field is enclosed perpendicularly to the skin surface. The conductivity is measured through planar structures of epidermis and dermis under the first electrode, across blood vessels to the second electrode and through next skin structures. Imperfection of these systems is obvious: there is a lack of local measurement and the total measured signal includes fluctuating parameters of human skin structure that are insensitive to nervous system.

First, it was assumed that increase in the skin conductivity during a stress stimulus is only caused by the skin perspiration. Later, an important factor of potential barrier existence near the stratum lucidum layer was discovered, analysed and proven. Its thickness changes due to the nervous system activity. The greatest degree of conductivity variations occurs in the skin of palms and bottom parts of fingers [5] and [6].Therefore, our sensors are designed particularly for local measurement of top layers of the skin (stratum corneum and stratum lucidum). For such design the ideal means are interdigital arrays microelectrodes (IDAE).

Figure 1 presents an equivalent substitute model for the measurement of impedance properties of the human skin. R_E and C_E characterize the electrical parameters of IDAE electrode system, R_C, C_C characterizes the electrical parameters of stratum corneum and R_{EQ}, C_{EQ} properties of equipotential area. Electrical parameters were determined and calculated with respect to the published values [7], [8] and [9]. Psychological conditions changes and the composition of the stratum corneum were simulated by R_E and C_E values.

By decreasing the distance between the electrodes (d) the sensitivity to simulated measurement system is increasing. This is due to reduction of the value of R_{C1} with decreasing electrode distance d. If the distance between the electrodes is smaller than the thickness of the upper laminar structures, the importance of the parameter R_{C1} increases, because its value becomes lower than RC and therefore it is more dominant because of parallel connection.

A great help for modelling of various kinds of electrode systems is the program QUICKFIELD. Isotropic or anisotropic properties of the study area may be defined by components ε_x and ε_y. The analysed structure (Fig. 2) is drawn and analysed as a system - substrate - conductive strip (of the material) - cutaneous environment. Individual areas are differentiated by their size, relative permittivity (Fig. 1(a)) and the energized or de-energized area. By using this program, not only an overview of the layout of the field can be obtained, but also the values at various points in a two-dimensional structure. Based on that computer analysis we can optimize the electrode system. In our conditions, we worked with two different planar electrodes:

- symmetric electrodes: $100 \times 100 \times 100$ μm (Fig 2(a)),

- asymmetric electrodes: $100 \times 50 \times 30$ μm (Fig 2(b)),

where 1st electrode width (w_1) × slit (d) × 2nd electrode width (w_2) [6].

Based on electrical model of interdigital arrays (IDAE) microelectrode/skin interface and simulations we have found that electric field distribution and depth of penetration into the outer skin layers depend mainly on the configuration and size of an electrode system.

This knowledge provides the possibility to examine different (separate) layers of epidermis by electrical impedance method. In short we can say that "electric field penetration depth" into human body is relatively close to "distance between the coplanar electrodes". It is perfectly valid for symmetric configuration, where the distance between electrodes is equal to electrodes width. The results of analysis also showed that in case of non-symmetric electrodes the electric field is more enclosed in outer layers of skin [10].

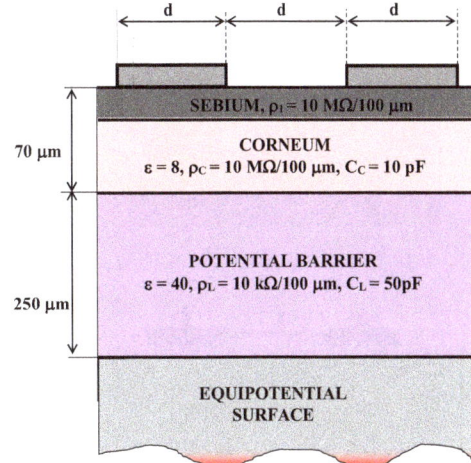

(a) Skin structure physical parameters and typical dimensions.

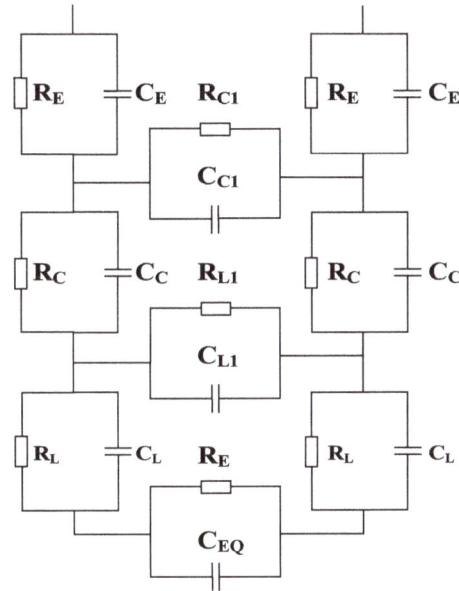

(b) Reciprocal electrical model.

Fig. 1: Model of human skin.

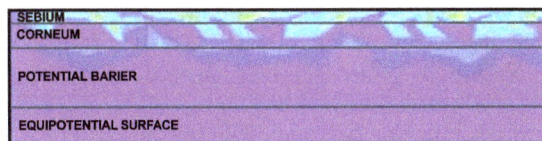

(a) Symmetric electrodes: $100 \times 100 \times 100$ μm.

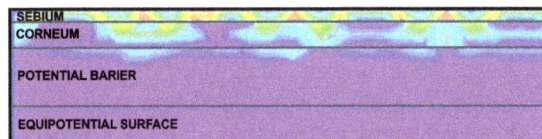

(b) Asymmetric electrodes: $100 \times 50 \times 30$ μm.

Fig. 2: IDA microsensor/human skin interface simulations - distribution of electric field intensity [6].

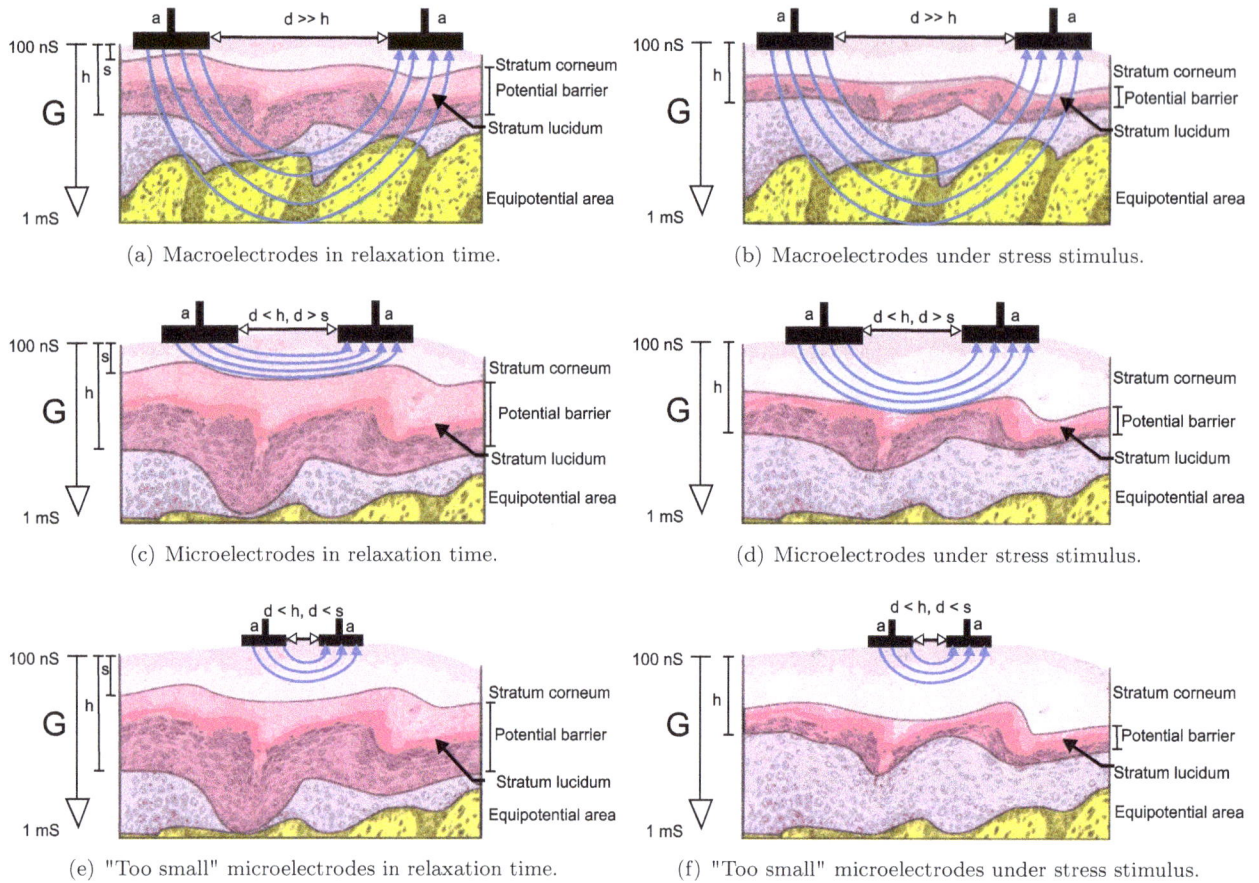

(a) Macroelectrodes in relaxation time.

(b) Macroelectrodes under stress stimulus.

(c) Microelectrodes in relaxation time.

(d) Microelectrodes under stress stimulus.

(e) "Too small" microelectrodes in relaxation time.

(f) "Too small" microelectrodes under stress stimulus.

Fig. 3: The dominant vector intensity lines of the electric field in human skin.

If there are different electrodes applied on human skin, various space distributions of electrical field in the skin can occur.

In case of using macroelectrodes, when the distance between the coupled electrodes is greater than the thickness of electric active layers of skin h (stratum corneum (the outermost layer of the skin) with potential barrier) $d \gg h$, the vector intensity lines of the electric field are enclosed perpendicularly to the skin surface across the planar skin structures through dermis with high conductance (Fig. 3(a) and Fig. 3(b)).

In case of use of microelectrode pairs the lines of electric field are enclosed in parallel direction relative to laminar skin structures of epidermis. This is because the distance between the electrodes is less than the thickness of electric active layers of skin: $d < h$ and higher then thickness of stratum corneum: $d > s$ (d - electrode distance, h - thickness of electric active layers of skin, s – thickness of stratum corneum). From inner layers of skin, the electric field intensity lines are embossed to the surface (to the area with a lower conductivity) by the influence of the potential barrier which is

generated by electrical double-layer around stratum lucidum (Fig. 3(c)).

Under a stress stimulus the potential barrier narrows down and the electric field can reach inner layers of human skin with higher conductivity, and therefore the total conductivity increases (Fig. 3(d)). Such configuration is ideal for the analysis of electrophysiological processes in human skin under stress.

In case of small sized microelectrodes the vector intensity lines of the electric field are enclosed in top layers of stratum corneum and the flow of electric lines is independent from thickness of potential barrier (Fig. 3(e) and Fig. 3(f)). Such electrodes are more ideal for surface analysis in cosmetics [10].

In our designed sensor system, we used IDA microelectrodes with utilized 200 μm × 200 μm × 200 μm (1st electrode width × slit × 2nd electrode width) dimensions. The obtained pre-experimental results proved that the optimal amplitude of driving signal should be selected from 1.5 V to 3 V. The driving signal frequency is not critical; however, an optimal value of several kHz has been proved.

3.   Sensor Systems

3.1.   Testing System

Several methods, applicable to continuous measurement of the human skin impedance, were considered, tested and analysed. As a result of this analysis we have chosen in the 1st setup the auto-balancing bridge method. It has high accuracy, short measurement time, high repeating rate, frequency and amplitude signal definition, possibility to measure both real and imaginary impedance components, controllability by a microprocessor, digital processing, etc. [11].

The proposed complex measurement system, offering these features, is composed of two main parts (Fig. 4). The core of the proposed portable monitoring system (Fig. 4(a)) is the integrated circuit AD5933 [12] that provides measurement of the human skin impedance sensed by the developed microsensor. The measurement process is controlled by the microprocessor nRF24E1 via I2C interface.

Using the RF wireless communication interface, the microprocessor sends the measured data to the receiver part (Fig. 4(b)) on the PC side. Consequently, the personal computer executes data storage and data post-processing. Additionally, the microcontroller also provides an initial configuration of integrated circuit AD5933 (setting the frequency and amplitude of the driving signal, measurement time slots, power management, etc.).

PCB has been realized on double layer FR4 board by SMT technology with minimum strip width of 0.2 mm and minimum clearance width of 0.2 mm. The total size of the measurement unit is 50×60 mm and 11×17 mm for the receiving module (Fig. 4(c)).

In order to verify the developed and realized system a comparison of our microelectrodes-based EDA method to the commercial macroelectrode approach [13], usually used in the laboratory medical or psychological experiments, was carried out. The $200 \ \mu m \times 200 \ \mu m \times 200 \ \mu m$ IDAE electrode microelectrode was placed on middle-finger of non-dominant hand, where macroelectrodes where connected between index and ring finger, so we could have expected that specific conductivities were similar. The comparison was performed by using the standard „Distraction" psychotests [14], where the signals were measured simultaneously. It shows that the responses given by both approaches were similar. Under stress stimulus the conductivity is increasing and in relax decreasing. However, the microelectrode signals are observed to be more stable with a shorter response time (Fig. 5).

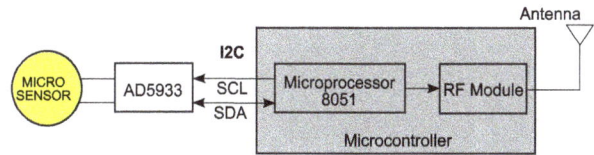

(a) Portable measurement block.

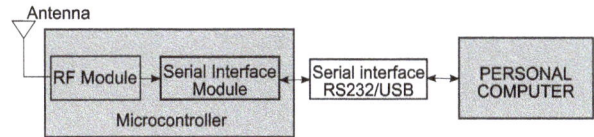

(b) Receiver module.

(c) PCB design of measurement and receiver module.

Fig. 4: Realized EDA monitoring system.

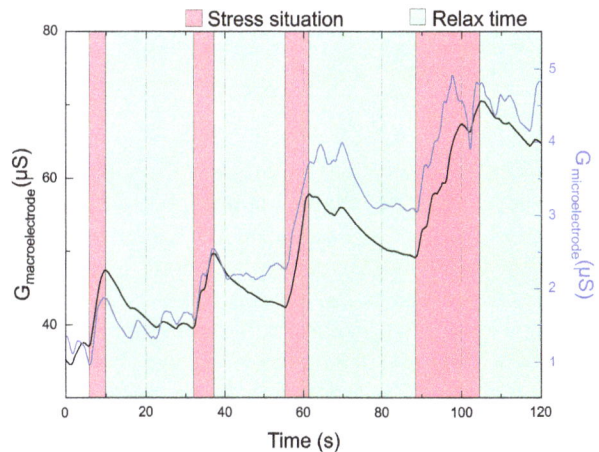

Fig. 5: Comparison of absolute conductivity values during a psychological experiment: standard macroelectrode versus IDA microelectrode.

3.2.   EDA Ring

After detailed testing of the 1st design we decided to construct a practical sensor system in the form of small ring (Fig. 6) that could be useful in practical life. In this system, the driving generator provides a sinus signal wave with amplitude of 1.6–3 V and frequency 1 kHz to gold-plated $200 \times 200 \times 200 \ \mu m$ IDAE structure, where the equivalent current is measured and skin conductivity is calculated. The sampling rate can be set in range from 0.33 to 33 SPS (Samples Per Second). Total dimensions of EDA-ring are $20 \times 20 \times 5$ mm and it is intended to be placed on the ring-finger of non-dominant hand.

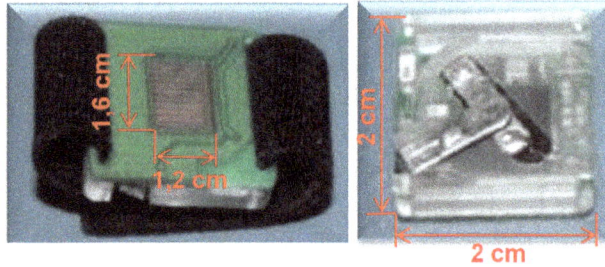

Fig. 6: EDA-ring: design.

Tab. 1: EDA-ring: technical parameters.

Technical parameters of sensor system	
Measuring signal Amplitude	1.6–3 V
Measuring signal Frequency	1 kHz (sinus)
ADC resolution	10 bit
Sample rate	0.33–33 SPS
Connectivity	ISM band 2.4 GHz, standard IEEE802.15.4
Transmission range	10 m at interior, 30 m at exterior
Supply voltage	1x CR 2032 (Lithium) 3 V, 12 - 24h stamina
Temperature range	0–70 °C
Weight	3 g
Dimensions	20 × 20 × 5 mm
Electrodes	IDAE 200 / 200 μm, Au galvanic plated on Cu

With this sensor system, a complex physiological research on group of 28 volunteers (mean age=23.5 years; SD=1.41; 13 males; 15 females) at Department of Psychology of Comenius University was done. The influence of body position and mental (Dual N-back [14]) or psychical activity (squatting) was evaluated. The experiment took about 30 minutes for each person.

In Fig. 7, there are shown exemplary EDA results from 3 persons. Each person provides individual signal dependent on his own characteristic features. The signals have similar tendencies – in relaxation, the conductivity is decreasing and under mental or physical burden (stress stimulus) it is increasing.

3.3.   Smart Clothes

Last sensor system is designed as multifunctional holter and in this case, it was built-in into specially designed smart clothes. Holter (Fig. 8) is based on analog font-end TI ADS1292R and microcontroller ATxmega 128A3 (Fig. 8(a)). The ADS1292R is two-channel, 24-bit, delta-sigma analog-to-digital converter with a built-in programmable gain amplifier, internal reference and an on-board oscillator [15]. The system has been extended by the gyroscope L3GD20, accelerometer with magnetometer LSM303D [16] and barometer with temperature sensor BMP180 [17]. The lat-

est design was reworked and improved, new firmware, controlling software and data transfer system were reprogrammed, acting now as USB flash device. The ADS1292R incorporates all features commonly required in portable, low-power medical electrocardiogram with sports and fitness applications. Power consumption of one channel is only 335 μW. Used version of analog-to-digital converter ADS1292R also includes a fully integrated impedance measurement function (Fig. 8(b)), where 32/64 kHz modulating square wave signal is driving the human body impedance trough known impedances Z_k. After demodulation and low pass (2–4 Hz) filtering we can obtain the EDA impedance Z_b (Fig. 3) [11]. Also in this version, compared to previous two systems, the impedance is measured at different frequency and so the relative sensitivity may be different.

Tab. 2: Smart clothes EDA monitoring system: technical parameters.

Technical parameters of sensor system	
Number of Channels	2
Programmable Gain	1, 2, 3, 4, 6, 8 or 12
Input-Referred Noise	8 mVPP(150-Hz BW, G = 6)
ADC resolution	24-bit, no data missing
Sample rate	125–8 kSPS
MCU	16 bit AVR, 32 MHz
SRAM	8 kbytes
Built-In	Impedance circuit (32, 64 kHz)
Acceleration sensor	3D ±2g/±4g/±8g/±16g (16 bit) 3–1600 SPS
Gyroscope	250, 500, 2000 dps (16 bit) 95–750 SPS
Magnetometer	2, 4 ,8 ,12 G (16 bit) 3–100 SPS
Barometer	300–1100 hPa (16 bit) 2–50 SPS
Data storage	Integrated 4 GB SD card
Output data format	CSV, EDF+
Power supply	1x Li-Pol 120 mAh
Connectivity	Micro USB
Optional	Bluetooth - 10 m range
Temperature range	0–70 °C
Electrodes	Disposable Ag/AgCl
Dimensions	37 × 25 × 15 mm
Weight	20 g
Next features: RGB LED and acoustic signalization, trigger button, real-time clock.	

In Fig. 9, there is our implementation of designed holter in smart shirt. Whole system, including electronics, fabric electrodes and T-shirt technology was prepared at our department. In comparison with the first two sensor systems, this design uses conductive fabric electrodes (MedTex P180, E130, Zel etc.). Therefore, the quality of the resulting signal is slightly lower.

The system was compared with standard macroelectrodes measurement systems [13]. Macroelectrodes were connected between index and ring finger and fabric electrodes of smart T-shirt measured signal in the area of brachioradialis muscle.

The comparison was performed by using the standard „Distraction" psychotests [14], where the signals were measured simultaneously. Results (Fig. 10) shows, that due to using different sensing method, the measured absolute values and sensitivities are different, but the responses, curve shapes, given by both approaches were similar. Under stress stimulus the conductivity is rapidly increasing and in relax it is rapidly decreasing. Also, the smart clothes offer lower sensitivity, mainly due to lower response of examined body areas to neural activity, but the suitability of smart clothes for long-term monitoring is incomparable with macroelectrodes.

4. Conclusion

Linear equations modelling and simplified power losses calculation have been used to predict approximate optimal DG size. Each location of DG at load buses has optimal size of DG for minimum power losses. It is noted that optimal size for optimal location is not necessary to be the same as optimal location for optimal size. The proposed method has acceptable accuracy with less time and memory consumption where it is crucial factor in real-time management of power grids. The loss reduction by properly placed and appropriate size of DG is one of the more significant findings to emerge from this study. With these benefits, control and assessment of large scales power grid will become easily predictable as more intermittent power sources, such as wind and solar, come online. Different IEEE test bus systems have been tested and results are validated with exact calculations.

Acknowledgment

This work was supported under grant VEGA 1/0739/16 and contracts APVV-14-0740, APVV-0496-12 and APVV-15-0763.

References

[1] BOQUETE, L., J. M. R. ASCARIZ, J. CANTOS, R. BAREA, J. M. MIGUEL, S. ORTEGA and N. PEIXOTO. A portable wireless biometric multi-channel system. *Measurement*. 2012, vol. 45, iss. 6, pp. 1587–1598. ISSN 0263-2241. DOI: 10.1016/j.measurement.2012.02.018.

[2] CARMO, J. P. and J. H. CORREIA. RF CMOS transceiver at 2.4 GHz in wearables for measuring the cardio-respiratory function. *Measurement*. 2011, vol. 44, iss. 1, pp. 65–73. ISSN 0263-2241. DOI: 10.1016/j.measurement.2010.09.027.

[3] BOUCSEIN, W., D. C. FOWLES, S. GRIMNES, G. BEN-SKAKHAR, W. T. ROTH, M. E. DAWSON and D. L. FILION. Committee Report - Publication recommendations for electrodermal measurements. *Psychophysiology*. 2012, vol. 49, iss. 8, pp. 1017–1034. ISSN 1469-8986. DOI: 10.1111/j.1469-8986.2012.01384.x.

[4] HANDLER, M., R. NELSON, D. KRAPOHL and C. R. HONTS. An EDA Primer for Polygraph Examiners. *Polygraph*. 2010, vol. 39, iss. 2, pp. 68–108. ISSN 1533-9793.

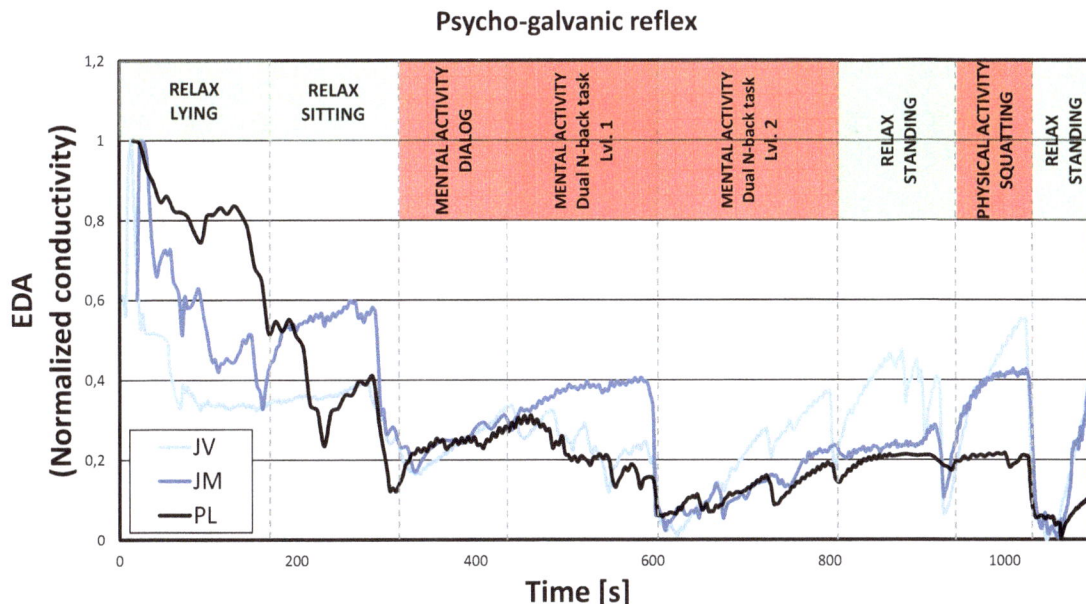

Fig. 7: Complex physiological monitoring: (3 probands).

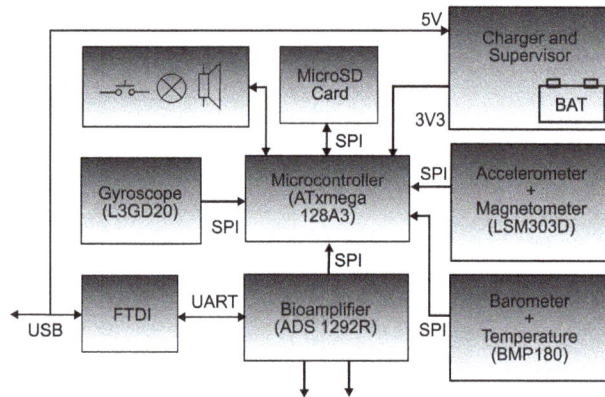

(a) Block diagram of holter module.

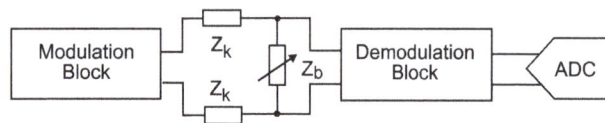

(b) Schematic of EDA monitoring circuit.

Fig. 8: Smart clothes EDA monitoring system.

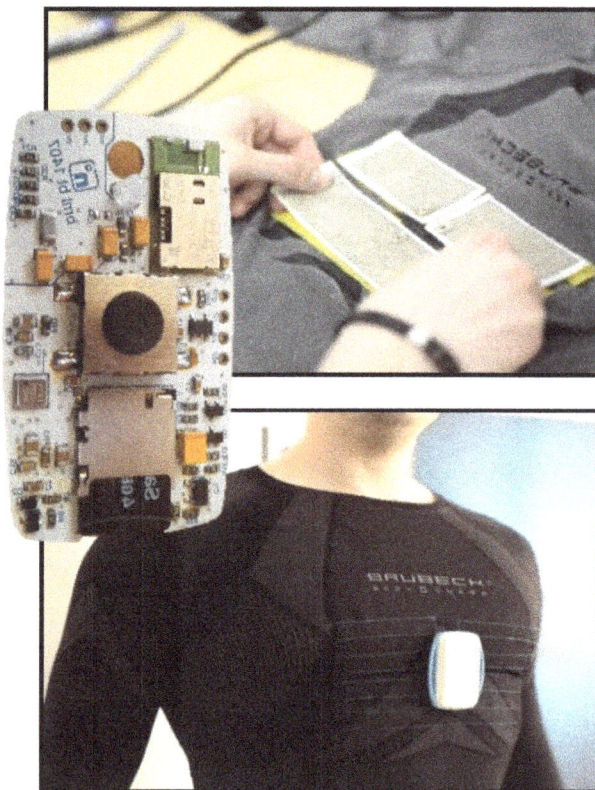

Fig. 9: Realization of smart clothes EDA monitoring system.

[5] WEIS, M., T. DANILLA, L. MATAY, P. HRKUT and J. KAKOS. Noninvasive Biomedical Sensors on the Biology - Interface of Human Skin. In: *7th International Conference on Measurement in Clinical Medicine*. Stara Lesna, 1995, pp. 89–91.

[6] IVANIC, R., I. NOVOTNY, V. REHACEK, V.

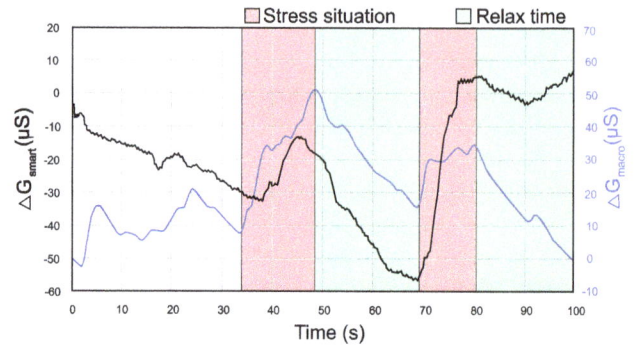

Fig. 10: Newton Raphson and linear method Comparison for 30 Bus.

TVAROZEK and M. WEIS. Thin film non-symmetric microelectrode array for impedance monitoring of human skin. *Thin Solid Films*. 2003, vol. 433, iss. 1–2, pp. 332–336. ISSN 0040-6090. DOI: 10.1016/S0040-6090(03)00389-4.

[7] BALIN, A. K. and A. M. KLIGMAN. *Aging and the Skin*. New York: Raven Press, 1989. ISBN 978-0890047781.

[8] SHEPPARD, Jr., N. F., R. C. TUCKER and C. WU. Electrical conductivity measurements using microfabricated interdigitated electrodes. *Analytical chemistry*. 1993, vol. 65, iss. 9, pp. 1199–1202. ISSN 0003-2700. DOI: 10.1021/ac00057a016.

[9] WILHELM, K. P. Possible Pitfalls in Hydration Measurements. *Skin Bioengineering: Techniques and Applications in Dermatology and Cosmetology*. Basel: Karger, 1998. ISBN 3-8055-6519-4.

[10] VAVRINSKY, E., V. STOPJAKOVA, I. BREZINA, L. MAJER, P. SOLARIKOVA and V. TVAROZEK. Electro Optical Monitoring and Analysis of Human Cognitive Processes. In: *Semiconductor Technologies*. Rijeka: InTech, 2010, pp. 465–490. ISBN 978-953-307-080-3. DOI: 10.5772/8558.

[11] MAJER, L., V. STOPJAKOVA and E. VAVRINSKY. Wireless Measurement System for Non-Invasive Biomedical Monitoring of Psycho-Physiological Processes. *Journal of Electrical Engineering*. 2009, vol. 60, iss. 2, pp. 57–68. ISSN 1335-3632.

[12] 1 MSPS, 12 Bit Impedance Converter Network Analyzer. *Analog Devices* [online]. 2008. Available at: http://www.analog.com.

[13] BOUCSEIN, W. *Electrodermal Activity*. New York: Springer, 2015. ISBN 978-1-461-41126-0.

[14] OWEN, A. M., K. M. MCMILLAN, A. R. LAIRD and E. BULLMORE. N-Back Working Memory

Paradigm: A Meta-Analysis of Normative Functional Neuroimaging Studies. *Human Brain Mapping*. 2005, vol. 25, iss. 1, pp. 46–59. ISSN 1097-0193. DOI: 10.1002/hbm.20131.

[15] *Texas Instruments*. [online]. 2012. Available at: http://www.ti.com.

[16] *STMicroelectronics*. [online]. 2010. Available at: http://www.st.com.

[17] *Bosch Sensortec*. [online]. 2014. Available at: http://www.bosch-sensortec.com.

About Authors

Erik VAVRINSKY received his M.Sc. and Ph.D. degree from the Faculty of Electrical Engineering and Information Technology, Slovak University of Technology in years 2002 and 2005. He has carried out postgraduate studies in area of thin-film microsensors and electrochemical biosensors. In present, he participates at the University education process as a lecturer. His current research interests include design, characterization and testing of sensor systems and bio-monitoring devices. Dominant part of his multidisciplinary research is focused on human physiology.

Viera STOPJAKOVA received her M.Sc. and Ph.D. degrees in Electronics from Slovak University of Technology (STU) in Bratislava, Slovakia, in 1992, and 1997, respectively. From October 1997 to September 2003 she was an assistant professor at Microelectronics Department, FEI STU in Bratislava. Since October 2003 she has been an associate and currently a full professor at the same department. She has been involved in several EU funded research projects such as Tempus, ESPRIT, Copernicus, Inco-Copernicus, 5th EU Framework project REASON, 7th FP project IDESA, ENIAC-JU projects END and MAS. Currently, she is a coordinator of 2 grants funded by the Slovak Ministry of Education and the European Social Fund project NANOSYS. She has published over 70 papers in various scientific journals and conference proceedings; and she is a co-inventor of two US patents. Her main research interests include IC design and test, on-chip current testing, design and test of mixed-signal circuits and systems, biomedical monitoring, and smart sensors.

Jozef MIHALOV studied at Slovak University of Technology in Bratislava where he received his Ph.D. degree in field of design and IC characterization. In present, he participates at the University research at the Department of Integrated Circuits Design and Test.

Martin DARICEK studied at Slovak University of Technology in Bratislava where he received his Ph.D. degree in field of device and IC characterization. He has carried out postgraduate studies in area of IC characterization methods development. Currently he focuses on off-chip innovative measuring, testing and bio-monitoring devices design, characterization, development and optimization. He is as well responsible for preparation of low-level firmware schemes. Besides this is he currently responsible for R&D projects management. In present, he participates at the University education process as a lecturer.

Martin DONOVAL received his master and Ph.D. degrees in Electronics from Slovak University of Technology in Bratislava. He has carried out postgraduate studies in area of IC design. Afterwards he continued working at the university as a research assistant in magnetic force affected IC sensors design and magnetic force IC sensors. He participated in the establishment of the Department of electronic systems with focus on reliability testing, design of smart electronic systems and off-chip innovative solutions. Today he is responsible for projects particularly in a field of bio-electronics. As a member of the team he is involved in development of new prototypes and dedicated integrable electronics and reliability testing.

Helena SVOBODOVA obtained her Master's degree in the field of molecular biology at Faculty of Natural Science, Comenius University, in 2014. In present, she is a Ph.D. student at the department of biophysics. Her research interests include metabolism of iron, neurodegenerative diseases, electroencephalography, sensor systems and bio-monitoring devices.

15

Reliable Detection of Rotor Faults in IM Using Frequency Tracking and Zero Sequence Voltage

Khalid DAHI, Soumia EL HANI, Ilias OUACHTOUK

Departement of Electrical Engineering, Ecole Normale Superieure de l'Enseignement Technique, Mohammed V University in Rabat, Avenue des Nations Unies, Agdal, Rabat, Morocco

khalid.dahi@um5s.net.ma, s.elhani@um5s.net.ma, ilias_ouachtouk@um5.ac.ma

Abstract. *The AC Alternating Current Induction Motor (IM) is the most commonly used AC motor in industrial applications because of its simplicity, robust construction, and relatively low manufacturing costs. To avoid expensive repairs in IM, early faults detection is needed. In this context, this paper presents a novel approach used to detect rotor asymmetries in induction motors. The Zero Sequence Voltage (ZSV) defined as the potential difference between the null point of the supply voltage system and the neutral of the star connection of IM stator winding is measured and employed for tracking the amplitude of the most sensitive harmonics in the spectrum of ZSV. This detection leads to make a criterion to take a decision about the state of the machine without a Prior knowledge. Simulation and experimental results obtained from real tests are presented to validate the study.*

Keywords

Frequency tracking, induction motors, rotor fault diagnosis, Zero Sequence Voltage.

1. Introduction

Induction Motors (IM) are used in many industrial processes and are frequently integrated in commercially available equipment. Robustness, cost advantage, high power capabilities, and performance are the major concerns of IM applications. Although IM are reliable, they are subjected to some failures. Therefore, monitoring and diagnosing of IM faults received considerable attention in recent years and it is motivated by objectives for reliability and serviceability in electrical drives [1].

Studies in the area of diagnosis and monitoring of electrical drives have shown that stator and rotor faults are assumed equal [3]. However, the vast majority of articles deal mainly with rotor fault (69 %) first and then with stator faults (30 %) and finally bearing faults [2].

Knowing that the topic is important, this paper deals with rotor fault detection in IMs. This fault that physically result from the short/open circuits or the increase of the rotor resistance. Among the rotor fault accrued in this type of machines we can cite the Broken Rotor Bars (BRBs) in case of machines with squirrel cage.

There are various methods that have been developed to detect BRBs in IMs such us vibration analysis, temperature analysis, acoustic measurement, neuronal and artificial intelligence based methods. However, the most used techniques are based on the monitoring of the stator current spectrum (known as Motor Current Signature Analysis (MCSA) [4], [5] and [6]). Based on that, MCSA is simple and effective in appropriate operating conditions.

Despite the advantages that this method provides, this technique has some significant practical limitations. MCSA is influenced by the operating conditions especially in machine working under very low load. The detection of BRBs is difficult if the IM is supplied by a power converter or if the IM operates in a system under time-varying conditions. An other limitation of MCSA is expressed with recent machines where the IM are frequently installed with inverters which provide some advantages but make the stator current inaccessible to diagnosis [3].

To overcome these limitations, this work proposes the use of Zero Sequence Voltage (ZSV), known also by the Line neutral voltage which is a potential difference between the null point of the supply voltage system and the neutral of the star connection of IM stator winding [8], [10] and [11]. Compared to MCSA, in case of motor with BRBs, if the slip is very close to zero the frequency that characterizes the fault will not be observable. which is not the case using the ZSV that gives better detection capabilities at no-load condition [9]. It will be shown in this paper that the ZSV contains much more information for IM diagnosis compared to other methods.

In this paper, a new algorithm is proposed aiming at the reliable fault detection of BRBs in IMs. The proposed method is based on the tracking of the amplitude of the most sensitive harmonics in the ZSV spectrum and then apply the proposed criteria to make decision about the IM state. The analysis is performed with the application of the FFT-based method. Experimentation and simulation using Finite Element Model (FEM) [12] and [13] are performed on different motor conditions (Load, number of BRBs, ...).

The structure of this paper is as follows. In Sec. 2. , some information are given about the detection of rotor fault in IM using the ZSV. In Sec. 3. , the proposed method is theoretically explained. In Sec. 4. we present a validation of the proposed method in both simulation and experimental results. Finally, in Sec. 5. we present some future works and the conclusions of this study.

2. Detection of Rotor Asymmetries Using the ZSV

In [11], the Zero sequence voltage is used to detect short circuit faults in IM by applying an experimental approach without a theoretical analysis. This work goes beyond, since develops a model to detect rotor fault in IM taking into account the ZSV. In Eq. (1), An analytical solution for the ZSV can be obtained for the case when the machine is excited with a balanced supply star connected machine.

$$V_0 = \frac{1}{3}(V_{an} + V_{bn} + V_{cn}). \qquad (1)$$

The measurement scheme of ZSV is given in Fig. 1. The ZSV is taking into account by the following mathematical relationship:

$$V_0 = R_a I_{sa} + L_a \frac{dI_{sa}}{dt} + \frac{L_a}{d\theta}\Omega I_{sa} - V_n, \qquad (2)$$

where: R_a represents the stator-phase resistance, L_a his inductance, I_{sa} the current passing through it, θ Rotation speed, the angular position of the rotor and V_n supply simple voltage generated by network supply. The presence of a fault rotor reveals additional components in the spectrum of ZSV at frequencies given by

the relation:

$$f_{hdef} = [3h - (3h \pm 1)s] f_s, \qquad (3)$$

where s is slip, f_s is supply frequency and h is harmonic order.

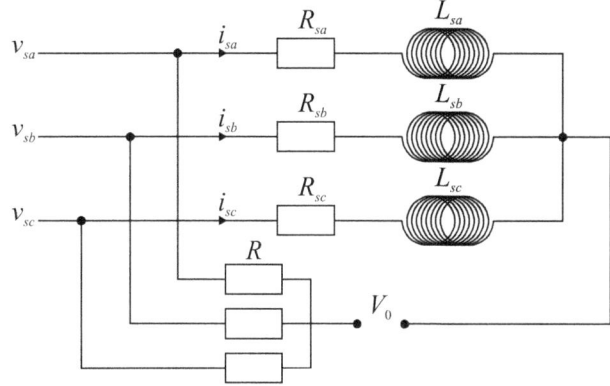

Fig. 1: Zero sequence voltage measurement.

These components computed as derivative of the sum of the three-phase linkage fluxes, appear in case of rotor asymmetry. Experiments are presented in Sec. 4. to correlate the amplitude of the ZSV components to the machine load. The amplitude of these components increases with slip and load conditions.

3. Suggested Methodology - ZSV and FT Combination

In this section, a fault detection criterion based on the combination between the ZSV and the tracking of the important frequencies that characterize the rotor fault in IM is presented. We show that the fault detection criterion can be easily used for this type of fault.

3.1. Problem Formulation

Many methods in the literature are based on the stator current analysis, torque analysis, power analysis, etc [7]. It was shown in [15] using a Finite Element Model, that the Zero Sequence Component (ZSC) can be used to detect and identify rotor fault in induction machine. Also, the amplitudes in ZSC are greater than those of the MCSA as shown in [9].

Most of diagnosis methods used up to now which deals with detection of rotor fault do not provide a direct fault detection criterion. These methods requires the knowledge of the healthy state to make a decision about the rotor.

For that, the proposed harmonic tracking approach is developed using the ZSV as electrical indicator to detect BRBs without the knowledge of a primary state of the motor. This approach is based on standard deviation calculations taken on two frequency ranges, the first standard deviation will be calculated on the first frequency range, this range identifies where the frequency $(3 - 4s)f_s$. The second standard deviation represent the picture of measurement noise present between frequencies located at $(3 - 4s)f_s$ and $(3 - 6s)f_s$ that is going to be mentioned in this paper as f_{tar1} and f_{tar2} respectively. Finally, using this approach, we are able to generate a decision about the rotor fault which is presented in the end of the proposed method section. It is important to note that The method is valid both for line connected as well as for inverter-fed machines.

Compared to recent works in area of induction machine fault diagnosis, the proposed method is effective and simple to implement because it uses the FFT-based method. Aslo, the slip of the motor is calculated in each test which is not the case in most recent works [6], [8] and [14].

3.2. Tracking Harmonics Module

As already mentioned, the proposed method is based on the analysis of the frequency $(3 - 4s)f_s$ which is function of the motor slip s. This is why it is necessary to calculate the slip of the IM. The easiest way to do it is the using of a speed sensor in case of experiment test. In this paper we focus on some harmonics to estimate the slip.

In [10], it is demonstrated that the equation given the principal RSH found on the spectrum of the line neutral voltage is given by:

$$f_{RSH} = f_s \left[\lambda \frac{N_r}{p} (1 - s) \pm 1 \right]. \tag{4}$$

The slip can be expressed as:

$$s = 1 - \frac{p}{\lambda N_r} \left[\frac{f_{RSH}}{f_s} \pm 1 \right]. \tag{5}$$

The frequency component $f_s \left[\lambda \frac{N_r}{p} (1 - s) + 1 \right]$ has a much more significant amplitude than the frequency component $f_s \left[\lambda \frac{N_r}{p} (1 - s) - 1 \right]$, which will facilitate its detection.

Practically, all IM have a slight asymmetry of construction induced, in the spectrum of ZSV, the appearance of the frequency component whose frequency equal to $(3-2s)f_s$. Therefore, the slip can be expressed from:

$$s = \frac{1}{2} \left[3 - \frac{f_{RSH}}{f_s} \right]. \tag{6}$$

A searching interval is defined because the frequency of the component $(3 - 2s)f_s$ changes according to the load motor; their boundaries depend on the max and min values of the slip s_{min} and s_{max}; these correspond to unloaded machine and full load machine respectively. Consequently, the searching frequency f_{SR} belongs to the following interval:

$$f_{SR} \in [(3 - 2_{s\,max})f_s \ldots (3 - 2_{s\,min})f_s]. \tag{7}$$

In our case, given that we know the fundamental frequency f_s, and as our machine is operating with a nominal speed of 2880 rpm which gives a minimum frequency f_{RS} equal to 143.6 Hz, therefore the range selected our detection of this jump will be $[140, 150]$ Hz.

The next step is to identify the value of the $(3-2s)f_s$ component and its amplitude in the spectrum of the ZSV, the best way to do this is define a frequency range corresponding to the wanted harmonic, this component which has the highest magnitude nearest the 3rd harmonic in the interval defined as:

$$R = [f_{tar} - i\Delta f; f_{tar} + j\Delta f], \tag{8}$$

where: f_{tar} is target harmonic obtained via the estimated slip, Δf is the frequency resolution ($\Delta f = f_s/N$), i and j are integers. Once the slip is determined.

Next, the idea is to compare the two standard deviation around the f_{tar1} and f_{tar2} frequencies.

$$\text{RANGE1} = \left[(3 - 4s)fs - \frac{\delta}{2}; (3 - 4s)fs + \frac{\delta}{2} \right],$$
$$\text{RANGE2} = \left[(3 - 4s)fs - \frac{\delta}{2}; (3 - 6s)fs + \frac{\delta}{2} \right]. \tag{9}$$

For an adequate understanding of the principle of calculation of these standard deviations, Fig. 2 shows

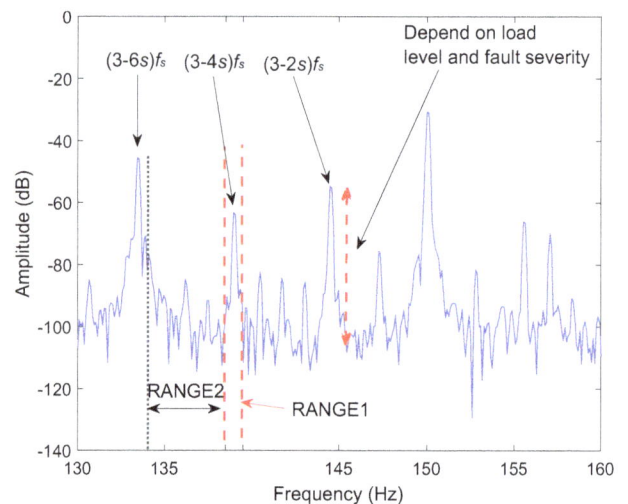

Fig. 2: Explication of the proposed method.

a representation where the standard deviation s_j is calculated on the gray frequency range while the standard deviation s_n is calculated on the black frequency range.

3.3. Design of a Threshold and Decision Making Strategy

In order to make our indicator more robust and to limit false alarms detection, a threshold has been introduced in the criterion that will symbolize with C_{th}. This threshold compares the variance s_j with the variance s_n. Therefore, the authors have defined the following criterion as given in Tab. 1.

Tab. 1: Rotor asymmetry decision module.

Criterion	Rotor State
$C_{RF} \leq C_{th}$	Healthy rotor
$C_{RF} \geq C_{th}$	Defective rotor

Where $C_{RF} = \dfrac{s_j}{s_n}$, and C_{th} is the sensitivity degree of our fault detection indicator and it is determined in function of the studied IM. Both, the rotor fault index C_{RF} and the corresponding threshold parameter C_{th} are determined from experimental results based on the signal detection theory.

The proposed method can be summarized as follows:

- After acquisition, the ZSV is sampled, and the slip is measured for each case (to be used just for comparison with the estimated slip).

- The slip estimation module is built using Eq. (5) and Eq. (6).

- The estimated slip is used to search the frequency component $(3 - 2s)f_s$ near the 3rd harmonic.

- Once this frequency is estimated, the f_{tar1} and f_{tar2} are estimated too, and the standard deviation in the two ranges is calculated to build the criterion. These values can be compared with predefined thresholds to evaluate the machine's condition.

4. Application to IM Rotor Faults Detection

In this section experimental and simulation results are compared. The proposed approach has been implemented in MATLAB-SIMULINK on a DELL Attitude PC, 2 GHz with 4 GB of RAM.

The proposed method will be illustrated using the case of an IM with rotor asymmetry. Nevertheless, the same procedure can be followed to the treatment of any other type of machine fault or working conditions.

4.1. Simulation Results using FE Model of IM

The IM modeling is briefly presented in this section, then the proposed approach is used to detect BRBs faults. IM model for both healthy and faulty motors has been developed based on the Finite Element theory [12], [13], [15] and [16]. The FE field-circuit Model of the IM takes into account the non-linearity of the magnetic materials and is suitable for a deep study of Squirrel Cage Induction Motor (SCIM). It is based on subdivision of the mathematical model into disjoint components of simple geometry called FE, and solving of Maxwell equations by considering the geometry of the structures and properties of the materials.

The basic principle of the FE method applied to electromagnetic calculation is introduced in our previous works [17], by presenting Maxwell's equations first and the dielectric behavior law. We infer from these equations some external parameters such as: the stator currents, Power, torque and the Joule losses.

The magnetic flux is defined as:

$$\varphi = \iint_s \vec{B} \cdot \mathrm{d}\vec{S} = \oint_{c(s)} A \cdot \mathrm{d}l, \tag{10}$$

where S is any surface, $C(S)$ the closed contour and A a vector potential. The electromagnetic torque is determined by:

$$\Gamma_{em} = R^2 L_z \iint_{2p} H_t B_n \mathrm{d}q, \tag{11}$$

where L_z is the axial length of the machine. The Joule losses density is defined as the scalar product of the electric field and the electric current density. Using the behavior laws used to determine total Joule losses in a volume:

$$pj(t) = \iiint_v \rho \frac{\mid Jz(t) \mid^2}{2} \mathrm{d}V. \tag{12}$$

Similarly, the density of magnetic energy is given by the product of the field and of the magnetic induction:

$$\begin{aligned} W_m &= \frac{1}{2} \iiint_v \mathrm{d}V \int_0^B h \mathrm{d}b, \\ &= \frac{1}{2} \iiint_{\text{conductor}} \mathrm{d}V \int_0^A j \mathrm{d}a. \end{aligned} \tag{13}$$

The assignment of elements of the stator and the rotor's electrical circuits to the magnetic circuit is performed interactively using the FLUX2D software. The geometry, the circuit model and the finite element mesh

(a) Geometry.

(b) Mesh zoom.

(c) Electrical circuit model of the IM.

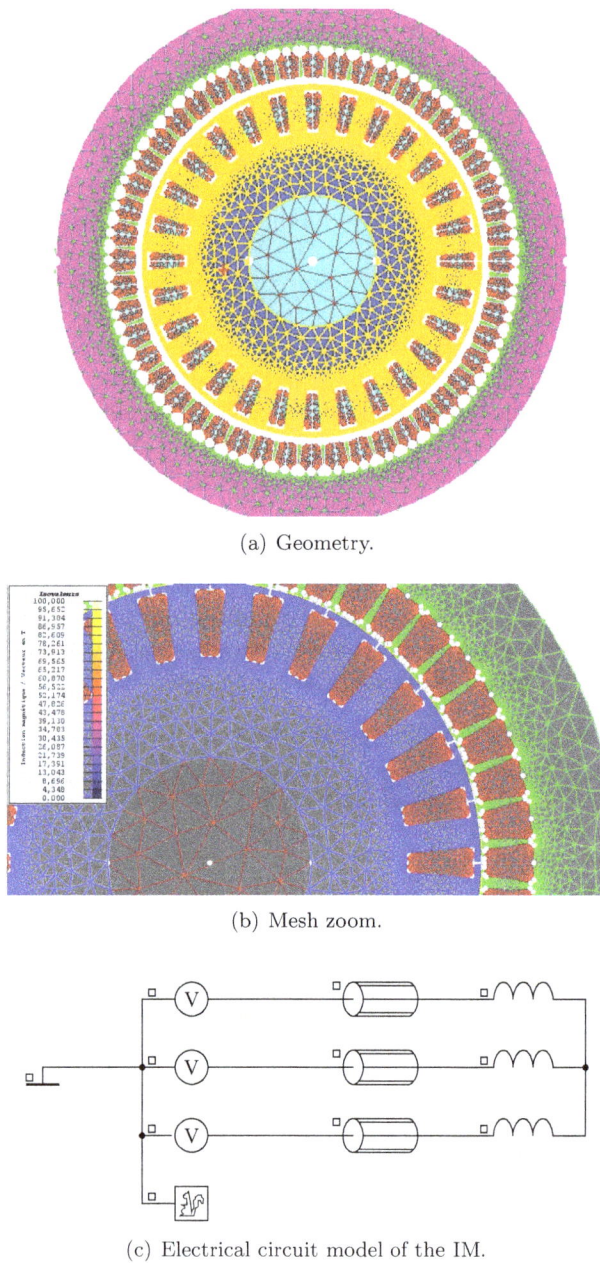

Fig. 3: Finite element model.

In Fig. 4 the rotor currents obtained by the FEM are presented, the results are more accurate, but require more computing time, up to 8 hours in our model.

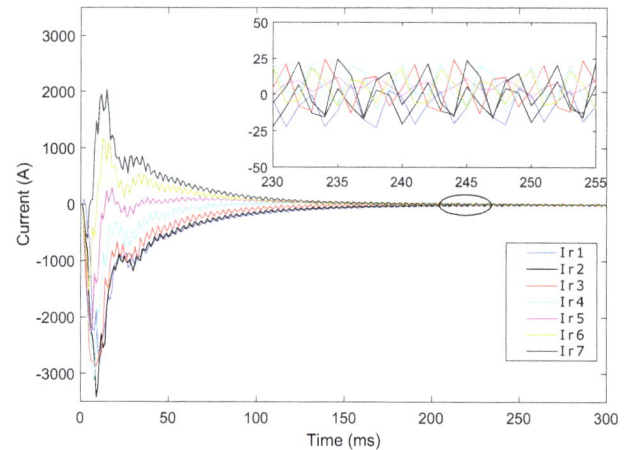

Fig. 4: The rotor currents on an area of 90 degree mechanical (one quarter of the machine).

Figure 5 shows the simulated spectral content of the ZSV – obtained from the FE model – of a healthy and faulty IM operating under rated load at 2889 rpm. In case of faulty IM, the amplitudes of the ZSV odd harmonics increase. Consequently, as seen in Fig. 5 the more appropriate harmonics to be analyzed to diagnose rotor faults is the third harmonic as mentioned in [9].

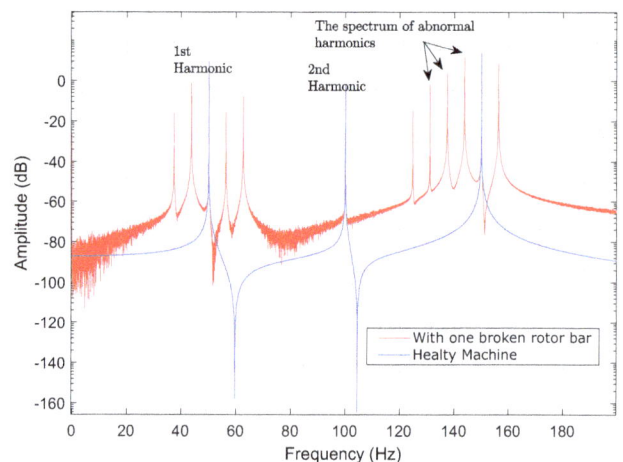

Fig. 5: Simulation: Spectral content of a healthy and faulty machine, operating under 75 % of load.

in Fig. 3, correspond to a $p = 2$ pole pairs motor of 3 kW, 400 V and $f = 50$ Hz supplied, which is used in the experimental study.

In reality inter bar currents exist in adjacent bars which lead to increase the joule losses. Consequently, in order to model this situation in FE Model, the value of the BB resistance should be selected such that it covers this condition. Furthermore, distribution of the flux lines around the broken bars differs from healthy bars. Since this is not the subject of this work, for more information on modeling using FE, Authors in [14] deals with this subject in details.

Table 2 gives the results for BRBs using the proposed technique. The broken bars are adjacent. The criterion has been evaluated for different fault degrees. We can notice that the fault criterion varies in proportion to the number of broken rotor bars and the load conditions. It is worthy to notice that the proposed approach leads to information about faults presence.

We will not present all tests (great execution time). but we focus on tests that will enable us the validation

Tab. 2: Fault detection result of the proposed method.

Supply	Rotor Condition	f_{tar1}	f_{tar1}	Meas. Slip (%)	Est. Slip (%)	s_j	s_n	C_{RF}	Decision
				Simulation Results					
	S-H0	145.23	139.19	×	4.77	0.019	0.008	2.44	Healthy
	S-H100	144.9	139.23	×	5.09	0.175	0.058	3.02	Healthy
	S-1bb50				No Max detected				
	S-1bb75	145.41	140.82	×	4.59	0.414	0.036	11.51	Defective
				Experimental Results					
	S-H0				No Max detected				
	S-H50	145.23	139.19	4.79	4.77	0.384	0.108	3.55	Healthy
	S-H100	144.9	139.23	5.14	5.09	0.105	0.073	1.45	Healthy
	S-1bb0				No Max detected				
Network	S-1bb25	145.41	140.82	4.63	4.59	0.610	0.067	9.11	Defective
	S-1bb50	144.40	138.81	5.71	5.60	0.15	0.007	21.42	Defective
	S-1bb100	142.33	134.66	3.78	7.67	0.29	0.003	79.30	Defective
	S-2bb0	148.56	147.06	1.47	1.44	0.0316	0.005	6.32	Defective
	S-2bb25	144.37	138.74	5.71	5.63	0.088	0.0082	10.74	Defective
	S-2bb50	143.8	138	6.25	6.2	0.302	0.0063	48.01	Defective
	S-2bb100	142.95	137.31	7.11	7.05	0.36	0.0062	56.3	Defective
	I-H0				No Max detected				
Inverter	I-H100	144.9	139.23	5.14	5.09	0.105	0.073	1.45	Healthy
	I-1bb0	147.40	144.80	2.54	2.61	0.015	0.007	4.89	Defective
	I-1bb75	145.34	140.68	4.72	4.66	2.218	0.093	23.86	Defective

of the proposed method especially test which are figured in the second column of table Tab. 2. Other tests can be seen in the next subsection.

Fig. 7 are compared in Tab. 2. The first column in this table corresponds to the machine supply (Line-fed or inverter-Fed), the rotor state is presented in the second column, the third and fourth gives the value of

4.2. Experimental Results

Figure 6 shows the structure of the laboratory setup. The motor under test is a 3 kW, 50 Hz, 220/380 V, 2-poles. This motor is directly coupled to a DC machine acting as a load. The sampling frequency was 26 kHz. Voltage and current signals are measured by using LEM Hall sensors. These sensors are connected to the data acquisition board, which is connected to a personal computer.

For motor conditions, three types of test motors were used: a healthy motor, a motor with one broken bar and a motor with two broken rotor bars. The motor consists also on inverter which is connected to the load motor in order to connect the load condition of the tested motor.

The measured signals were analyzed by using Fast Fourier Transform (FFT) using a Hanning window to minimize the frequency leakage. For the tested motor, the experiments were performed in the steady state condition to obtain accurate information about the broken rotor bars.

Figure 7 shows the experimental spectrum of the ZSV in healthy and faulty case rated load. It is clear that the 3rd harmonic is appropriate to detect BRBs in IM using ZSV. This supports the simulations results obtained from FE model.

To highlight the sensitivity of the proposed method described in Sec. 3. , results shown in Fig. 5 and

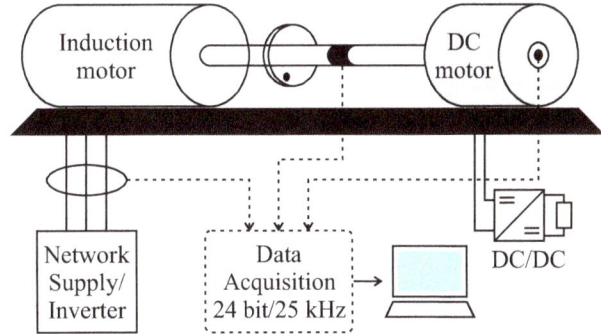

(a) Block diagram of the studied machine.

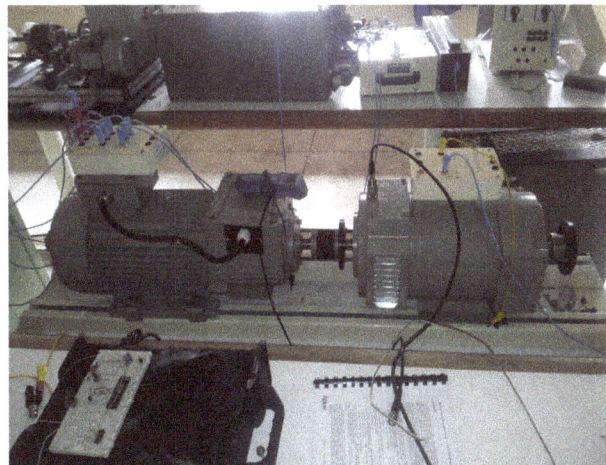

(b) The used Induction Machine.

Fig. 6: Experimental setup.

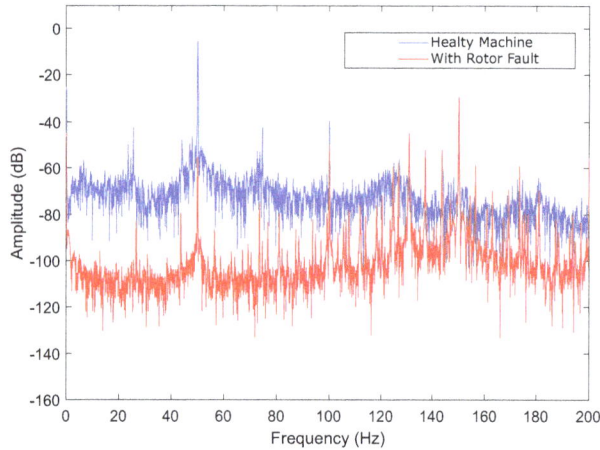

Fig. 7: Experiment: Spectral content of a healthy and faulty machine, operating under 75 % of load.

Tab. 3: Experiment details.

Des.	Supply	Details
S-H0		Healthy, unloaded
S-H50		Healthy, under 50% of load
S-H100		Healthy, under 100% of load
S-1bb0		One BRB, unloaded
S-1bb25		One BRB, 25% of load
S-1bb50	Network	One BRB, 50% of load
S-1bb100		One BRB, 100% of load
S-2bb0		Two BRB, unloaded
S-2bb25		Two BRB, 25% of load
S-2bb50		Two BRB, 50% of load
S-2bb100		Two BRB, 100% of load
I-H0		Healthy, unloaded
I-H100	Inverter	Healthy, under 100% of load
I-1bb0		One BRB, unloaded
I-1bb75		One BRB, 75% of load

the frequencies f_{tar1} and f_{tar2} respectively, fifth and sixth column gives the calculated and estimated slip. The values of s_j and s_n calculated on the frequency ranges RANGE1 and RANGE2 are presented in the seventh and eighth column respectively. And then the module decision is presented in the two last columns.

Figure 8 shows an example of the application of the frequency tracking module in the **S-1bb50** case. In the first plot of this figure, Fig. 8(a), the amplitude of target frequencies f_{tar}, f_{tar1} and f_{tar2} can be observed in the ZSV spectrum. In Fig. 8(b), frequency tracking module is shown. It has been built by keeping only the amplitudes of the components of the searched harmonics. For the rest, the same process is done for all tests in Tab. 3.

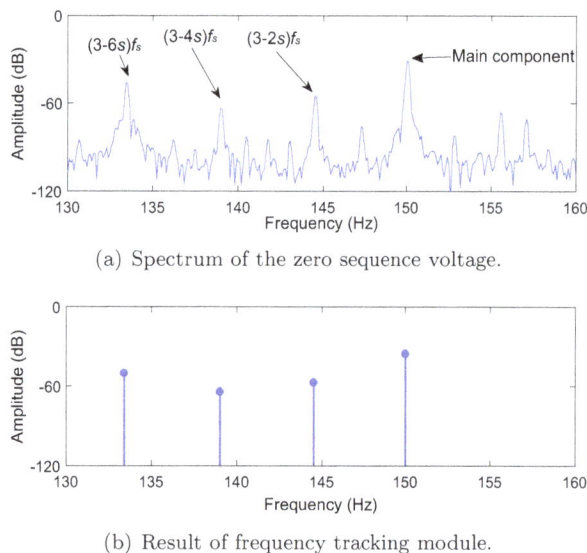

(a) Spectrum of the zero sequence voltage.

(b) Result of frequency tracking module.

Fig. 8: Experiment: Example of application of the frequency tracking module.

The method described above (Sec. 3.) is applied on the ZSV when the machine is directly star-connected to the three-phase network. According to the column giving C_{RF}, we note that is low for a machine operating with a healthy rotor (**S-H100** and **S-H50**), then we perceive that for some healthy functioning we do not detect the frequency component $(3 - 2s)f_s$. In this case, the rotor is considered in good condition (**S-H0**).

The appearance of a partial rotor fault does not induce a significant increase of s_j relative to s_n, which does not allow to conclude on such a failure. For an important rotor fault (**S-1bb100**), we note that this report is greater 5 times that in tests where the machine is healthy. From these results, it can be concluded that the proposed approach is validated, even if C_{RF} in tests **S-1bb0** and **S-H0** is less pronounced as seen in Tab. 2, but the results are satisfactory.

It is noted that the C_{RF} criterion vary too much despite the variation in the load level. In the defective case a notable variation between fault conditions is seen. From the Tab. 2, C_{RF} does not exceed 5 for a healthy machine and it is greater than 5 for a defective machine.

This conclusion led us to make an IM diagnosis method without reference (this reference usually obtained from a healthy functioning). In other words, if the report s_j/s_n is less than 5 then the machine is healthy, and defective if greater than 5 (as indicated in the last column of Tab. 2). which validate the Tab. 1 where $C_{th} = 5$ for the studied motor.

Results from Tab. 2 show a similar behavior when comparing the simulated results obtained based on FE model proposed in this work with experimental results obtained from real tests. After analyzing these results, the proposed criterion provides sufficiently good sensitivity allowing the detection of rotor asymmetries in IMs.

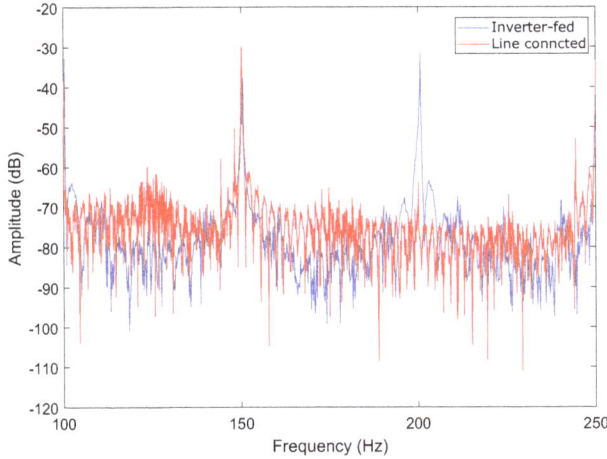

Fig. 9: ZSV in Healthy case with Inverter-fed and line connected machine, operating under 25 % of load.

4.3. Case Study: Induction Machine Supplied by Inverter Fed

The theoretical development of the proposed method, and its comparison with previous approaches, is illustrated with a case study, a 3 kW star-connected induction motor supplied with an inverter, which drives a variable load.

It is well known that in inverted-fed the spectral content of the currents is influenced by the inverter. in [18], it is demonstrated that is difficult to extract useful information from the study of stator currents of motor in case of inverter-fed because it generate additional harmonics in line current spectrum. On the other hand, the Zero Sequence Voltage based method presented in this paper allows minimizing the additional harmonics on the V_0 spectrum.

After tests, the results are shown in the second part of Tab. 2. If the motor operates at a very low slip (**I-H0**), the diagnosis of this type of fault is especially challenging because the fault harmonics are very close to the mains component and also because the harmonic generated by the inverter as shown in Fig. 9. Compared to other works from the literature, the proposed method gives better result even the machine is supplied with an inverter.

5. Future Research

As mentioned in this paper, the fault index C_{RF} and the threshold criterion C_{th} are determined just from the obtained experimental results. The aim in the next works is to combine the proposed method with the detection probability and the false alarm probability in order to detect rotor faults in IM in order to make a robust condition on C_{th}. Basis of Neyman-Pearson

approach (NPA), see [19] and [20], is that for decision of accepting or rejecting hypothesis in binary case is necessary to consider cost of accepting H_1 and rejection H_0 hypothesis. Using the NPA, the index C_{RF} and the optimal threshold C_{th} in fault decision module will be obtained by:

$$\{C_{RF}, C_{th}\} = \arg_{\{C_{RF}, C_{th}\}} \max \xi_{C_{RF}}, \quad (14)$$

where $\xi_{C_{RF}} = P_{D,C_{RF}} - P_{FA,C_{RF}}$ and:

$$\begin{aligned} P_{D,C_{RF}} &= Pr\{C_{RF} > C_{th}; H1\}, \\ P_{FA,C_{RF}} &= Pr\{C_{RF} < C_{th}; H0\}. \end{aligned} \quad (15)$$

Another extension of the proposed method is the detection of multiple faults in IMs is currently under development (eccentricity, bearing fault,...) and will be presented in a future papers.

6. Conclusion

In this paper, an affective approach to detect rotor fault in IM has been proposed, it is based on the analysis of a new fault indicator that uses both the harmonic tracking and the Zero sequence voltage. The presented indicator allows to have a knowledge about the rotor state, the fault severity and the corresponding slip for the data acquisition. With the proposed method, the decision making is done. It was shown that this approach -in addition that it is easy to implement- provides some advantages over existing methods in the literature:

- Robustness and efficiency under time-varying operating conditions.

- The proposed method has a high sensitivity to detect rotor fault.

The description of the proposed method, its theoretical justification, Simulation and the experimental validation are fully confirmed under a wide variety of supply types and working conditions.

References

[1] CAPOLINO, G. A., J. A. ANTONINO-DAVIU and M. RIERA-GUASP. Modern Diagnostics Techniques for Electrical Machines, Power Electronics, and Drives. *IEEE Transactions on Industrial Electronics.* 2015, vol. 62, iss. 3, pp. 1738–1745. ISSN 0278-0046. DOI: 10.1109/TIE.2015.2391186.

[2] NAHA, A., A. K. SAMANTA, A. ROUTRAY and A. K. DEB. A Method for Detecting

Half-Broken Rotor Bar in Lightly Loaded Induction Motors Using Current. *IEEE Transactions on Instrumentation and Measurement*. 2016, vol. 65, iss. 7, pp. 1614–1625. ISSN 0018-9456. DOI: 10.1109/TIM.2016.2540941.

[3] FILIPPETTI, F., A. BELLINI and G. A. CAPOLINO. Condition monitoring and diagnosis of rotor faults in induction machines: State of art and future perspectives. In: *IEEE Workshop on Electrical Machines Design Control and Diagnosis (WEMDCD)*. Paris: IEEE, 2013, pp. 196–209. ISBN 978-1-4673-5656-5. DOI: 10.1109/WEMDCD.2013.6525180.

[4] THOMSON, W. T. and M. FENGER. Current signature analysis to detect induction motor faults. *IEEE Industry Applications Magazine*. 2001, vol. 7, iss. 4, pp. 26–34. ISSN 1077-2618. DOI: 10.1109/2943.930988

[5] JUNG, J.-H., J.-J. LEE and B.-H. KWON. Online Diagnosis of Induction Motors Using MCSA. *IEEE Transactions on Industrial Electronics*. 2006, vol. 53, iss. 6, pp. 1842–1852. ISSN 1557-9948. DOI: 10.1109/TIE.2006.885131.

[6] LEE, S. B., D. HYUN, T.-J. KANG, C. YANG, S. SHIN, H. KIM, S. PARK, T.-S. KONG and H.-D. KIM. Identification of False Rotor Fault Indications Produced by Online MCSA for Medium-Voltage Induction Machines. *IEEE Transactions on Industry Applications*. 2016, vol. 52, iss. 1, pp. 729–739. ISSN 0093-9994. DOI: 10.1109/TIA.2015.2464301.

[7] DAHI, K., S. EL HANI, S. GUEDIRA, I. OUACHTOUK and A. ECHCHAACHOUAI. Diagnonis of rotor asymmetries in Induction motor using ESA. In: *2016 International Conference on Electrical and Information Technologies (ICEIT)*. Tangiers: IEEE, 2016, pp. 1–8. ISBN 978-1-4673-8469-8. DOI: 10.1109/EITech.2016.7519564.

[8] GYFTAKIS, K. N., J. A. ANTONINO-DAVIU, R. GARCIA-HERNANDEZ, M. D. MCCULLOCH, D. A. HOWEY and A. J. MARQUES CARDOSO. Comparative Experimental Investigation of Broken Bar Fault Detectability in Induction Motors. *IEEE Transactions on Industry Applications*. 2016, vol. 52, iss. 2, pp. 1452–1459. ISSN 0093-9994. DOI: 10.1109/TIA.2015.2505663.

[9] GARCIA, P., F. BRIZ, M. W. DEGNER and A. B. DIEZ. Diagnostics of induction machines using the zero sequence voltage. In: *IEEE Industry Applications Conference*. Seattle: IEEE, 2004, pp. 735–742. ISBN 0-7803-8486-5. DOI: 10.1109/IAS.2004.1348496.

[10] OUMAAMAR, M. E. K., A. KHEZZAR, M. BOUCHERMA, H. RAZIK, R. N. ANDRIAMALALA and L. BAGHLI. Neutral Voltage Analysis for Broken Rotor Bars Detection in Induction Motors Using Hilbert Transform Phase. In: *42nd Annual Meeting of the IEEE-Industry-Applications-Society*. New Orleans: IEEE, 2007, pp. 1940–1947. ISBN 978-1-4244-1259-4. DOI: 10.1109/07IAS.2007.295.

[11] CASH, M. A., T. G. HABETLER and G. B. KLIMAN. Insulation failure prediction in AC machines using line-neutral voltages. *IEEE Transactions on Industry Applications*. 1998, vol. 34, iss. 6, pp. 1234–1239. ISSN 0093-9994. DOI: 10.1109/28.738983

[12] FERKOVA, Z. Comparison between 2D and 3D Modelling of Induction Machine Using Finite Element Method. *Advances in Electrical and Electronic Engineering*. 2015, vol. 13, no. 2, pp. 120–126. ISSN 1804-3119. DOI: 10.15598/aeee.v13i2.1346.

[13] WALLACE, A. K. and A. WRIGHT. Novel Simulation of Cage Windings Based on Mesh Circuit Model. *IEEE Transactions on Power Apparatus and Systems*. 1974, vol. PAS-93, iss. 1, pp. 377–382. ISSN 0018-9510. DOI: 10.1109/TPAS.1974.293957.

[14] BOUGHRARA, K., N. TAKORABET, R. IBTIOUEN, O. TOUHAMI and F. DUBAS. Analytical Analysis of Cage Rotor Induction Motors in Healthy, Defective, and Broken Bars Conditions. *IEEE Transactions on Magnetics*. 2015, vol. 51, iss. 2, pp. 1–17. ISSN 0018-9464. DOI: 10.1109/TMAG.2014.2349480.

[15] GYFTAKIS, K. N. and J. C. KAPPATOU. The Zero-Sequence Current as a Generalized Diagnostic Mean in Delta-Connected Three-Phase Induction Motors. *IEEE Transactions on Energy Conversion*. 2014, vol. 29, no. 1, pp. 138–148. ISSN 0885-8969. DOI: 10.1109/TEC.2013.2292505.

[16] SINGH, A., B. GRANT, R. DEFOUR, C. SHARMA and S. BAHADOORSINGH. A review of induction motor fault modeling. *Electric Power Systems Research*. 2016, vol. 133, iss. 1, pp. 191–197. ISSN 0378-7796. DOI: 10.1016/j.epsr.2015.12.017.

[17] OUACHTOUK, I., S. E. HANI, S. GUEDIRA, L. SADIKI and K. DAHI. Modeling of squirrel cage induction motor a view to detecting broken rotor bars faults. In: *1st International Conference on Electrical and Information Technologies*. Marrakech: IEEE, 2015, pp. 347–352. ISBN 978-1-4799-7479-5. DOI: 10.1109/EITech.2015.7163001.

[18] ROUX, W. L., R. G. HARLEY and T. G. HABETLER. Detecting faults in rotors of PM drives. *IEEE Industry Applications Magazine*. 2008, vol. 14, iss. 2, pp. 23–31. ISSN 1077-2618. DOI: 10.1109/MIA.2007.915789.

[19] CHARFI, F., S. LESECQ and F. SELLAMI. Fault diagnosis using SWT and Neyman Pearson detection tests. In: *IEEE International Symposium on Diagnostics for Electric Machines, Power Electronics and Drives*. Cargese: IEEE, 2009, pp. 1–6. ISBN 978-1-4244-3440-4. DOI: 10.1109/DEMPED.2009.5292788.

[20] MUSTAFA, M. O., D. VARAGNOLO, G. NIKO-LAKOPOULOS and T. GUSTAFSSON. Detecting broken rotor bars in induction motors with model-based support vector classifiers. *Control Engineering Practice*. 2016, vol. 52, iss. 1, pp. 15–23. ISSN 0967-0661. DOI: 10.1016/j.conengprac.2016.03.019.

About Authors

Khalid DAHI was born in Errachidia, Morocco, in 1988. He received the M.Sc. degree in electrical engineering in 2012 from the Mohammed V University in Rabat- Morocco. Where he is currently working toward the Ph.D. degree in the department of electrical engineering. Since 2012, His research interests are related to electrical machines and drives, diagnostics of induction motors. His current activities include monitoring and diagnosis of induction machines in wind motor.

Soumia EL HANI has been Professor at the ENSET (Ecole Normale Superieure de l'Enseignement Technqique - Rabat, Morroco) since October 1992. She is a Research Engineer at the Mohammed V University in Rabat Morocco, in charge of the research team electromechanical, control and diagnosis, IEEE member, member of the research laboratory in electrical engineering at ENSET - Rabat. Author of several publications in the field of electrical engineering, including robust control systems, diagnosis and control systems of wind electric conversion. She has been general co-Chair the two editions of "the International Conference on Electrical and Information Technologies", held in Marrakech, March 2015 and Tangier, May 2016 respectively.

Ilias OUACHTOUK was born in Foum Zguid-Tata, Morocco, in 1991. He received his M.Sc. degree in electrical engineering from the Mohammed V University in Rabat, Morocco, in 2014. Where he is currently working toward the Ph.D. degree, in the department of electrical engineering. Since 2015. His research interests include modeling and diagnosis of electrical drives, in particular, synchronous and asynchronous motors.

Appendix A
Induction Motor Parameters

- $P_N = 3$ kW,
- $U_{1N} = 220$ V,
- $f_s = 50$ Hz,
- $p = 2$,
- $N_r = 28$,
- $R_s = 1.5$ Ω,
- $R_b = 96.9 \cdot 10^{-6}$ Ω,
- $L_b = 0.28 \cdot 10^{-6}$ H,
- $L_e = 0.036 \cdot 10^{-6}$ H,
- $j = 4\pi \cdot 10^{-7} \dfrac{W_b}{\text{A} \cdot \text{m}}$.

16

Load Insensitive, Low Voltage Quadrature Oscillator Using Single Active Element

Jitendra MOHAN, Bhartendu CHATURVEDI

Department of Electronics and Communication Engineering, Jaypee Institute of Information Technology,
Sector-62, 201304 Noida, Uttar Pradesh, India

jitendramv2000@rediffmail.com, bhartendu.prof@gmail.com

Abstract. *In this paper, a load insensitive quadrature oscillator using single differential voltage dual-X second generation current conveyor operated at low voltage is proposed. The proposed circuit employs single active element, three grounded resistors and two grounded capacitors. The proposed oscillator offers two load insensitive quadrature current outputs and three quadrature voltage outputs simultaneously. Effects of non-idealities along with the effects of parasitic are further studied. The proposed circuit enjoys the feature of low active and passive sensitivities. Additionally, a resistorless realization of the proposed quadrature oscillator is also explored. Simulation results using PSPICE program on cadence tool using 90 nm Complementary Metal Oxide Semiconductor (CMOS) process parameters confirm the validity and practical utility of the proposed circuit.*

Keywords

Load insensitive, quadrature oscillator, resistorless.

1. Introduction

Nowadays, research in analog signal processing has gone in the direction of low-voltage and low-power design. In addition, many new applications continue to emerge where new analog topologies have to be designed to ensure the trade-off between power and speed requirements. Finally, the modern development towards miniaturized circuits has given a strong and significant enhancement towards the implementation of low-voltage and low-power analog circuits. Ever since the introduction of analog signal processing, the need of new active devices has always been very significant. Recently, with the increasing demand of low-voltage and low-power circuits, current conveyors have achieved popularity [1], [2] and [9]. With the use of current conveyors, a number of applications can be realized such as: differentiators, integrators, impedance simulators, impedance converters, biquadratic filters, instrumentation amplifiers, oscillators, etc. The realizations of oscillators using different variation of current conveyors have received significant attention due to their numerous advantages. In the literature a number of quadrature oscillators based on different active elements are also reported. The quadrature oscillators in [3], [4], [5] and [6] produced voltage-mode signals and the ones in [7], [8], [9], [10] and [11] produced current-mode signals. Although some of the quadrature oscillators in [12], [13], [14] and [15] generated both voltage-mode signals as well as current-mode signals. Moreover, few of them are based on single active element [7], [8], [10] and [15].

The idea behind this paper is to propose a new load insensitive, low voltage quadrature oscillator using a single Differential Voltage Dual-X Second generation Current Conveyor (DV-DXCCII) along with five grounded passive components (three grounded resistors and two grounded capacitors). The proposed circuit design offers many advantages such as the use of single active element, the use of all grounded passive components, operated at low voltage, simultaneous availability of quadrature voltage and load insensitive quadrature current outputs, low active and passive sensitivities and resistorless realization.

2. Proposed Circuit

Differential Voltage Dual-X second generation Current Conveyor (DV-DXCCII) is an analogue building block [16] which is characterized by Eq. (1).

(a)

(b)

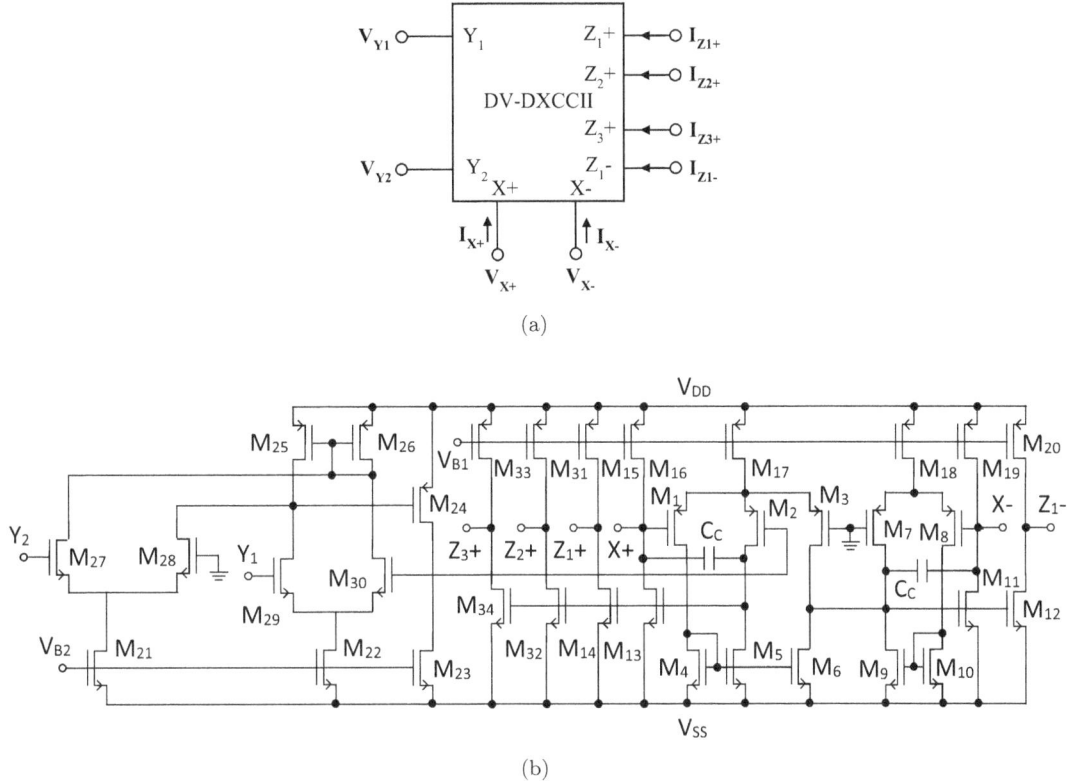

Fig. 1: (a) Schematic Symbol of DV-DXCCII (b) CMOS realization of DV-DXCCII [16].

The schematic symbol and CMOS realization of DV-DXCCII are shown in Fig. 1, where the CMOS realization comprises of Differential Voltage Current Conveyor (DVCC), i.e. M_{21} - M_{30}, with unused Z stages and dual-X second generation current conveyor (DXCCII), i.e. M_1 – M_{20}. In the realization of DV-DXCCII, X terminal of DVCC will drive the Y terminal of the DXCCII. The output terminals Z_{1+} and Z_{1-} are realized from the joint drains of M_{14}, M_{15} and M_{12}, M_{20} transistors, respectively. The features of both DVCC [2] and DXCCII [9] are combined together in this single active element. However, additional Z+ stages (Z_{2+} and Z_{3+}) have been implemented by taking the extra cascaded structures of transistors (M_{31} – M_{32} and M_{33} – M_{34}).

proposed oscillator provides two quadrature current outputs and three quadrature voltage outputs simultaneously. Since both the current outputs (I_{O1} and I_{O2}) are directly available at the high output impedance terminals (Z_{1-} and Z_{3+}) without any load thus the proposed circuit enjoys the benefit of load insensitive current outputs. The characteristic equation of the proposed quadrature oscillator using Eq. (1) is given as:

$$s^2 + s\left[\frac{1}{C_1 R_1} + \frac{1}{C_2 R_2} - \frac{1}{C_1 R_2}\right] + \frac{1}{C_1 C_2 R_1 R_2} = 0. \quad (2)$$

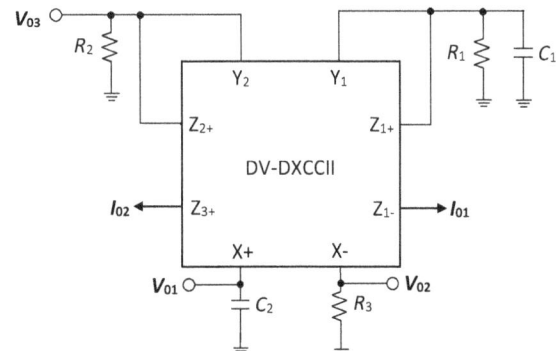

$$\begin{bmatrix} \mathbf{I}_{Y1} \\ \mathbf{I}_{Y2} \\ \mathbf{V}_{X+} \\ \mathbf{V}_{X-} \\ \mathbf{I}_{Z1+} \\ \mathbf{I}_{Z2+} \\ \mathbf{I}_{Z3+} \\ \mathbf{I}_{Z1-} \end{bmatrix} = \begin{bmatrix} 0 & 0 & 0 & 0 \\ 0 & 0 & 0 & 0 \\ 1 & -1 & 0 & 0 \\ -1 & 1 & 0 & 0 \\ 0 & 0 & 1 & 0 \\ 0 & 0 & 1 & 0 \\ 0 & 0 & 1 & 0 \\ 0 & 0 & 0 & 1 \end{bmatrix} \begin{bmatrix} \mathbf{V}_{Y1} \\ \mathbf{V}_{Y2} \\ \mathbf{I}_{X+} \\ \mathbf{I}_{X-} \end{bmatrix}. \quad (1)$$

The proposed circuit of load insensitive, low voltage quadrature oscillator is shown in Fig. 2. The proposed circuit employs a single DV-DXCCII, three grounded resistors, and two grounded capacitors. The

Fig. 2: Proposed load insensitive, low voltage quadrature oscillator.

Condition of Oscillation (CO) is given as:

$$CO: \frac{1}{C_1 R_1} + \frac{1}{C_2 R_2} \geq \frac{1}{C_1 R_2}, \qquad (3)$$

and the expression of Frequency of Oscillation (FO) is given as:

$$FO: \omega_0 = \sqrt{\frac{1}{C_1 C_2 R_1 R_2}}. \qquad (4)$$

The phasor diagrams for the three quadrature voltage outputs (V_{O1}, V_{O2}, and V_{O3}) and two quadrature current outputs (I_{O1} and I_{O2}) are shown in Fig. 3(a) and Fig. 3(b), respectively. The three voltage and two current outputs of Fig. 3 are related as:

$$V_{O3} = j\omega C_2 R_2 V_{O1}, V_{O1} = -V_{O2}, \qquad (5)$$

$$I_{O2} = -j\omega C_2 R_3 I_{O1}. \qquad (6)$$

It is apparent from Eq. (5) and Eq. (6) that all the voltage and current outputs are in quadrature relationship.

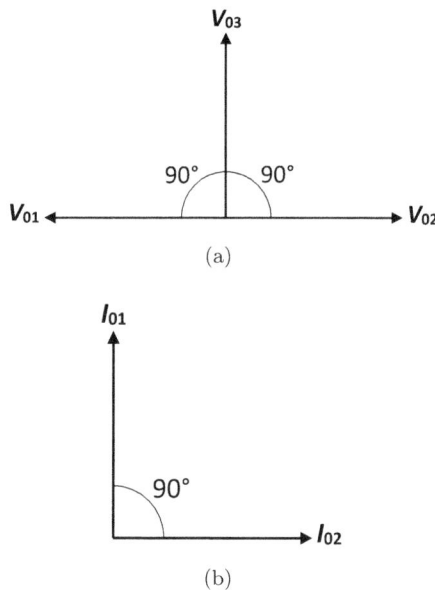

(a)

(b)

Fig. 3: Phasor diagram depicting (a) quadrature voltage relationship (b) quadrature current relationship.

3.   Non-Ideal Analysis

By considering the non-ideal voltage and current transfer gains (α_i, where $i = 1, 2$ and β_j, where $j = 1, 2, 3, 4$) of DV-DXCCII, the modified voltage and current terminals relationship can be rewritten as:

$$\begin{bmatrix} \mathbf{I}_{Y1} \\ \mathbf{I}_{Y2} \\ \mathbf{V}_{X+} \\ \mathbf{V}_{X-} \\ \mathbf{I}_{Z1+} \\ \mathbf{I}_{Z2+} \\ \mathbf{I}_{Z3+} \\ \mathbf{I}_{Z1-} \end{bmatrix} = \begin{bmatrix} 0 & 0 & 0 & 0 \\ 0 & 0 & 0 & 0 \\ \beta_1 & -\beta_2 & 0 & 0 \\ -\beta_3 & \beta_4 & 0 & 0 \\ 0 & 0 & \alpha_1 & 0 \\ 0 & 0 & \alpha_2 & 0 \\ 0 & 0 & \alpha_3 & 0 \\ 0 & 0 & 0 & \alpha_4 \end{bmatrix} \begin{bmatrix} \mathbf{V}_{Y1} \\ \mathbf{V}_{Y2} \\ \mathbf{I}_{X+} \\ \mathbf{I}_{X-} \end{bmatrix}. \qquad (7)$$

Here, β_1 and β_2 are the voltage transfer gains from Y_1 and Y_2 terminals, respectively to the $X+$ terminal. Similarly β_3 and β_4 are the voltage transfer gains from Y_1 and Y_2 terminals, respectively to the $X-$ terminal. α_1, α_2, α_3 are the current transfer gains from $X+$ terminal to Z_{1+}, Z_{2+}, Z_{3+} terminals, respectively. α_4 is the current transfer gain from $X-$ terminal to $Z1-$ terminal. The ideal value of these voltage and current transfer gains is unity depending upon selected operating frequency.

Using Eq. (7), the proposed quadrature oscillator is reanalyzed. The non-ideal characteristic equation is obtained as:

$$s^2 + s \left[\frac{1}{C_1 R_1} + \frac{1}{\alpha_2 \beta_2 C_2 R_2} - \frac{\alpha_1 \beta_1}{\alpha_2 \beta_2 C_1 R_2} \right] + \frac{1}{\alpha_2 \beta_2 C_1 C_2 R_1 R_2} = 0. \qquad (8)$$

The non-ideal CO and FO are given as:

$$CO: \frac{1}{C_1 R_1} + \frac{1}{\alpha_2 \beta_2 C_2 R_2} \geq \frac{\alpha_1 \beta_1}{\alpha_2 \beta_2 C_1 R_2}, \qquad (9)$$

$$FO: \omega_0 = \sqrt{\frac{1}{\alpha_2 \beta_2 C_1 C_2 R_1 R_2}}. \qquad (10)$$

The active and passive sensitivities with respect to ω_0 are given below

$$S^{\omega_0}_{\alpha_2, \beta_2} = S^{\omega_0}_{C_1, C_2, R_1, R_2} = -\frac{1}{2}, \quad S^{\omega_0}_{\alpha_1, \beta_1} = 0. \qquad (11)$$

Equation (11) shows that the active and passive sensitivities with respect to frequency of oscillation are within unity in magnitude. Therefore, the proposed quadrature oscillator enjoys good active and passive sensitivity performance.

4.   Parasitic Study

The performance of the proposed quadrature oscillator by considering the effects of various parasitic of DV-DXCCII is explored in this section. The parasitic model of DV-DXCCII is shown in Fig. 4, which

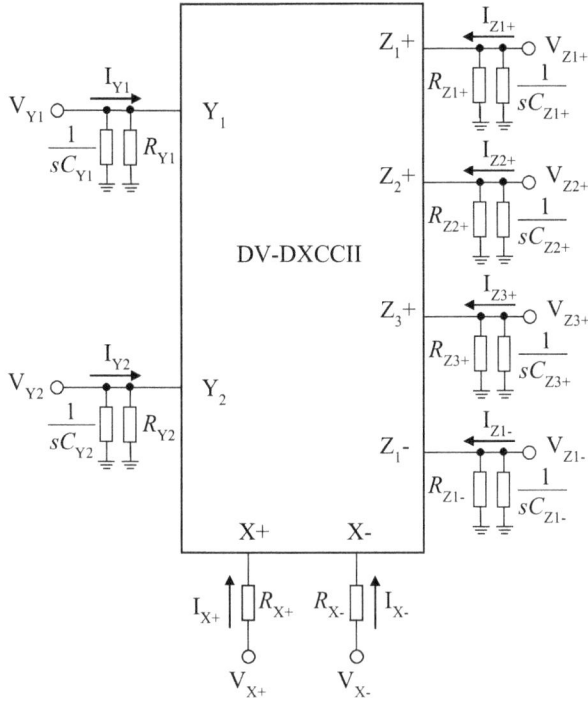

Fig. 4: The parasitic model of DV-DXCCII.

shows various ports parasitic. These various parasitic are port Y parasitic in the form of $R_Y//1/(sC_Y)$, port Z parasitic in the form of $R_Z//1/(sC_Z)$, and port X parasitic in the form of R_X. It is also worth mentioning that in the non-ideal case the parasitic resistances and capacitances appearing at the high input impedance terminal (Y) and high output impedance terminal (Z) are absorbed into the external grounded resistors and grounded capacitors as they are shunt with them. Thus, all grounded passive components based circuits are more suitable for monolithic integration. Moreover, the parasitic resistances at low impedance terminals (X+ and X−) are negligible (ideally they are zero). The parasitic impedances appearing at the X terminals would be connected between virtual grounds and actual ground and thereby eliminating their effect. However, the circuit with only capacitor at X terminal would show performance deterioration at higher frequencies [17]. In practice, to alleviate the effects of the parasitic impedances, the impedances should be chosen as:

$$Z_1 = \frac{1}{sC_1'}//R_1',$$

where:

$$R_1' = R_1//R_{Y1}//R_{Z1+},$$
$$\frac{1}{sC_1'} = \frac{1}{sC_1}//\frac{1}{sC_{Y1}}//\frac{1}{sC_{Z1+}},$$
$$Z_2 = \frac{1}{sC_Y}//R_2', \tag{12}$$

where:

$$R_2' = R_2//R_{Y2}//R_{Z2+},$$
$$\frac{1}{sC_Y} = \frac{1}{sC_{Y2}}//\frac{1}{sC_{Z2+}},$$
$$Z_3 = \frac{1}{sC_2} + R_{X+},$$
$$R_3' = R_3 + R_{X-}.$$

where, R_{Y1}, and R_{Y2} are the parasitic resistances and C_{Y1}, and C_{Y2} are the parasitic capacitances at the Y_1 and Y_2 terminals, respectively, R_{Z1+} and R_{Z2+} are the parasitic resistances and C_{Z1+} and C_{Z2+} are the parasitic capacitances at the Z_{1+} and Z_{2+} terminals, respectively, and R_{X+} and R_{X-} represent the parasitic resistances appearing at the X+ and X− terminals, respectively.

5. Simulation Results

PSPICE simulations of the proposed circuit of quadrature oscillator of Fig. 2 have been performed using the CMOS realization of DV-DXCCII of Fig. 1(b) with 90 nm CMOS parameters as given in Tab. 1. The dimensions of MOS transistors used in DV-DXCCII are given in Tab. 2. The supply voltages and bias voltage are taken as ± 1 V and $V_{BB} = -0.4$ V, respectively. Therefore, the proposed circuit is benifitted with the feature of low operating voltage. The proposed circuit of quadrature oscillator is designed at frequency of oscillation of 39.8 MHz by choosing the passive component values as $C_1 = 1$ pF, $C_2 = 2$ pF, $R_1 = 4$ kΩ, $R_2 = 2$ kΩ and $R_3 = 2$ kΩ. The frequency of oscillation as obtained from simulation is 39.63 MHz which is 0.42 % in error with the designed value. The three voltage outputs and two current outputs along with their Fourier spectrums are shown in Fig. 5 and Fig. 6, respectively. Total Harmonic Distortion (THD) is found to be within 1 %.

Tab. 1: 90 nm CMOS Model parameters.

NMOS: LEVEL= 7 BINUNIT= 1 MOBMOD= 2 RDSMOD= 0 CAPMOD= 2 EPSROX= 3.9 TOXE= 2.2E-9 NGATE= 1E20 RSH= 7.2 VTH0= 0.2637059 K1= 0.5459903 K2= -0.106347 K3= 1.00071E-3 K3B= 8.2559564 W0= 1E-10 LPE0= 8.09095E-8 LPEB= 6.541354E-8 DVT0= 0.0241091 DVT1= 0.0354959 DVT= -5.083204E-5 DVTP0= 0 DVTP1= 0 DVT0W=0 DVT1W= 0 DVT2W= -0.032 U0= 180.5863743 UA = 5.007749E-15 UB= 1E-26 UC = -1E-10 EU= 0.1098817 VSAT= 7.526087E5 A0= 2 AGS = 0 B0= 1.248816E-6 B1=1E-7 KETA= 0.05 A1= 0 A2= 1 WINT= 4.513582E-15 LINT=3.004861E-12 DWG= -2.358907E-8 DWB = 4.286229E-8ETA0 = 2.9476E-3 VOFF= -0.0349398 NFACTOR= 1.9128092CDSC= 2.4E-4 CDSCB= 0 CDSCD= 0 ETAB = -0.0117687DSUB= 0.1841044 CIT= 0 PCLM= 0 RDSWMIN= 100PDIBLC2= 0.014020828 PDIBLC1= 0.6278172 PSCBE1=5.552668E8 RSW= 100 PDIBLCB= -1E-3 DROUT= 0.6771686 PVSAT= 1.838993E3 RDWMIN= 0 PSCBE2= 3.09264E-6 PVAG= 0.0491186 DELTA= 1.741612E-3 PRWG= 3 FPROUT= 1.689547E-4 RDSW= 100 RDW= 100 PRWB= 0.0996955 RSWMIN= 0 WR= 1 XPART= 0.5

CGSO= 1E-10 CGBO= 0CF= 0 CGDO= 1E-10 WKETA= 0.039914 PKETA= -1.059394E-3 PETA0= 0 CJS = 8.93E-4 CJD= 8.93E-4 MJS= 0.3003 MJSWGD= 0.1757671 MJD= 0.3003 MJSWS=0.2357 MJSWD= 0.2357 PBSWD= 0.4 CJSWS= 1.59E-10CJSWD= 1.59E-10 CJSWGS= 3.065074E-11 PBSWGS=0.429054 CJSWGD= 3.065074E-11 MJSWGS= 0.1757671PBSWGD= 0.429054 TNOM= 27 PB= 0.4697817 PBSWS= 0.4

PMOS: LEVEL= 7 BINUNIT= 1 MOBMOD= 2 RDSMOD= 0 CAPMOD= 2 EPSROX= 3.9 TOXE= 2.4E-9 NGATE= 1E20RSH= 7.6 VTH0= -0.1828674 K= 0.5378175 K2= -0.100053 K3= 87.186991 K3B= 10 W0= 1.982838E-6 LPE0= 3.624927E-8 LPEB= -1.472555E-8 DVT0= 9.268739E-3 DVT1= 0.050834DVT2= -5.288937E-5 DVTP0= 0 DVTP1= 0 DVT0W= 0DVT1W= 0 DVT2W= -0.032 U0= 100 UA= 2.065422E-15 UB= 1E-23 UC= -1.83813E-17 EU= 1.6379893 VSAT=8.399744E4 A0= 2 AGS= 2.0870724 B0= 1.733997E-7 B1= 1E-7 KETA= 0.05 A1= 0 A2= 1 WINT= 1.540463E-8 LINT=1.382131E-10 DWG= -7.761435E-8 DWB= 2.989342E-8VOFF= -0.0422107 NFACTOR= 3.480545E-3 ETA0= 1E-5ETAB= 0 DSUB= 1 CIT= 0 CDSC= 2.4E-4 CDSCB= 0CDSCD= 0 PCLM= 0.5340493 PDIBLC1= 0.5061971PDIBLC2= 4.729943E-3 PDIBLCB= 0 DROUT= 0.9793851PSCBE1= 1.695917E8 PSCBE2= 3E-6 PVAG= 0.4223462DELTA= 0.0135647 FPROUT= 1.439073E-5 RDSW=872.5912948 RDSWMIN= 100 RDW= 100 RDWMIN= 0RSW= 100 RSWMIN= 0 PRWG= 0.1222031 PRWB= 0.1 WR= 1 XPART= 0.5 CGSO= 1E-10 CGDO= 1E-10 CGBO= 0 CF= 0 CJS= 6.4337E-9 CJD= 6.4337E-9 MJS= 0.475 MJD= 0.475MJSWS= 0.432 MJSWD= 0.432 CJSWS= 1.084E-9CJSWD= 1.084E-9 CJSWGS= 6.3E-11 CJSWGD= 6.3E-11MJSWGS= 0.2169436 MJSWGD= 0.2169436 PB= 0.9646063PBSWS= 0.965 PBSWD= 0.965 PBSWGS= 0.9009362 PBSWGD= 0.9009362 TNOM= 27 WKETA= -0.0364142PETA0= 0 PVSAT= -100

Tab. 2: Dimensions of MOS transistors used in DV-DXCCII.

MOS Transistors	W(μm)	L(μm)
M1-M2, M27-M30	0.18	0.09
M3, M7-M8	0.36	0.09
M4-M5	0.3	0.09
M6, M9-M10	0.6	0.09
M11-M20, M31-M34	1.2	0.09
M21-M22	2.61	0.09
M23	5.4	0.09
M24	1.8	0.09
M25	0.72	0.09
M26	0.76	0.09

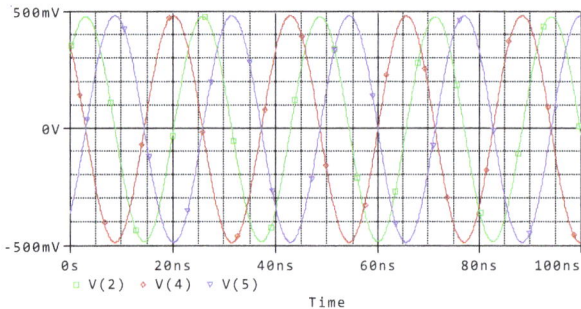

(a)

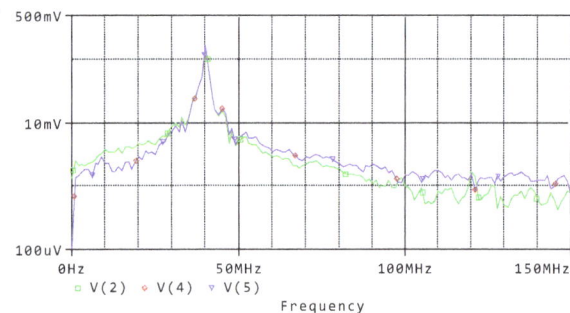

(b)

Fig. 6: (a) Two quadrature current outputs (b) Fourier spectrum at 39.63 MHz.

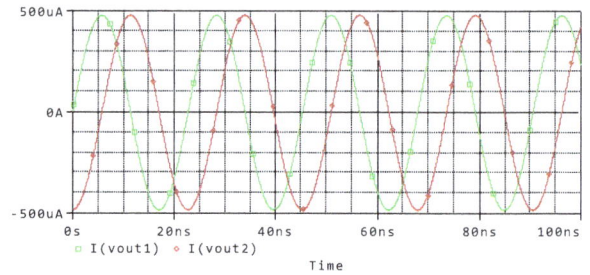

(a)

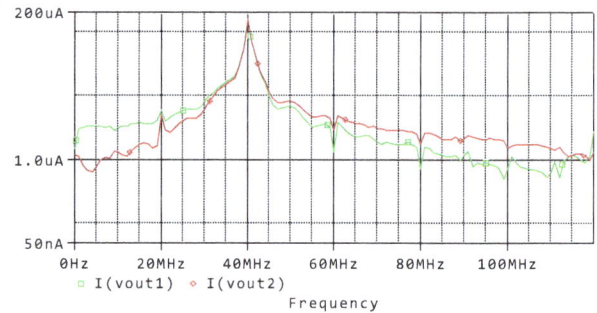

(b)

Fig. 5: (a) Three quadrature voltage outputs (b) Fourier spectrum at 39.63 MHz.

6. Resistorless Load Insensitive, Low Voltage Quadrature Oscillator

In this section, the integration and tuning aspects of the proposed circuit of Fig. 2 have been explored in the form of resistorless realization of load insensitive, low voltage quadrature oscillator. As far as active elements are concerned, its realization in CMOS technology is available. The passive components in form of resistors and capacitors can also be made compatible in CMOS technology [18] and [19]. The resistors can be replaced by active-MOS resistors with added advantage of tunability through external voltage [20]. The resistorless realization of the proposed circuit of load insensitive,

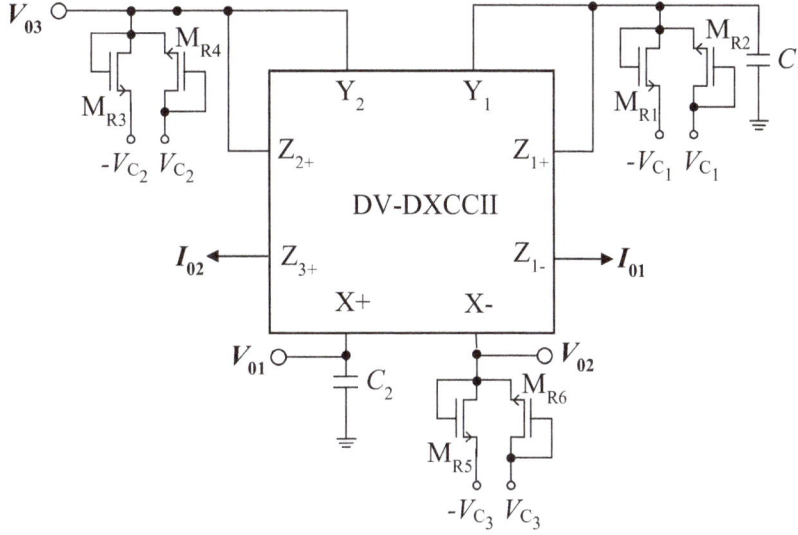

Fig. 7: Proposed resistorless load insensitive, low voltage quadrature oscillator.

low voltage quadrature oscillator is shown in Fig. 7. The resistorless load insensitive, low voltage quadrature oscillator is realized by replacing all the resistors with the two n-MOS transistors based active resistors [20].

The characteristic equation for the resistorless load insensitive, low voltage quadrature oscillator is given as:

$$s^2 + s\left[\frac{1}{C_1 R_{MOS1}} + \frac{1}{C_2 R_{MOS2}} - \frac{1}{C_1 R_{MOS2}}\right] + \frac{1}{C_1 C_2 R_{MOS1} R_{MOS2}} = 0. \quad (13)$$

The CO and FO are found as:

$$CO : \frac{1}{C_1 R_{MOS1}} + \frac{1}{C_2 R_{MOS2}} \geq \frac{1}{C_1 R_{MOS2}}, \quad (14)$$

$$FO : \omega_0 = \left(\frac{1}{C_1 C_2 R_{MOS1} R_{MOS2}}\right)^{\frac{1}{2}}. \quad (15)$$

All active resistors, R_{MOSk} (where, $k = 1, 2, 3$) is the equivalent resistance of the n-MOS transistors which is defined as:

$$R_{MOSk} = \left[2\mu C_{OX}\left(\frac{W}{L}\right)(V_{Cl} - V_t)\right]^{-1}, \quad (16)$$

where $k, l = 1, 2, 3$,

where, μ, C_{OX}, W/L and V_t are the carrier mobility, gate capacitance per unit area, aspect ratio of n-MOS and threshold voltage.

Next, the resistorless quadrature oscillator is designed for frequency 55 MHz. The transistor aspect ratios for the MOS based active resistors are selected as $(W/L)_{MR1} = (W/L)_{MR2} = (W/L)_{MR3} =$

7.2 μm/0.09 μm and capacitors values are selected as $C_1 = 1$ pF and $C_2 = 2$ pF. The FO for the proposed resistorless circuit is tuned to 55 MHz by selecting the control voltages $V_{C1} = -V_{C1} = 0.53$ V, $V_{C2} = -V_{C2} = V_{C3} = -V_{C3} = 0.86$ V. The three quadrature voltage outputs and two quadrature current outputs are shown in Fig. 8 and Fig. 9, respectively. The Fourier spectrums of the outputs are shown in Fig. 10 and Fig. 11. The THD is again found to be within 1 %. The obtained results verify the realization of resistorless load insensitive, low voltage quadrature oscillator.

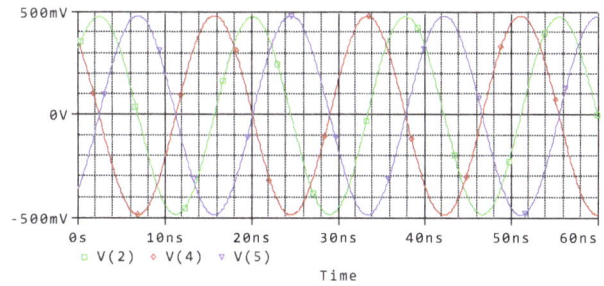

Fig. 8: Three quadrature voltage outputs at 55 MHz.

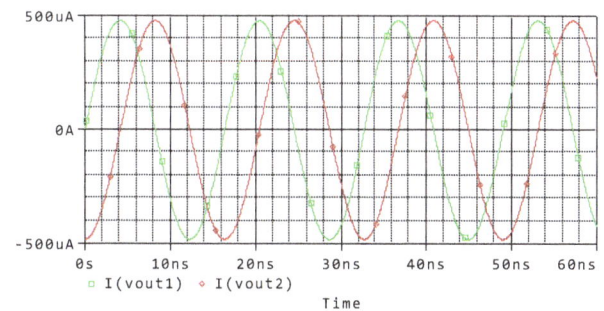

Fig. 9: Two quadrature current outputs at 55 MHz.

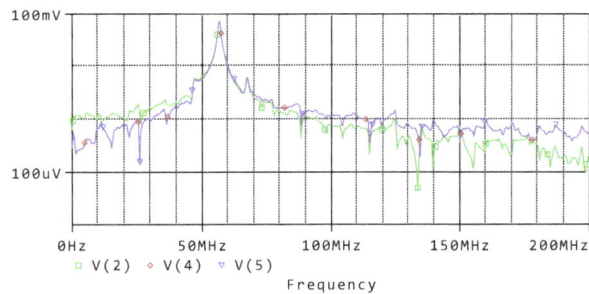

Fig. 10: Fourier spectrums of voltage outputs.

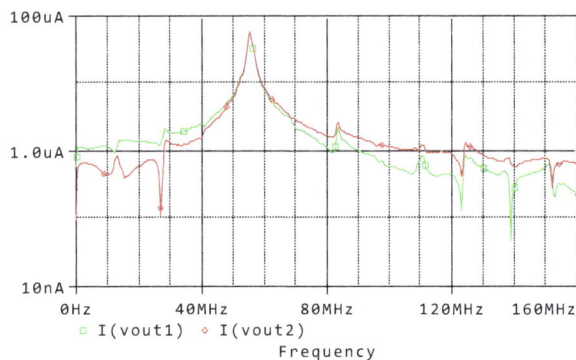

Fig. 11: Fourier spectrums of current outputs.

7.   Conclusion

A novel load insensitive, low voltage quadrature oscillator has been presented in this paper. The proposed circuit consists of single DV-DXCCII as active element and all grounded passive components, which is ideal for IC implementation. The proposed circuit of quadrature oscillator provides three quadrature voltage outputs and two quadrature current outputs simultaneously from the same configuration. The proposed quadrature oscillator also offers good active and passive sensitivities. Furthermore, the resistorless realization of the proposed circuit is also explored. Simulations results are given to support the presented theory.

References

[1] WILSON, B. Recent developments in current conveyors and current mode circuits. *IEE Proceedings G (Circuits, Devices and Systems)*. 1990, vol. 137, iss. 2, pp. 61–77. ISSN 0956-3768. DOI: 10.1049/ip-g-2.1990.0014.

[2] ELWAN, H. O. and A. M. SOLIMAN. Novel CMOS differential voltage current conveyor and its applications. *IEE Proceeding Circuits Devices and Systems*. 1997, vol. 144, iss. 3, pp. 195–200. ISSN 1350-2409. DOI: 10.1049/ip-cds:19971081.

[3] SOLIMAN, A. M. Simple sinusoidal RC oscillators using current conveyors. *International Journal of Electronics*. 1975, vol. 42, iss. 4, pp. 309–311. ISSN 0020-7217. DOI: 10.1080/00207217508920504.

[4] AHMED, M. T., I. A. KHAN and N. MINHAJ. On transconductance-C quadrature oscillators. *International Journal of Electronics*. 1997, vol. 83, iss. 2, pp. 201–208. ISSN 0020-7217. DOI: 10.1080/002072197135526.

[5] KHAN, I. A. and S. KHWAJA. An Integrable Gm-C Quadrature Oscillator. *International Journal of Electronics*. 2000, vol. 87, iss. 11, pp. 1353–1357. ISSN 0020-7217. DOI: 10.1080/002072100750000150.

[6] PROMMEE, P. and K. DEJHAN. An Integrable Electronic-Controlled Quadrature Sinusoidal Oscillator Using CMOS Operational Transconductance Amplifier. *International Journal of Electronics*. 2002, vol. 89, iss. 5, pp. 365–379. ISSN 0020-7217. DOI: 10.1080/713810385.

[7] ABUELMAATTI, M. T. and H. A. AL-ZAHER. Current-mode sinusoidal oscillators using single FTFN. *IEEE Transactions on Circuits and Systems-II*. 1999, vol. 46, iss. 1, pp. 69–74. ISSN 1057-7130. DOI: 10.1109/82.749100.

[8] CAM, U., A. TOKER, O. CICEKOGLU and H. KUNTMAN. Current-mode high output impedance sinusoidal oscillator configuration employing single FTFN. *Analog Integrated Circuits and Signal Processing*. 2000, vol. 24, iss. 3, pp. 231–238. ISSN 1573-1979. DOI: 10.1023/A:1008365726144.

[9] ZEKI, A. and A. TOKER. The dual-X current conveyor (DXCCII): a new active device for tunable continuous-time filters. *International Journal of Electronics*. 2002, vol. 89, iss. 12, pp. 913–923. ISSN 0020-7217. DOI: 10.1080/00207210031000120461.

[10] BIOLEK, D., A. U. KESKIN and V. BIOLKOVA. Grounded capacitor current mode single resistance-controlled oscillator using single modified current differencing transconductance amplifier. *IET Circuits, Devices and Systems*. 2010, vol. 4, iss. 6, pp. 496–502. ISSN 1751-8598. DOI: 10.1049/iet-cds.2009.0330.

[11] MAHESHWARI, S. and B. CHATURVEDI. High output impedance CMQO using DVCCs and grounded components. *International Journal of Circuit Theory and Application*. 2011, vol. 39, iss. 4, pp. 427–435. ISSN 1097-007X. DOI: 10.1002/cta.636.

[12] MAHESHWARI, S. and I. A. KHAN. Mixed mode quadrature oscillator using translinear conveyors and grounded components. In: *International Conference on Multimedia Signal Processing and Communication Technologies*. Aligarh: IEEE, 2011, pp. 153–155. ISBN 978-1-4577-1107-7. DOI: 10.1109/MSPCT.2011.6150462.

[13] BEG, P., M. S. ANSARI and M. A. SIDDIQI. DXCC-II based mixed-mode three phase oscillator. In: *International Conference on Multimedia Signal Processing and Communication Technologies*. Aligarh: IEEE, 2011, pp. 9–11. ISBN 978-1-4577-1107-7. DOI: 10.1109/MSPCT.2011.6150507.

[14] BIOLEK, D., A. LAHIRI, W. JAIKLA, M. SIRIPRUCHYANUN and J. BAJER. Realization of electronically tunable voltage-mode/current-mode quadrature sinusoidal oscillator using ZC-CG-CDBA. *Microelectronics Journal*. 2011, vol. 42, iss. 10, pp. 1116–1123. ISSN 0026-2692. DOI: 10.1016/j.mejo.2011.07.004.

[15] MOHAN, J., S. MAHESHWARI and I. A. KHAN. Mixed-mode quadrature oscillators using single FDCCII. *Journal of Active and Passive Electronic Devices*. 2007, vol. 2, iss. 1, pp. 227–234. ISSN 1555-0281.

[16] CHATURVEDI, B. and J. MOHAN. Single DV-DXCCII based voltage controlled first order all-pass filter with inverting and non-inverting responses. *Iranian Journal of Electrical and Electronic Engineering*. 2015, vol. 11, no. 4, pp. 301–309. DOI: 10.22068/IJEEE.11.4.301.

[17] FABRE, A., O. SAAID and H. BARTHELEMY. On the frequency limitations of the circuits based on second generation current conveyors. *Analog Integrated Circuits Signal Processing*. 1995, vol. 7, iss. 2, pp. 113–129. ISSN 1573-1979. DOI: 10.1007/BF01239166.

[18] GEIGER, R. L., P. E. ALLEN and N. R. STRADER. *VLSI Design Techniques for Analog and Digital Circuits*. New York: McGraw-Hill Publishing Company, 1990. ISBN 0070232539.

[19] RAZAVI, B. *Design of analog CMOS integrated circuits*. Boston, MA: McGraw Hill Edition, 2001. ISBN 0072380322.

[20] WANG, Z. 2-MOSFET transresistor with extremely low distortion for output reaching supply voltages. *Electronics Letters*. 1990, vol. 26, iss. 13, pp. 951–952. ISSN 0013-5194. DOI: 10.1049/el:19900620.

About Authors

Jitendra MOHAN obtained his B.Tech. degree from S. R. T. M. University, Nanded. He got his M.Tech degree from Aligarh Muslim University and Ph.D. degree from Uttarakhand Technical University. He is currently working as an Associate Professor in the Department of Electronics and Communication Engineering at Jaypee Institute of Information Technology, Noida (India). His main areas of interest are current-mode circuits and analog signal processing applications. He has guided 01 Master's Dissertation and 13 projects. He has more than 45 International Journal and conference papers.

Bhartendu CHATURVEDI received B.Tech. degree in Electronics and Communication Engineering and M.Tech. degree in Electronics Engineering, with specialization in Electronic Circuits and System Design He has completed his Ph.D. in Electronics Engineering from Department of Electronics Engineering of Aligarh Muslim University, Aligarh, India. He is currently working as Assistant Professor in the Department of Electronics and Communication Engineering of Jaypee Institute of Information Technology, Noida, India. His research interests include Analog Signal Processing, Circuits and Systems. He has published around 40 research papers in reputed international journals and conferences and also authored 1 book chapter.

17

Continuous Wavelet Transform Analysis of Surface Electromyography for Muscle Fatigue Assessment on the Elbow Joint Motion

TRIWIYANTO[1,3], Oyas WAHYUNGGORO[1], Hanung ADI NUGROHO[1], HERIANTO[2]

[1]Department of Electrical Engineering & Information Technology, Faculty of Engineering,
Universitas Gadjah Mada, Grafika No. 2, Yogyakarta, Indonesia
[2]Department of Mechanical & Industrial Engineering, Faculty of Engineering,
Universitas Gadjah Mada, Grafika No. 2, Yogyakarta, Indonesia
[3]Department of Electromedical Engineering, Politeknik Kesehatan Surabaya,
Pucang Jajar Timur No. 10, Surabaya, Indonesia

triwiyanto123@gmail.com, oyas@ugm.ac.id, adinugroho@ugm.ac.id, herianto@ugm.ac.id

Abstract. *Studying muscle fatigue plays an important role in preventing the risks associated with musculoskeletal disorders. The effect of elbow-joint angle on time-frequency parameters during a repetitive motion provides valuable information in finding the most accurate position of the angle causing muscle fatigue. Therefore, the purpose of this study is to analyze the effect of muscle fatigue on the spectral and time-frequency domain parameters derived from electromyography (EMG) signals using the Continuous Wavelet Transform (CWT). Four male participants were recruited to perform a repetitive motion (flexion and extension movements) from a non-fatigue to fatigue condition. EMG signals were recorded from the biceps muscle. The recorded EMG signals were then analyzed offline using the complex Morlet wavelet. The time-frequency domain data were analyzed using the time-averaged wavelet spectrum (TAWS) and the Scale-Average Wavelet Power (SAWP) parameters. The spectral domain data were analyzed using the Instantaneous Mean Frequency (IMNF) and the Instantaneous Mean Power Spectrum (IMNP) parameters. The index of muscle fatigue was observed by calculating the increase of the IMNP and the decrease of the IMNF parameters. After performing a repetitive motion from non-fatigue to fatigue condition, the average of the IMNF value decreased by 15.69 % and the average of the IMNP values increased by 84 %, respectively. This study suggests that the reliable frequency band to detect muscle fatigue is 31.10–36.19 Hz with linear regression parameters of 0.979 $mV^2 \cdot Hz^{-1}$ and 0.0095 $mV^2 \cdot Hz^{-1}$ for R^2 and slope, respectively.*

Keywords

CWT, elbow joint angle, EMG, muscle fatigue, wavelet.

1. Introduction

In everyday life, when the limb performs an intensive repetitive motion, the muscle can experience muscle fatigue. The muscle fatigue is a condition in which the muscle cannot sustain the force on the given certain task. Muscle fatigue can provide a useful information which needs to be considered in the area of ergonomic, robotic exoskeleton based on EMG control, and sport. Furthermore, the muscle fatigue can be used to prevent the muscle disorder. Several techniques have been used to determine the muscle fatigue, and analyzing the EMG signals are widely used to indicate the muscle fatigue [1]. When the muscle is in the fatigue condition, it is proved that the spectral parameters (frequency and amplitude) of the EMG signal will change [1]. Basmajian and De Luca had reported the results of their study, where during constant force contraction, the amplitude of the EMG signal increased and both mean and median frequency shifted to the lower values [1]. Generally, a conventional method to measure the spectral parameters of the EMG signal is by utilizing the Fast Fourier Transform (FFT) method. In this case, the EMG signal, however, is assumed to be in the stationary condition which the muscle fatigue is determined by performing constant force or isometric contraction [2] on the subject's limb. Determination of the muscle fatigue in the dynamic motion of the limb is closely related to daily activities. During the dynamic motion, the muscle length changes [1] in accordance with the limb joint angle, and for this issue, the non-stationary characteristic of the EMG signal increases [3].

The studies that addressed the muscle fatigue in the dynamic contraction have been conducted by several previous researchers. Gonzales et al. determined the muscle fatigue during a repetitive motion on the knee using the mean and variance of instantaneous frequency based on Choi–William distribution [4]. Chowdhury et al. used Discrete Wavelet Transform (DWT) to determine the muscle fatigue on the neck and shoulder during a repetitive motion [5]. In their study, the changes in the spectral parameter were observed by calculating the DWT coefficients. Karthick and Ramakrishnan proposed a method to observe the progression of the muscle fatigue during the elbow motion in the flexion and extension [6] by utilizing the time-frequency distribution. Triwiyanto et al. proposed the DWT analysis of the EMG signal to conclude which level of decomposition is mostly determined for the muscle fatigue in the dynamic motion [7]. The models used in the previous studies have not discussed the relationship between the elbow joint angle and the time-frequency parameters when the muscle was in the fatigue condition.

It is obvious that the EMG signals have the non-stationary characteristic which means that the frequency of the EMG signal changes by the time. In this case, the EMG signals analysis using the CWT becomes the most suitable method compared to the FFT. Therefore, to address the limitations that have been mentioned in the previous studies, a new method needs to be presented for investigating the relationship between the elbow joint angle, and the spectral and time-frequency parameters of the EMG signal when the muscle is in the non-fatigue and fatigue condition. The purpose of this study is to analyze the effect of the muscle fatigue on the spectral and time-frequency parameters using the CWT. The specific objectives of the study are:

- to calculate the linear regression parameters,

- to investigate the effect of the elbow joint angle on the time-frequency parameters in the non-fatigue and fatigue condition, and

- to test the significant difference of the power spectrum density between non-fatigue and fatigue conditions.

2. Theoretical Background

Wavelet analysis is a method to decompose the signal into several parts of the signal based on wavelet basis function. A wavelet function, $\psi_{\tau,s}(t)$, is built based on a mother wavelet function composed of scaling and translation parameters. In this case, s refers to scaling parameter related to the frequency of the signal and τ is translation parameter. The wavelet function is written as follows [8]:

$$\psi_{\tau,s}(t) = \frac{1}{\sqrt{|s|}} \psi\left(\frac{t-\tau}{s}\right), \quad s, \tau \in \mathbb{R}, s \neq 0. \quad (1)$$

The Continuous Wavelet Transform (CWT) of the signal, $x(t)$, is written as follows [8]:

$$\psi_{\tau,s}(t) = \int_{-\infty}^{\infty} x(t) \frac{1}{\sqrt{|s|}} \psi^*\left(\frac{t-\tau}{s}\right) \mathrm{dt}. \quad (2)$$

The CWT function on Eq. (2) is composed of the scaling, translation, wavelet function $\psi\left(\frac{t-\tau}{s}\right)$, and the signal $x(t)$. In order to analyze the EMG signal, the function $x(t)$ in Eq. (2) can be substituted by the EMG signals. In this study, the Morlet mother wavelet was used to implement the CWT. The complex-Morlet mother wavelet is defined as follows [8]:

$$\psi = \pi^{-\frac{1}{4}} \left(e^{i2\pi f_0 t} - e^{-\frac{(2\pi f_0)^2}{2}}\right) e^{-\frac{t^2}{2}}, \quad (3)$$

where f_0 indicates the Morlet frequency constant ($f_0 = 0.849$).

The local Power Spectrum Density (PSD) of the CWT for a certain scale range is measured using the Scaled-Average Wavelet Power (SAWP), as shown in Eq. (4) [9]:

$$\overline{W}_n^2 = \frac{\delta_s \delta_\tau}{C_\delta} \sum_{s=s_1}^{s_2} \frac{|W_n(s)|^2}{s}, \quad (4)$$

where s_1 and s_2 indicate the ranges of the CWT scale, C_δ indicates the coefficient of the mother wavelet, δ_s is the increment of the scale, δ_τ is the time translation of the mother wavelet and W_n is the CWT coefficient. The global scale PSD is measured for all ranges of the scale. A local PSD for the short period of time is measured using the Time-Averaged Wavelet Spectrum (TAWS) described as follows [9]:

$$\overline{W}^2(s) = \frac{1}{N} \sum_{n=0}^{N-1} |W_n(s)|^2, \quad (5)$$

where W_n is the CWT coefficient, and n and N are the specific time range to average the CWT coefficients.

3.　Materials and Method

3.1.　Participants

Four healthy male volunteers (age: 22.4 ± 3.2 years old, weight: 65.4 ± 5.6 kg, height: 169 ± 4.2 cm) who had no history of the muscular disorder were recruited in this study. Before the data collection, the subjects were recommended not to do any hard work that could harm the elbow joint. Furthermore, the subjects were given the explanation how to do the movements of flexion and extension and told any potential risks that might occur during the experiment.

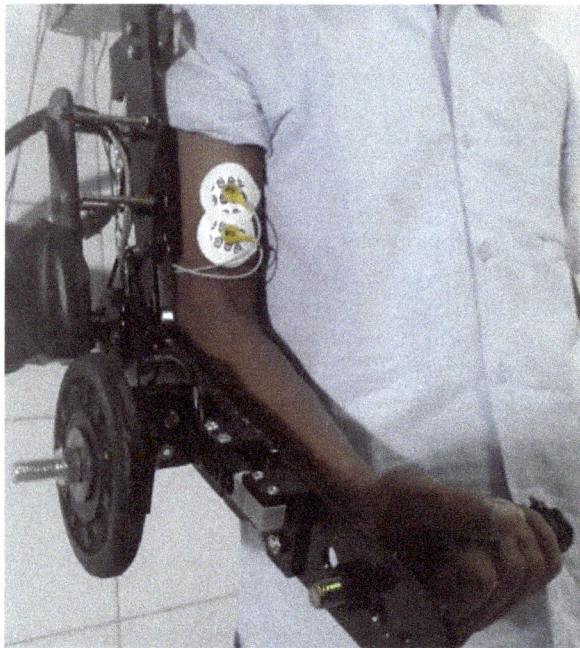

Fig. 1: The exoskeleton frame to synchronize the elbow joint motion (flexion and extension).

3.2.　Equipment

In this study, the EMG signal was collected using one channel EMG system consists of pre-amplifier, band pass filter with the cut-off frequency of 20 and 500 Hz, adjustable gain amplifier and summing amplifier. The EMG signal was collected using three surface electrodes (Ag/AgCl, size: 57×48 mm, Ambu, Bluesensor R, Malaysia). Two electrodes were placed on biceps muscle and one electrode was placed on hand as a common ground electrode. An exoskeleton frame was used to synchronize the elbow joint motion (Fig. 1). The elbow joint angle was collected using a linear potentiometer. One kilogram of the load was placed on the edge of the exoskeleton frame.

3.3.　Data Collection

The EMG signal covers frequency in the range of 0 to 500 Hz, while the range of the dominant frequency falls between 50 and 150 Hz [10]. The EMG signal and elbow joint position were collected with a sampling frequency of 1000 Hz [11] and [12]. The choice of this sampling frequency was in accordance with the Nyquist rule [13]. Furthermore, the application program developed using Borland Delphi Professional (Version 7.0, Borland Software Corporation, Scotts Valley, California, USA) was used to acquire the EMG signal and CWT analysis.

During the flexion and extension motion from 0 to 120 degrees, the EMG signal and the elbow joint angle were recorded for off-line data analysis. In the data collection process, subjects were instructed to perform repetitive motion until they were perceived in fatigue conditions. The condition was indicated by the subject's elbow that could not perform the flexion and extension movements.

3.4.　Data Processing

The EMG signal and angle data were processed off-line using the Borland Delphi Professional (Version 7.0, Borland Software Corporation, Scotts Valley, California, USA), see Fig. 2. Four cycles of flexion and extension motion assigned by T1, T5, T10, and T15 were selected from the EMG data. The T1, T5, T10, and T15 were the cycles that were measured in the first minute, fifth minute, tenth minute, and last minute, respectively. T1 would be assumed as a time location of the non-fatigue condition and T15 denoted as a time location of the fatigue condition. Each cycle of motion was analysed using the CWT. In this study, the CWT implementation used the complex Morlet wavelet which was widely accepted in the EMG analysis [14], [15] and [16]. The coefficients of the CWT were calculated with the scale of 100 with the delta scale (δ_s) of 0.002. The mother wavelet was shifted with the total translation (τ_{total}) of 360. The delta translation (δ_τ) is calculated as follows:

$$\delta_\tau = \frac{(N-1)}{f_s \tau_{\text{total}}}, \qquad (6)$$

where f_s is the frequency sampling in Hz, N is the number of the sample point and τ_{total} is the number of total translation. The following time-frequency and spectral parameters were calculated:

- Scale-Average Wavelet Power (SAWP) is the average of PSD for the specific range of the scale band. In this study, the scales range to be analysed was from 0.002 to 0.052. The scales were divided into 10 bands in which each band has the scale length of 0.005. These ten bands were denoted by the

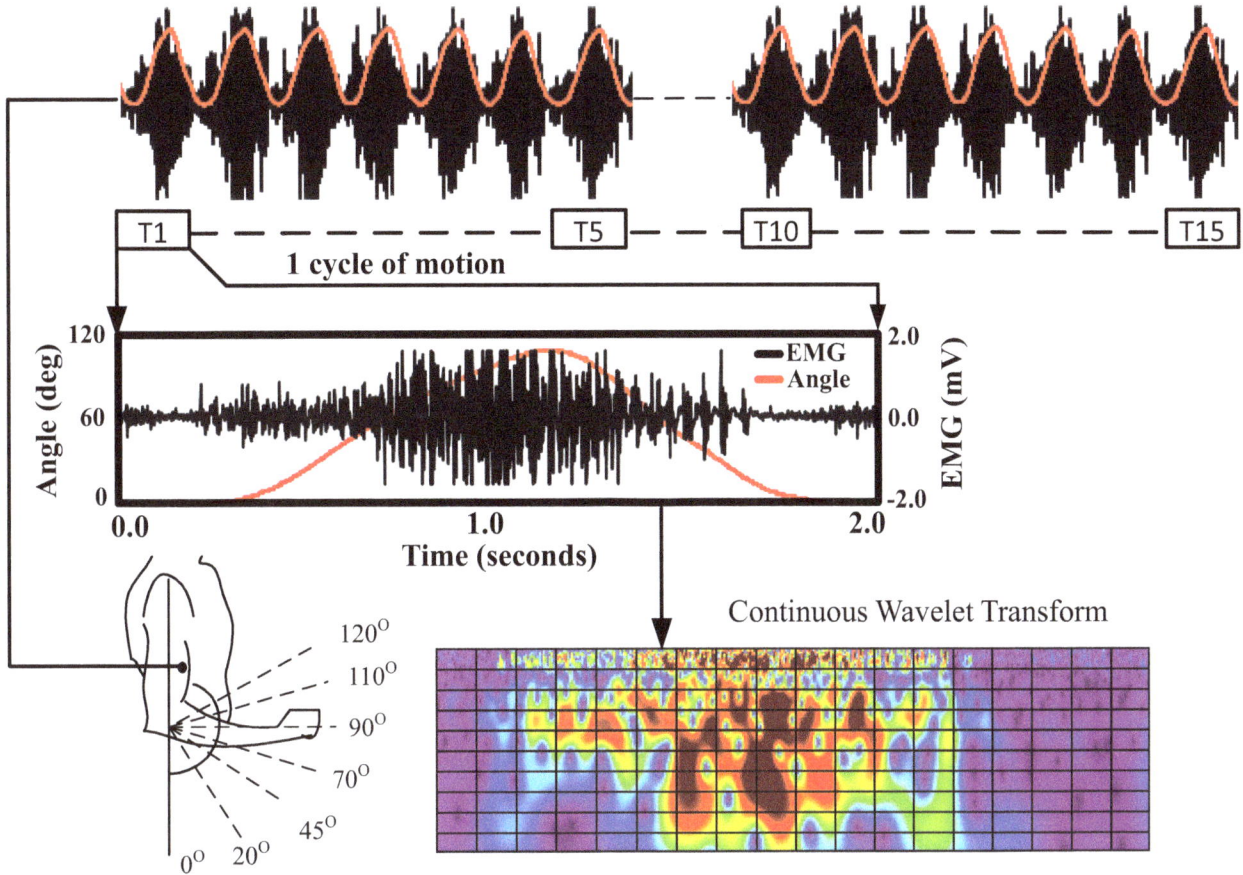

Fig. 2: The illustration of the EMG data processing. Four cycles of the EMG signal indicated by T1, T5, T10, and T15 was analyzed using the CWT. The black line is the EMG signal and the red line is the elbow joint angle.

1st to 10th bands (for example, the 1st band had the scale range of 0.002 to 0.007). The SAWS was calculated based on Eq. (4).

- Time-Average Wavelet Spectrum (TAWS) is the average of PSD for the specific length of the translation. In this study, the number of translation was 360. If the translation length was 20, the number of the TAWS was 18. The TAWS is calculated based on Eq. (5).

- Instantaneous Mean Frequency (IMNF) [6] is calculated to obtain the mean power of frequency at an instant time. The IMNF is formulated as follows:

$$IMNF(\tau) = \frac{\sum\limits_{f_1}^{f_2} f \, |CWT(s,\tau)|^2}{\sum\limits_{f_1}^{f_2} |CWT(s,\tau)|^2}, \qquad (7)$$

where f_1 denotes the lowest frequency and f_2 denotes the highest frequency and $|CWT|$ is the absolute of the CWT coefficients. The f_1 and f_2 were determined based on the frequency range of the processed EMG signal ($f_1 = 20$ Hz and $f_2 = 500$ Hz).

- Instantaneous Mean Power (IMNP) is the mean of PSD at an instant time [6].

$$IMNP = \frac{1}{N} \sum |CWT(s,\tau)|^2, \qquad (8)$$

where N indicates the number of scale and $|CWT|$ indicates the absolute of the CWT coefficients.

3.5. Statistical Analysis

Muscle fatigue affected the spectral and time-frequency parameters of the EMG signal. The significant difference of the parameters was examined using the one way single factor ANOVA with the confidence level of 95 % for T1, T5, T10 and T15 cycle. The effect of the muscle fatigue was significantly indicated by p-value. If the p-value is less than 0.05, then it indicates that the muscle fatigue has significantly affected.

4. Results and Discussion

In consideration to the purpose of the study, the change in the spectral parameters (the frequency and magni-

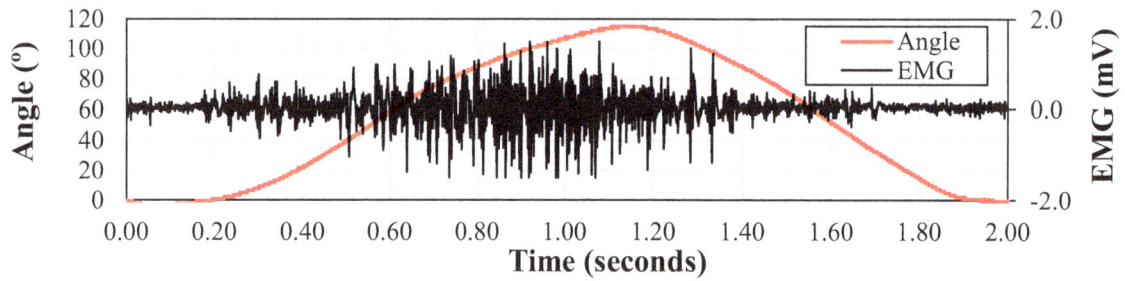

(a) The EMG signal during the flexion and extension motion with the motion period of 2 seconds (the black line is the EMG signal and the red line is the elbow joint angle).

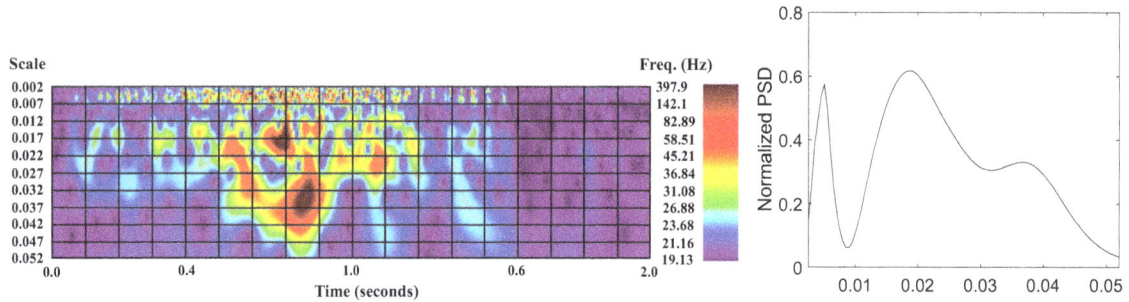

(b) The contour of the CWT coefficients.

(c) The global wavelet spectrum.

Fig. 3: The typical example of one cycle of the flexion and extension motion which was measured at the first cycle (T1) in the non-fatigue condition.

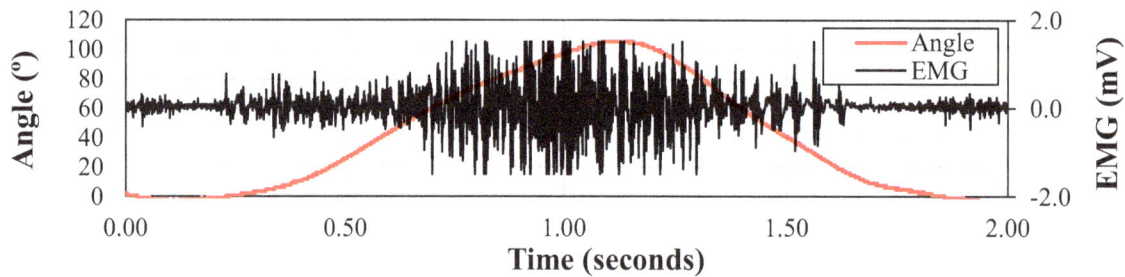

(a) The EMG signal during the flexion and extension motion with the motion period of 2 seconds (the black line indicates the EMG signal and the red line indicates the elbow joint angle).

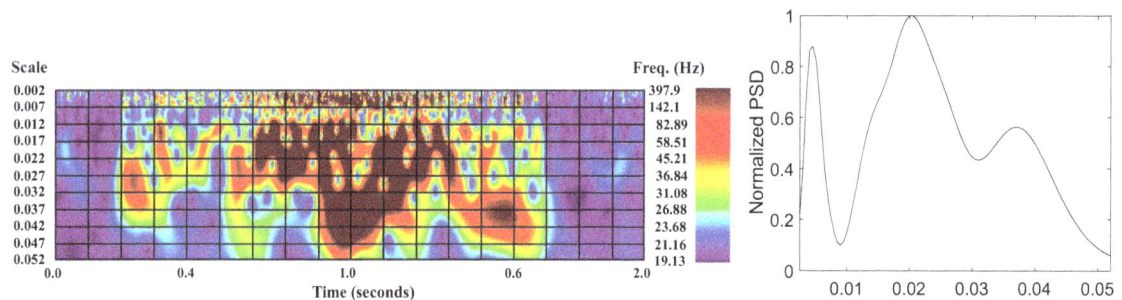

(b) The contour of the CWT coefficients.

(c) The global wavelet spectrum.

Fig. 4: The typical cycle of flexion and extension motion which was measured in the last cycle (T15) in the fatigue condition.

tude of power spectrum) was analyzed using the CWT. In the non-fatigue condition (Fig. 3), the averages of the IMNF and IMNP for all of the scale and time are 70.37 ± 10.16 Hz and 0.0165 ± 0.0094 mV^2, respec-

tively. After a repeated flexion and extension movements, the muscle will be in the fatigue condition and the magnitude of the power spectrum increased significantly. Thus, the contour of the wavelet power is wider compared to that in the non-fatigue condition, see Fig. 4(b). Therefore, the averages of the spectral parameters also changed. The averages of the IMNF and IMNP for all of the scale and time in the fatigue condition are 60.83 ± 13.95 Hz and 0.1039 ± 0.084 mV^2, respectively. The average of the IMNF decreased by 15.69 % and the average of the IMNP increased by 84.14 %. These results, i.e. the decrease of the frequency and the increase of the power, are in line with the previous study conducted by Basmajian and De Luca [1]. Similar results were also observed by Karthick and Krishnan which showed that the IMNF and IMDF of EMG signal decrease from non-fatigue to fatigue condition [6].

The Global Wavelet Spectrum (GWS) was calculated based on the average of the PSD for the total time range of 2 seconds as shown in Fig. 5.

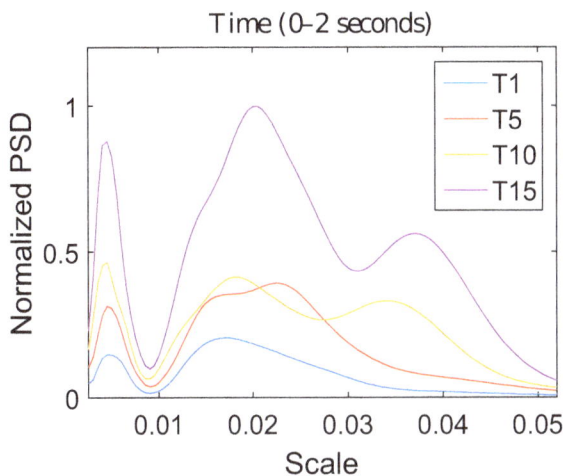

Fig. 5: The global wavelet spectrum for the 2 second duration.

Figure 5 shows that the highest magnitude of the global wavelet spectrum was in the center of scales: 0.017 (58.54 Hz), 0.0175 (56.87 Hz), 0.0185 (53.8 Hz), and 0.020 (49.76 Hz) for the time of T1, T5, T10 and T15, respectively. The GWS in the scale range of 0.002 to 0.008 showed a consistent increase from non-fatigue (T1) to fatigue condition (T15). The time of T1 shows the lowest magnitude of the GWS. It was the first cycle of the EMG measurement (at first minute) and was assumed as a non-fatigue condition. It was followed by T5, T10, and T15. For all of the scale, T15 had the highest GWS. The average of PSD from T1 to T15 increased significantly ($p < 0.05$) about 84.17 %. It was also observed in Fig. 5 in which the frequency changed to lower frequency by 10.48 %, from non-fatigue to fatigue condition. This phenomenon was in line with the Basmajian and de Luca's report. In their study on the assessment of the muscle fatigue, the frequency changed to the lower value and the amplitude increased significantly [1].

The Time-Averaged Wavelet Spectrum (TAWS) and the Scale-Averaged Wavelet Spectrum (SAWS) was calculated to find the specific time location of the spectral parameters.

4.1. Time-Averaged Wavelet Spectrum

Time-Averaged Wavelet Spectrum (TAWS) is calculated according to Eq. (5). The TAWS was calculated for a specific range of time, which was aimed to find the specific time and frequency location related to the elbow joint angle and the muscle fatigue condition. In order to perform the TAWS calculation, the 2 seconds period of the cycle was divided into 18 ranges of time.

Figure 6 shows that the magnitude of the PSD in the time ranges of 0.006–0.111 seconds, 0.117–0.222 seconds, 0.228–0.333 seconds, 0.339–0.444 seconds, 0.450–0.556 seconds, 1.672–1.778 seconds, 1.783–1.889 seconds, and 1.894–2 seconds are smaller compared to the others time range. In the time range of 0–0.556 seconds (Fig. 6(a), Fig. 6(b), Fig. 6(c), Fig. 6(d) and Fig. 6(e)), the TAWS of T1, T5, T10, and T15 showed small PSD ($< \sim 0.2$), except in the time range of 0.561–0.667 sec-

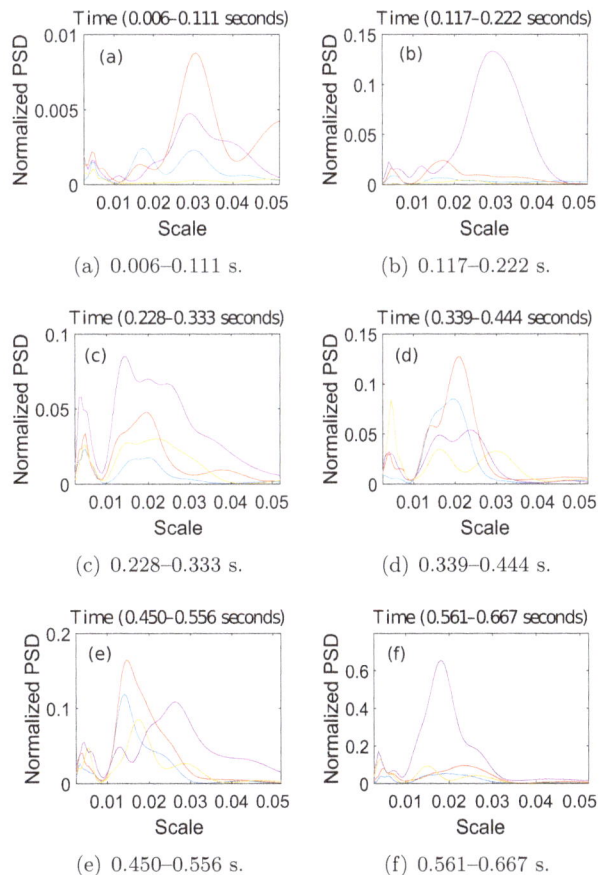

(a) 0.006–0.111 s. (b) 0.117–0.222 s.

(c) 0.228–0.333 s. (d) 0.339–0.444 s.

(e) 0.450–0.556 s. (f) 0.561–0.667 s.

Fig. 6: The TAWS for the time range.

onds. This was in accordance with the angle of the elbow joint at 50–70°. In the fatigue condition (T15), the TAWS tended to show a higher PSD than T1, T5 and T10 (Fig. 6(b), Fig. 6(c) and Fig. 6(f)).

Figure 7 shows the shift of the PSD to the higher scale value, from 0.017 (58.54 Hz) to 0.022 (45.24 Hz), from non-fatigue to fatigue condition. The TAWS increased significantly (p-value< 0.05) from T1 to T5, T1 to T10 and T1 to T15. Among the Fig. 7, Fig. 8, Fig. 9 and Fig. 10, Fig. 10 shows the highest PSD at the scale of 0.022 (equal to the frequency of 46.29 Hz).

The TAWS increased significantly (p-value < 0.05) from T1 to T5, T1 to T10 and T1 to T15. This highest PSD was related to the elbow-joint angle at 108° to 110°. The similar finding has also been reported by Chowdhury and Nimbarte. They observed the significant increase of PSD in the frequency band of 23–46 Hz and 46–93 Hz [17]. Figure 7, Fig. 8, Fig. 9 and Fig. 10, obviously, show that the position of the elbow joint affects the PSD. On the contrary, Doheny et al. reported

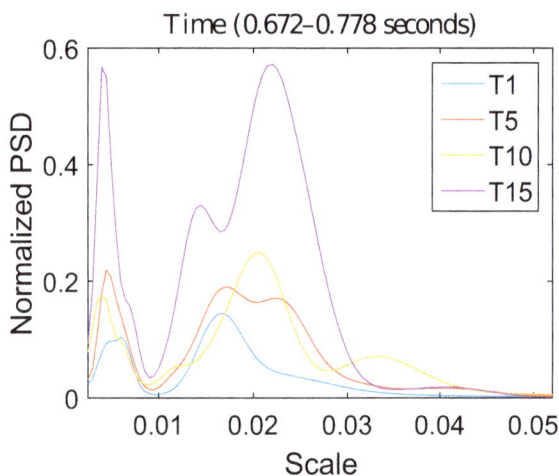

Fig. 7: The TAWS for the time range of 0.672–0.778 seconds.

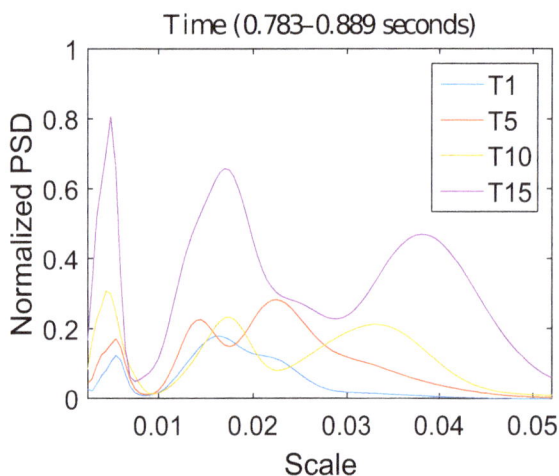

Fig. 8: The TAWS for the time range of 0.783–0.889 seconds.

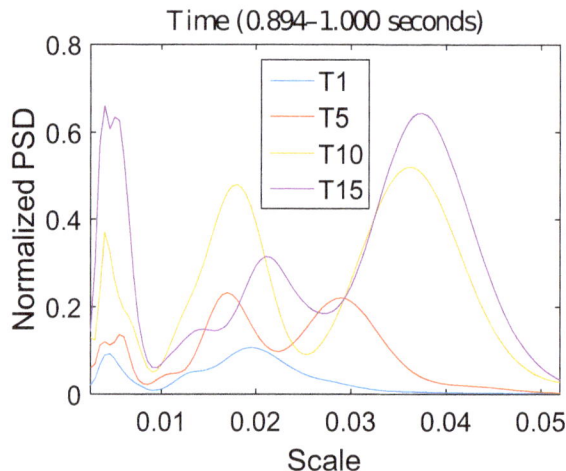

Fig. 9: The TAWS for the time range of 0.894–1 seconds.

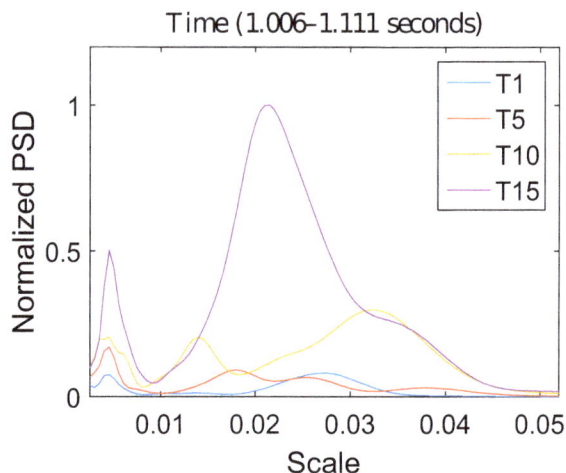

Fig. 10: The TAWS for the time range of 1.006–1.111 seconds.

that there was no relationship between the elbow joint angle and power or EMG amplitude [3]. This difference was due to the different technique in analyzing the EMG signals. They observed the relationship between the elbow joint angle and EMG amplitude using the amplitude-based parameters.

4.2. Scale-Averaged Wavelet Power

The Scale-Averaged Wavelet Power (SAWP) feature is calculated using Eq. (4). SAWP is used to test the fluctuation of PSD in the time series for specific frequency bands. By using this feature, the dominant PSD can be found in the specific time and elbow joint angle. In order to perform this calculation, the frequency was divided into ten frequency bands including 1st band (19.4–20.95 Hz), 2nd band (21.17–23.42 Hz), 3rd band (23.7–26.54 Hz), 4th band (26.9–30.62 Hz), 5th band (31.10–36.19 Hz), 6th band (36.86–44.23 Hz), 7th band (45.24–56.87 Hz), 8th band (58.54–79.62 Hz), 9th band (82.9–132 Hz) and 10th band (142–497 Hz).

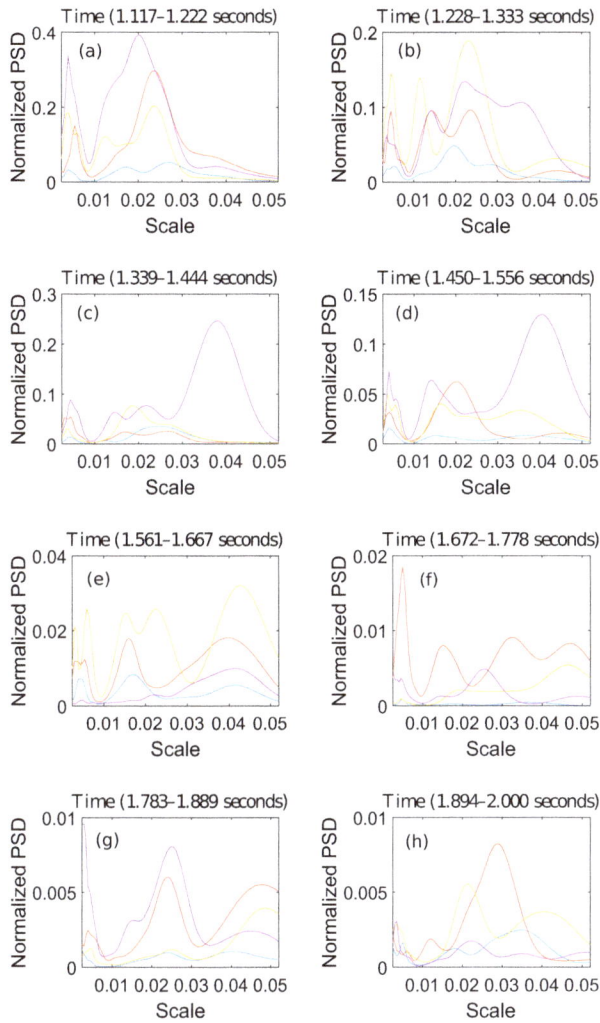

Fig. 11: The TAWS for the time range of 1.117–2 seconds.

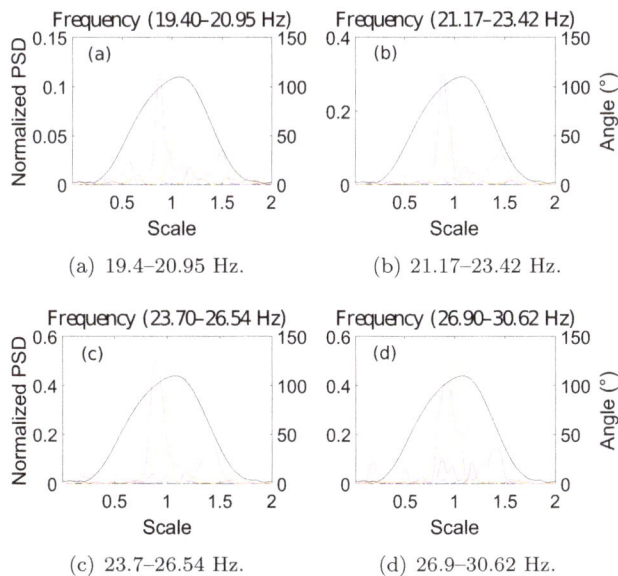

(a) 19.4–20.95 Hz.

(b) 21.17–23.42 Hz.

(c) 23.7–26.54 Hz.

(d) 26.9–30.62 Hz.

Fig. 12: The SAWP for the frequency band.

Figure 12, Fig. 13, Fig. 14, Fig. 15 and Fig. 17 show that the SAWP with the maximum PSD, in the fatigue condition (T15), was mostly found in the flexion motion at the time range of 0.8 to 1.03 seconds (at the

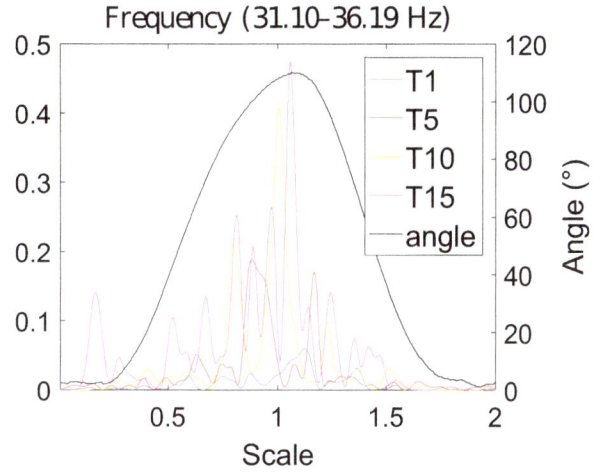

Fig. 13: The SAWP for the frequency band of 31.10–36.19 Hz.

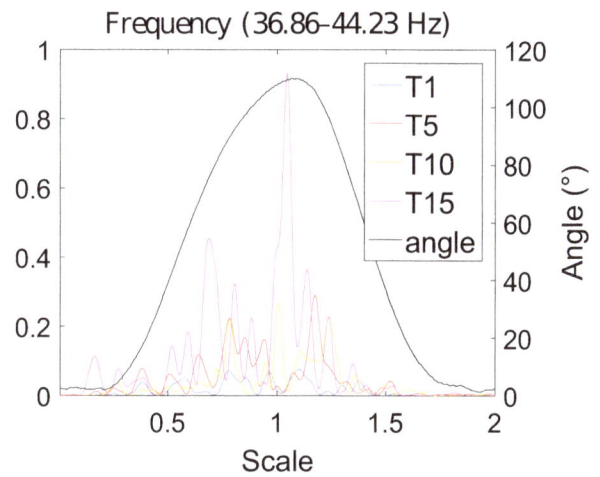

Fig. 14: The SAWP for the frequency band of 36.86–44.23 Hz.

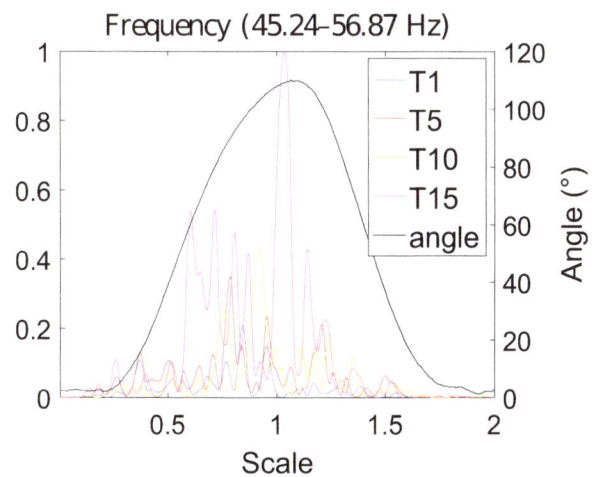

Fig. 15: The SAWP for the frequency band of 45.24–56.87 Hz.

angle of 97°–110°). Figure 16 shows that the PSD of the SAWP, in the fatigue condition, were found at several locations from 0.6 to 1.2 seconds. In this case, we could not use this band (58.54–79.62 Hz) to localize the muscle fatigue in the certain time and angle.

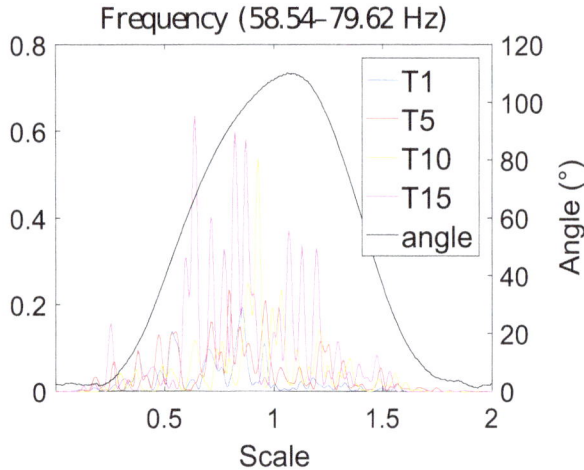

Fig. 16: The SAWP for the frequency band of 58.54–79.62 Hz.

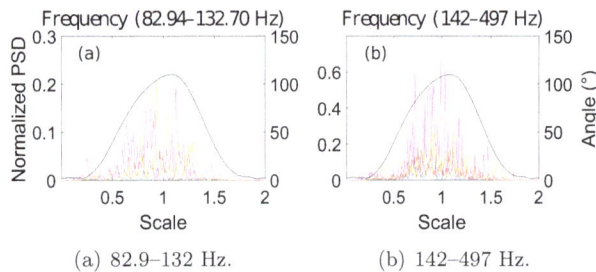

(a) 82.9–132 Hz.   (b) 142–497 Hz.

Fig. 17: The SAWP for the frequency band.

As shown in Fig. 12, Fig. 13, Fig. 14, Fig. 15, Fig. 16 and Fig. 17, each frequency band has different SAWP. The reliable frequency band, that could be effectively used to detect the muscle fatigue, was determined using the determination coefficient (R^2) of linear regression calculated from the average of SAWP for all the time. The linear regressions parameters (Slope, R^2, and Intercept) were calculated using the average of the SAWP for T1, T5, T10, and T15. In the frequency band of 31.10–36.19 Hz, the R^2 shows the highest value (0.979), see Tab. 1. The positive value in the R^2 indicated that the PSD magnitude increased linearly by the time. These results are similar to Karthick and Ramakrishnan's finding [6]. In their study, they used the IMNP to observe the effect of the muscle fatigue and obtained $R^2 = 0.57$ and slope = 0.0039 mV^2· Hz^{-1}. The increase of the PSD magnitude in this band was also similar to Chowdury and Nimbrate's finding. They found that in the frequency range of 23–46 Hz and 46–93 Hz, the magnitude of the power spectrum increased significantly.

Tab. 1: The summary of linear regression parameters for all the frequency bands.

Frequency Band (Hz)	R^2	Slope (mV^2·Hz^{-1})	Intercept (mV^2·Hz^{-1})
19.40–20.95	0.919	0.0020	0.0016
21.17–23.42	0.876	0.0058	0.0066
23.70–26.54	0.910	0.0116	0.0139
26.90–30.62	0.964	0.0123	0.0127
31.10–36.19	0.979	0.0095	0.0048
36.86–44.23	0.783	0.0135	0.0056
45.24–56.87	0.815	0.0169	0.0076
58.54–79.62	0.864	0.0106	0.0020
82.90–132.0	0.975	0.0039	0.0016
142.0–497.0	0.948	0.0111	0.0053

5.   Conclusion

This study investigated the effect of muscle fatigue quantitatively on the spectral and time-frequency parameters of the EMG signal using CWT. It was found that when the muscle was in the fatigue condition, the spectral parameters of the EMG signal changed. The average of the IMNF decreased by 15.69 % and the average of the IMNP increased by 84.14 %. The optimum fatigue detection was located at the time range of 1.006 to 1.111 seconds related to the elbow joint angle in the range of 108° to 110°. These findings suggest that the CWT analysis with SAWP and TAWS features can determine the specific frequency range, time location, and elbow joint angle that is most affected when the muscle is in fatigue condition.

References

[1] BASMAJIAN, J. V. and C. J. DE LUCA. Muscle Fatigue and Time-Dependent Parameters of the Surface EMG Signal. In: *Muscles alive: their functions revealed by electromyography.* Baltimore: Williams & Wilkins, 1985, pp. 201–222. ISBN 04-716-7580-6.

[2] MERLETTI, R. and P. PARKER. *Electromyography: Physiology, Engineering, and Non-Invasive Applications.* Hoboken, NJ: John Wiley & Sons, Inc., 2004. ISBN 0-471-67580-6.

[3] DOHENY, E. P., M. M. LOWERY, D. P. FITZ PATRICK and M. J. O'MALLEY. Effect of elbow joint angle on force-EMG relationships in human elbow flexor and extensor muscles. *Journal of Electromyography and Kinesiology.* 2008, vol. 18, no. 5, pp. 760–770. ISSN 1050-6411. DOI: 10.1016/j.jelekin.2007.03.006.

[4] GONZALEZ-IZAL, M., A. MALANDA, I. N. AMEZQUETA, E. M. GOROSTIAGA, F. MALLOR, J. IBANEZ and M. IZQUIERDO.

EMG spectral indices and muscle power fatigue during dynamic contractions. *Journal of Electromyography and Kinesiology*. 2010, vol. 20, no. 2, pp. 233–240. ISSN 1050-6411. DOI: 10.1016/j.jelekin.2009.03.011.

[5] CHOWDHURY, S. K., A. D. NIMBARTE, M. JARIDI and R. C. CREESE. Discrete wavelet transform analysis of surface electromyography for the fatigue assessment of neck and shoulder muscles. *Journal of Electromyography and Kinesiology*. 2013, vol. 23, no. 5, pp. 995–1003. ISSN 1050-6411. DOI: 10.1016/j.jelekin.2013.05.001.

[6] KARTHICK, P. A. and S. RAMAKRISHNAN. Surface electromyography based muscle fatigue progression analysis using modified B distribution time-frequency features. *Biomedical Signal Processing and Control*. 2016, vol. 26, iss. 1, pp. 42–51. ISSN 1746-8108. DOI: 10.1016/j.bspc.2015.12.007.

[7] TRIWIYANTO, O. WAHYUNGGORO, H. A. NUGROHO and HERIANTO. DWT Analysis of sEMG for Muscle Fatigue Assessment of Dynamic Motion Flexion-Extension of Elbow Joint. In: *8th International Conference on Information Technology and Electrical Engineering (ICITEE)*. Yogyakarta: IEEE, 2016, pp. 1–6. ISBN 97-815-0904-139-8. DOI: 10.1109/ICITEED.2016.7863300.

[8] ADDISON, P. S. *The Illustrated Wavelet Transform Handbook: Introductory Theory and Applications in Science, Engineering, Medicine and Finance*. New York: Taylor & Francis, 2002. ISBN 97-807-5030-692-8.

[9] TORRENCE, C. and G. P. COMPO. A Practical Guide to Wavelet Analysis. *Bulletin of the American Meteorological Society*. 1998, vol. 79, no. 1, pp. 61–78. ISSN 1520-0477. DOI: 10.1175/1520-0477(1998)079<0061:APGTWA>2.0.CO;2.

[10] DE LUCA, G. Fundamental Concepts in EMG Signal Acquisition. In: *Delsys* [online]. 2003, pp. 1–31. Available at: http://delsys.com/Attachments_pdf.

[11] ROGERS, D. R. and D. T. MACISAAC. EMG-based muscle fatigue assessment during dynamic contractions using principal component analysis. *Journal of Electromyography and Kinesiology*. 2011, vol. 21, no. 5. pp. 811–818. ISSN 1873-5711. DOI: 10.1016/j.jelekin.2011.05.002.

[12] WINSLOW, J., P. L. JACOBS and D. TEPAVAC. Fatigue compensation during FES using surface EMG. *Journal of Electromyography and Kinesiology*. 2003, vol. 6, no. 6, pp. 555–568. ISSN 1050-6411. DOI: 10.1016/S1050-6411(03)00055-5.

[13] TAN, L. and J. JIANG. *Digital Signal Processing: Fundamental and Applications*. Boston: Academic Press. ISBN 978-0123740908.

[14] LEAO, R. N. and J. A. BURNE. Continuous wavelet transform in the evaluation of stretch reflex responses from surface EMG. *Journal of Neuroscience Methods*. 2004, vol. 133, no. 1–2, pp. 115–125. ISSN 0165-0270. DOI: 10.1016/j.jneumeth.2003.10.003.

[15] BASTIAENSEN, Y., T. SCHAEPS and J. P. BAEYENS. Analyzing an sEMG signal using wavelets. In: *4th European Conference of the International Federation for Medical and Biological Engineering*. Berlin: Springer, 2008, pp. 156–159. ISBN 978-35-4089-207-6. DOI: 10.1007/978-3-540-89208-3_39.

[16] DE MICHELE, G., S. SELLO, M. C. CARBONCINI, B. ROSSI and S. K. STRAMBI. Cross-correlation time-frequency analysis for multiple EMG signals in Parkinson's disease: A wavelet approach. *Medical Engineering & Physics*. 2003, vol. 25, no. 5, pp. 361–369. ISSN 1350-4533. DOI: 10.1016/S1350-4533(03)00034-1.

[17] CHOWDHURY, S. K. and A. D. NIMBARTE. Comparison of Fourier and Wavelet Analysis for Fatigue Assessment During Repetitive Dynamic Exertion. *Journal of Electromyography and Kinesiology*. 2015, vol. 25, no. 2. pp. 5–13. ISSN 1050-6411. DOI: 10.1016/j.jelekin.2014.11.005.

About Authors

TRIWIYANTO was born in Surabaya, Indonesia. He received his M.Sc. degree in Electronic Engineering in Institute of Technology Sepuluh Nopember in 2004, Surabaya, Indonesia. He is currently a Ph.D. candidate in Electrical Engineering at Gadjah Mada University, Yogyakarta, Indonesia. His research interests include biomedical signal analysis, embedded system, electronic instrumentation, assistive and rehabilitation devices.

Oyas WAHYUNGGORO was born in Jogjakarta, Indonesia. He received his Ph.D. degree in Electrical and Electronic Engineering from the Universiti Teknologi Petronas, Malaysia in 2011. His research interests include biomedical signal processing, intelligent system, and control system.

Hanung ADI NUGROHO was born in Jogjakarta, Indonesia. He received his Ph.D. degree in Electrical and Electronic Engineering from the Universiti Teknologi Petronas, Malaysia in 2012. His research interests include biomedical signal processing and image processing.

HERIANTO was born in Jogjakarta, Indonesia. He received his D.Eng. in the Department of Mechanical and Control Engineering, Tokyo Institute of Technology, Japan, in 2009. His research interests include robotics and manufacture. His current research is product design and development especially in rehabilitation robot.

18

An Investigation Into Time Domain Features of Surface Electromyography to Estimate the Elbow Joint Angle

TRIWIYANTO[1,3], Oyas WAHYUNGGORO[1], Hanung Adi NUGROHO[1], HERIANTO[2]

[1]Department of Electrical Engineering & Information Technology, Faculty of Engineering,
Universitas Gadjah Mada, Jl. Grafika 2, 55281 Yogyakarta, Indonesia
[2]Department of Mechanical & Industrial Engineering, Faculty of Engineering,
Universitas Gadjah Mada, Jl. Grafika 2, 55281 Yogyakarta, Indonesia
[3]Department of Electromedical Engineering, Politeknik Kesehatan Surabaya,
Ministry of Health Indonesia Pucang Jajar Timur 10, 60282 Surabaya, Indonesia

triwiyanto123@gmail.com, oyas@ugm.ac.id, adinugroho@ugm.ac.id, herianto@ugm.ac.id

Abstract. *In literature, it is well established that feature extraction and pattern classification algorithms play essential roles in accurate estimation of the elbow joint angle. The problem with these algorithms, however, is that they require a learning stage to recognize the pattern as well as capture the variability associated with every subject when estimating the elbow joint angle. As EMG signals can be used to represent motion, we developed a non-pattern recognition method to estimate the elbow joint angle based on twelve time-domain features extracted from EMG signals recorded from bicep muscles alone. The extracted features were smoothed using a second order Butterworth low pass filter to produce the estimation. The accuracy of the estimated angles was evaluated by using the Pearson's Correlation Coefficient (PCC) and Root Mean Square Error (RMSE). The regression parameters (Euclidean distance, R^2 and slope) were then calculated to observe the effect of the features on elbow joint angle estimation. In this investigation, we found that for a 10-second long recording period, the MyoPulse Percentage (MYOP) Rate produced the best accuracy: with PCC of 0.97 ± 0.02 (Mean±SD) and RMSE of $11.37 \pm 3.04°$ (Mean±SD), respectively. The MYOP feature also showed the highest R^2 and slope value of 0.986 ± 0.0083 (Mean's) and 0.746 ± 0.17 (Mean's), respectively for flexion and extension motions during all recorded periods.*

Keywords

EMG, feature extraction, non-pattern recognition, time domain features.

1. Introduction

Surface ElectroMyoGraphy (EMG) is often used to control an assist device such as the upper and lower limb exoskeletons with the function to support human life [1]. It is obvious that the EMG signal can be related to the human limb motion. Several efforts on EMG signal detection have been made to investigate the relationships between muscle groups and limb movement [2] and [3]. In the EMG detection stage, Tang et al. [4] collected EMG signal from four muscle groups located at biceps brachii, brachioradialis, triceps, and anconeus to estimate the elbow joint angle. Benitez et al. [5] recorded the EMG signals from two muscle groups located at biceps and triceps to develop an orthotic system. The methods that utilize more muscle groups in estimating the elbow joint angle, however, would require more computational complexities in data processing.

In order to get information related to elbow joint motion, the recorded EMG signal should be processed by using time, frequency, or time-frequency domain methods to produce informative features. After the feature extraction stage, the EMG features can represent useful information related to the joint angle, force, and torque. Choosing an appropriate feature is essential because it determines the accuracy of the estimation. Some previous studies have preferred to use time domain features over those extracted from frequency and time-frequency domains to predict joint angle [6] and [7] and torque [8]. This preference is due to reduced complexity in data processing and the application of a simple algorithm to be implemented in the real time control. Generally, after the feature extraction process, the joint angle or torque is estimated using a machine learning algorithm or a classifier to improve the accuracy. The methods used in human-machine interaction based on EMG control, are divided into two categories: pattern recognition and non-pattern recognition [1] methods. In the pattern recognition methods, some previous studies used artificial neural networks [4], fuzzy controllers [1], and support vector machines [10] as their classifiers. The limitation in the pattern recognition methods, however, is that the system needs to be trained for each different subject due to the variability in the EMG signal. Therefore, in some cases, this method is not practically applicable. In the non-pattern recognition methods, some previous studies used onset analysis, proportional control, and threshold control [11] and [12]. These methods are simple to be implemented but their accuracies tend to be low. There is also limited literature on elbow-joint angle estimation using non-pattern recognition methods.

Although some efforts have been dedicated to pattern recognition and non-pattern recognition methods for elbow-joint angle estimation, there are still some limitations that should be addressed in furthering this research. Therefore, the purpose of this study is to develop a non-pattern recognition method for estimating the elbow joint angle using a single muscle group (biceps). To implement the proposed method, twelve time-domain features were investigated and a second-order Butterworth low pass filter was applied to filter the features. The specific objectives of the study are to:

- evaluate the accuracy of EMG features in estimating the elbow joint angle using the Pearson's Correlation Coefficient (PCC) and the Root Mean Square Error (RMSE),

- evaluate the regression parameters (Euclidean distance, R-squared, and slope) that relate to the elbow joint angle.

2.   Theoretical Background

2.1.   Time Domain Features

The recorded EMG signal was extracted to get the features that related to the human elbow-joint angle during flexion and extension motions. In this study, twelve Time-Domain (TD) features were extracted to estimate the elbow joint angle. These features were classified into three categories (based on energy, complexities, and frequency information) [13]. The energy-based features were as follows: the Root Mean Square

(RMS), Integrated EMG (IEMG), Variance (VAR), and Mean Absolute Value (MAV). The complexity of the EMG signal could be quantified by using the Waveform Length (WL), Average Amplitude Change (AAC), and Difference Absolute Standard Deviation Value (DASDV) features. The calculated frequency-based informative features were as follows: Zero Crossing (ZC), Sign Slope Change (SSC), Wilson Amplitude (WAMP), and MYOPulse Percentage (MYOP) Rate.

1)   RMS

The Root Mean Square (RMS) value represents the mean power of a signal over a window length of EMG samples. The mathematical equation to describe this feature is as follows [14]:

$$RMS = \sqrt{\frac{1}{N} \sum_{i=1}^{N} x_i^2}, \tag{1}$$

where x_i indicates the ith EMG signal and N indicates the length of the EMG signal.

2)   IEMG

The Integrated EMG (IEMG) value is an absolute summation of the EMG signal over a window length of EMG samples. The mathematical equation is described as follows [14]:

$$IEMG = \sum_{i=1}^{N} |x_i|. \tag{2}$$

3)   VAR

The Variance of the EMG signal, EMG (VAR), is the average value of the power of the EMG signal. VAR is formulated as follows [14]:

$$VAR = \frac{1}{N-1} \sum_{i=1}^{N} |x_i^2|. \tag{3}$$

4)   MAV

The Mean Absolute Value (MAV) is the average of the absolute value of the EMG signal for a window length N. The MAV is formulated as [14]:

$$MAV = \frac{1}{N} \sum_{i=1}^{N} |x_i|. \tag{4}$$

5) LOG

The Logarithm (LOG) parameter is a measure of the non-linear characteristic of the EMG signal. The LOG value is calculated based on the average of the logarithm of the EMG signal. The LOG value is defined as follows [14]:

$$LOG = \exp\left(\frac{1}{N}\sum_{i=1}^{N}\log(|x_i|)\right). \tag{5}$$

6) WL

The Waveform Length (WL) is used to measure the length of the signal between two consecutive samples x_{i+1} and x_i. WL is formulated as follows [14]:

$$WL = \sum_{i=1}^{N-1}|x_{i+1} - x_i|. \tag{6}$$

7) AAC

The Average Amplitude Change (AAC) is an the mean value of the waveform length within a window of length N. AAC is written as follows [14]:

$$ACC = \frac{1}{N}\sum_{i=1}^{N-1}|x_{i+1} - x_i|. \tag{7}$$

8) DASDV

The Difference Absolute Standard Deviation Value (DASDV) is calculated based on the standard deviation between x_{i+1} and x_i. DASDV is defined as follows [14]:

$$DASDV = \sqrt{\frac{1}{N-1}\sum_{i=1}^{N-1}(x_{i+1} - x_i)^2}. \tag{8}$$

9) ZC

The Zero Crossing (ZC) value is the number of time that the signal crosses a certain threshold value. ZC is calculated as [14]:

$$ZC = \sum_{i=1}^{N-1}[f(x_i \cdot x_{i+1}) \cap |x_i - x_{i+1}|] \geq \text{threshold},$$
$$f(x) = \begin{cases} 1, & \text{if} \rightarrow x \geq \text{threshold}, \\ 0, & \text{otherwise}. \end{cases} \tag{9}$$

10) SSC

The Sign Slope Change (SSC) is the number of times the slope of the signal changes its sign within a window of length N. It is formulated as follows [14]:

$$SSC = \sum_{i=1}^{N-1}[f[(x_i - x_{i+1}) \cdot (x_i - x_{i+1})]],$$
$$f(x) = \begin{cases} 1, & \text{if} \rightarrow x \geq \text{threshold}, \\ 0, & \text{otherwise}. \end{cases} \tag{10}$$

11) WAMP

The Wilson Amplitude (WAMP) is the number of times that the absolute value of the difference between two consecutive samples (x_{i+1} and x_i) exceeds a threshold value. It is defined as follows [14]:

$$WAMP = \sum_{i=1}^{N-1}[f(|x_i - x_{i+1}|)],$$
$$f(x) = \begin{cases} 1, & \text{if} \rightarrow x \geq \text{threshold}, \\ 0, & \text{otherwise}. \end{cases} \tag{11}$$

12) MYOP

The MyoPulse Percentage (MYOP) Rate is the average of the number of times that the EMG signal exceeds a predefined threshold. MYOP can be expressed as [15]:

$$MYOP = \frac{1}{N}\sum_{i=1}^{N}[f(x_i)],$$
$$f(x) = \begin{cases} 1, & \text{if} \rightarrow x \geq \text{threshold}, \\ 0, & \text{otherwise}. \end{cases} \tag{12}$$

2.2. Infinite Impulse Response

It is obvious that the EMG signal has random and stochastic characteristics in nature [16]. Therefore, in order to smooth and reduce the noise contaminating this signal, filtering is required. Commonly, the filtering stage, as it has been performed in previous studies [12] and [17], is conducted by applying a digital Low-Pass Filtered (LPF) to process the EMG signal. In this study, an Infinite Impulse Response (IIR) LPF was designed and implemented. The LPF was constructed using a 2nd order Butterworth filter with cutoff frequencies set between 80 Hz and 100 Hz, respectively. The IIR filter was implemented using a cascade bi quad filter. This digital filter was then implemented by using the following difference equation [18]:

$$y[n] = b_0 x[n] + b_1 x[n-1] + \cdots + b_P x[n-P] - $$
$$+ a_1 y[n-1] - a_2 y[n-2] - \cdots - a_Q y[n-Q], \tag{13}$$

where $x[n]$ indicates the nth input sample, $y[n]$ indicates the nth output sample, b_0, b_1, b_P, a_1, a_2, and a_Q are the filter coefficients, and $P = Q$ is the filter order.

3. Materials and Method

3.1. Participants

To implement the proposed method, four healthy male participants with no history of muscular disorder (age: 22.4 ± 3.2 years old, weight: 65.45 ± 5.67 kg) were recruited for this study after giving informed consent. Before the data collection process, the participants were recommended not to do any hard work especially anything that could potentially harm the elbow joint. The participants were instructed on how to perform the flexion and extension movements and were informed about any potential risk that could be involved in carrying out these motions.

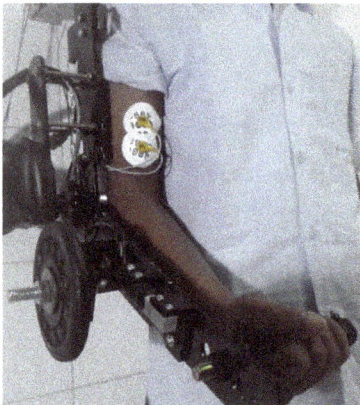

Fig. 1: The Exoskeleton frame to synchronize the elbow-joint motion.

3.2. Equipment

A one-channel EMG system comprised of: a preamplifier, a band pass filter (with cut-off frequencies of 20 to 500 Hz, respectively), a summing amplifier, and an adjustable gain amplifier, was built. EMG signals were collected using three disposable surface (pregelled Ag/AgCl) bioelectrodes. Two bioelectrodes were positioned on the biceps muscle with the third one placed on the hand as a common ground electrode. The participants held an exoskeleton frame which was used to synchronize the elbow joint motion (see Fig. 1). The elbow joint angle of the exoskeleton was collected using a linear potentiometer which was located at the joint between the arm and forearm of the exoskeleton. A one kilogram (1 kg) load was placed on the forearm of the exoskeleton.

3.3. Data Collection

Before the data collection process, the participants were instructed to follow some specific steps. EMG signals were recorded while the subject's arm held the exoskeleton and moved it in flexion and extension motions within the range of 0 to $140°$. As mentioned above, the exoskeleton was loaded with a 1 kg load (see Fig. 1). The motion periods were guided using a metronome program so that the flexion and extension movements could be regulated for 2 seconds, 4 seconds, 8 seconds and 10 seconds periods. EMG signals were recorded using a sampling frequency of 1000 Hz. For each period of motion, the participants performed flexion and extension motions for eight cycles (designated by C1, C2, C3, C4, C5, C6, C7, and C8) so that the total dataset comprised of 128 data points (4 participants × 4 periods × 8 cycles).

3.4. Data Processing

Figure 2 shows the processing of EMG signals to estimate the elbow joint angle. The collected EMG signals from biceps were processed to extract twelve Time-Domain (TD) features with a length of window of 200 milliseconds. The feature extraction process was conducted for each cycle of motion with the total of eight cycles. All of the extracted TD features such as: EMG_F (RMS, IEMG, VAR, MAV, LOG, WL, AAC, DASDV, ZC, SSC, WAMP, and MYOP) were calculated for each cycle and motion period. In order to obtain the estimated angle, the second order Butterworth low pass filter was then applied to smooth the features. As mentioned before, this IIR low pass filter was designed using the cut-off frequencies specified above to smooth out the EMG signals. The filtered feature Melbas then assumed as the estimated elbow joint angle. To evaluate the performance of the proposed method, the estimated elbow joint angle was analyzed using the Pearson's Correlation Coefficient (PCC) and Root Mean Squared Error (RMSE). The PCC was used to evaluate the relationship between the extracted TD features and the elbow joint angle. The RMSE value was used to evaluate the deviation between the estimated angle and the measured angle. The linearity of the estimated angle was also evaluated using linear regression parameters namely R^2, Slope and the Euclidian Distance.

3.5. Statistical Analysis

The statistical Analysis of Variance (ANOVA) was performed to observe if there was any statistical difference in performance and the regression parameters between the periods of motion (10 seconds, 8 seconds, 4 seconds,

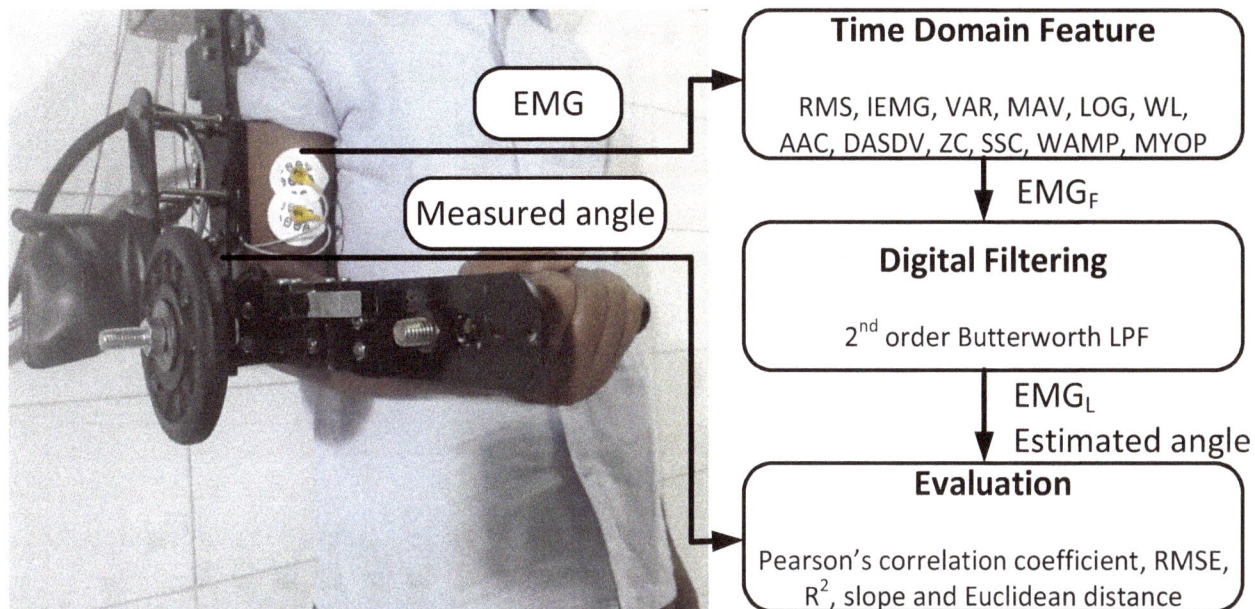

Fig. 2: The processing of EMG signas for flexion and extension movements to estimate the elbow joint angle. EMG signals were collected from biceps; time domain features were extracted and smoothed using a second order Butterworth low pass filter.

and 2 seconds). The significance test was established with confidence level of 95 % (alpha = 0.05).

4.   Results and Discussion

The recorded EMG signals and the measured angles acquired from four participants were processed offline for feature extraction and evaluation. A predefined threshold was required for ZC, SSC, WAMP, and MYOP features. The cut-off frequency of the LPF was also essential which determined the smoothness of the estimated angle. In this work, the threshold and cut-off frequencies were chosen such that elbow joint angle estimation could be made at the maximum performance. The detailed results of this study are explained and discussed in the following subsection.

4.1.   Accuracy of the Elbow Joint Angle Estimation

In this work, a relationship between the estimated angle and the measured angle was indicated by the PCC. A coefficient score approaching 1 indicates that there is a strong relationship and a score approaching 0 shows that there is a weak relationship. In the motion period of 10 seconds Fig. 3(a) and Fig. 3(b), the results show that the estimated angles based on the MYOP feature have the highest correlation coefficient (0.97 ± 0.02) (Mean±SD) and the lowest RMSE ($11.37 \pm 3.04°$)

(Mean±SD) value. In the motion period of 8 seconds, as shown in Fig. 3(c) and Fig. 3(d), the estimated angle based on the MYOP feature shows the highest correlation coefficient (0.97 ± 0.01) and the lowest RMSE ($11.25 \pm 2.44°$) value. Figure 3(e) and Fig. 3(f) show that the estimated angle from the MYOP feature has the highest accuracy (correlation coefficient = 0.91 ± 0.04 and RMSE = $17.58 \pm 3.08°$). The highest accuracies of the estimated angle are also found from the estimated angle based on the MYOP feature in the motion period of 2 seconds (0.88 ± 0.05 and $20.13 \pm 2.69°$ for correlation coefficient and RMSE, respectively). Over all periods of motion, there is a minimum of RMSE of $6.07°$ and a maximum correlation of 0.99 that occurred in the 10 second period of motion. Among the other features, the correlation coefficient of the estimated angle from Zero Crossing (ZC) feature showed the widest variance (Fig. 3(a), Fig. 3(b), Fig. 3(c), Fig. 3(d), Fig. 3(e) and Fig. 3(f)). The estimated angle based on the VAR feature showed wider variance of RMSE compared to the other features. The ANOVA tests showed that there was a significant difference (p-value < 0.05) in accuracy between groups of periods (10 seconds, 8 seconds, 4 seconds and 2 seconds) for all features except for the MYOP feature. In the period of motion of 8 seconds and 10 seconds, the MYOP feature showed that there was no significant difference in the RMSE value (p-value > 0.05). This indicates that the estimated angle using the MYOP feature is more consistent and produces higher accuracy to estimate the elbow joint angle for different motion periods compared to the other features.

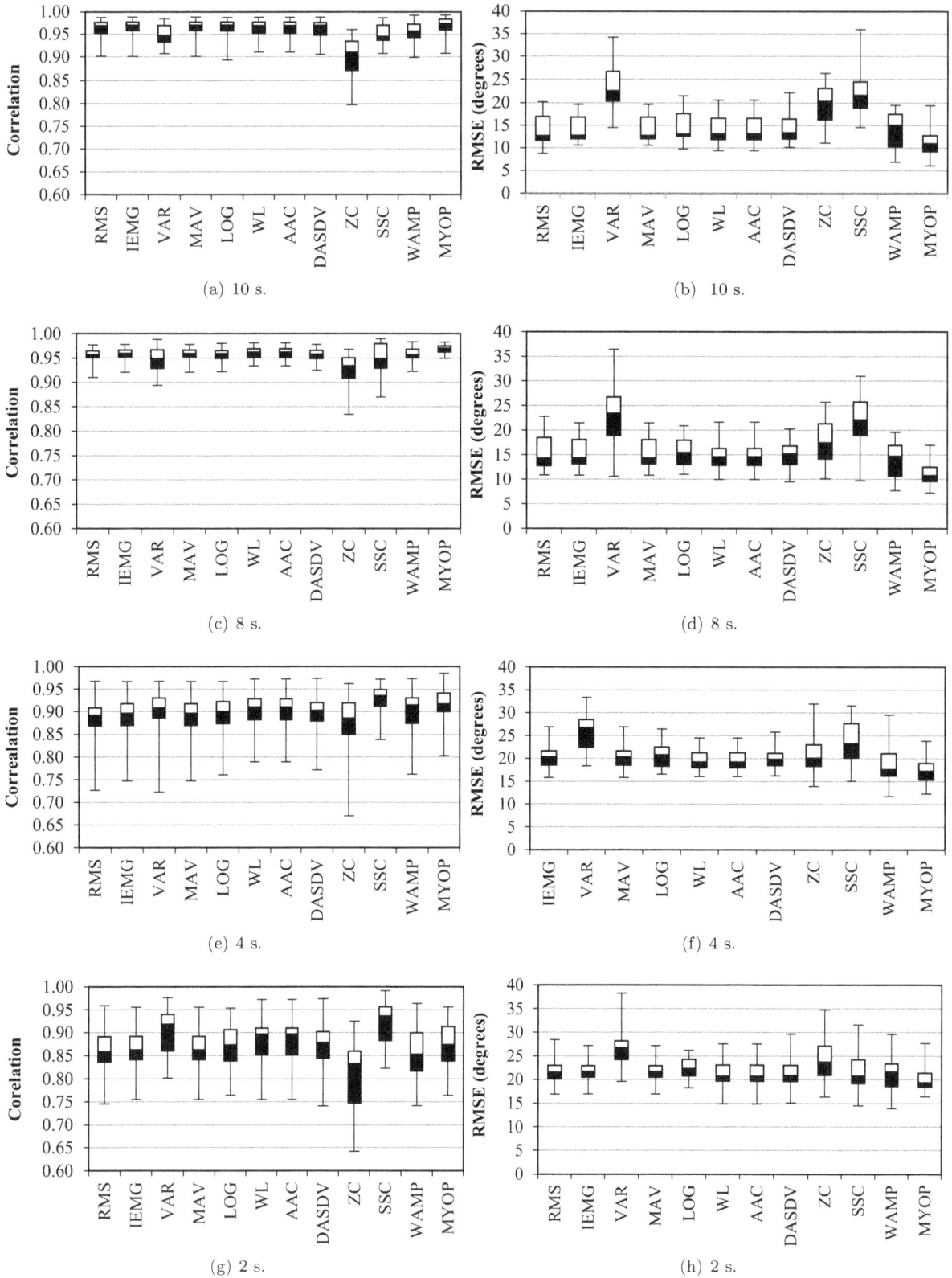

Fig. 3: The effect of TD features on the accuracy of the elbow joint angle estimation. The box plot of Pearson's Correlation Coefficient were calculated for the following periods of motion: (a) 10 s, (c) 8 s, (e) 4 s and (g) 2 s. The box plot of the RMSE value for periods of motion: (b) 10 s, (d) 8 s, (f) 4 s and (h) 2 s.

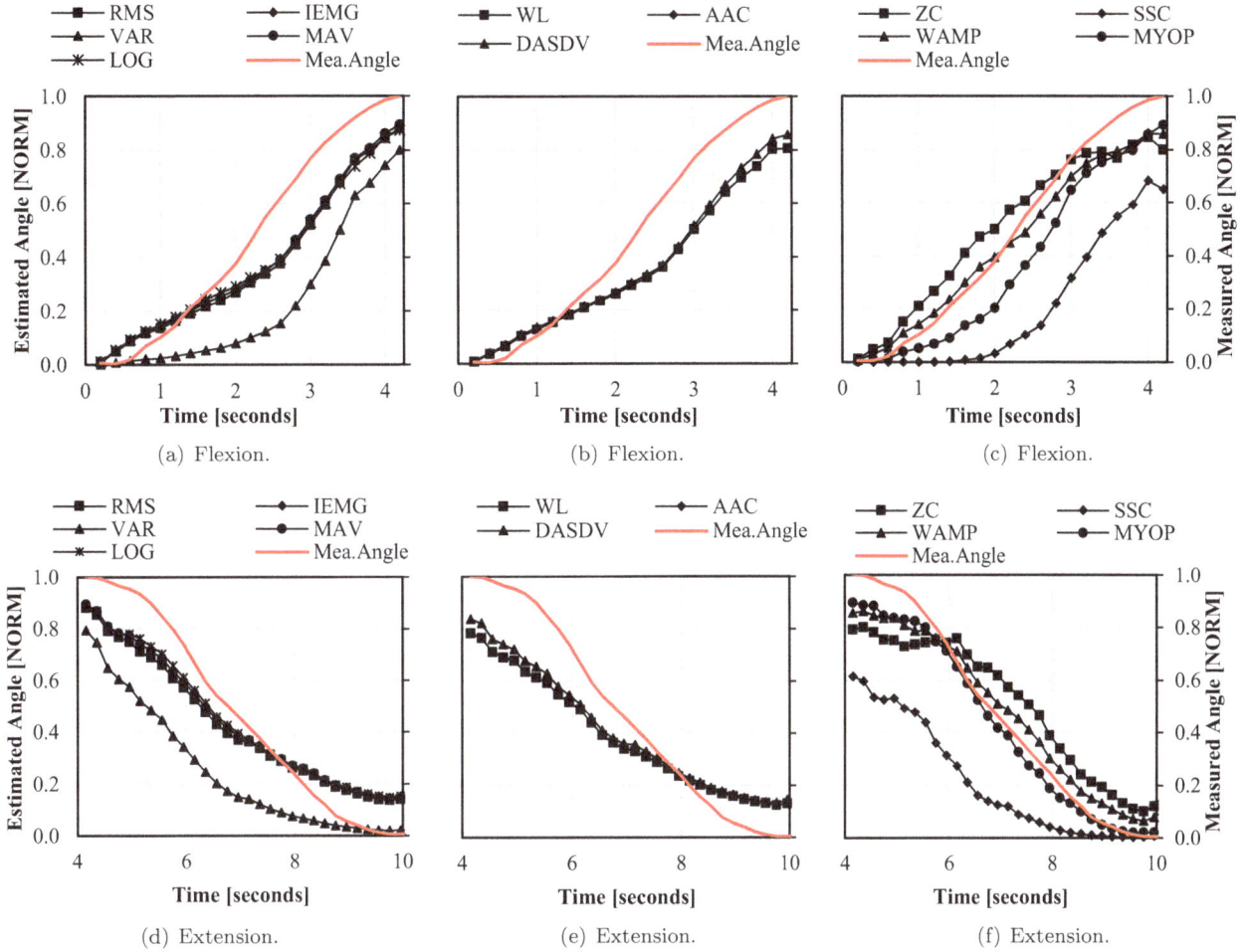

Fig. 4: Typical time response for the normalized estimated angle for a motion period of 10 seconds. The estimated angle based on RMS, IEMG, VAR, MAV, and LOG features for (a) flexion and (d) extension motions. The estimated angle from WL, AAC, and DASDV features for (b) flexion and (e) extension motions. The estimated angle from ZC, SSC, WAMP and MYOP features for (c) flexion and (f) extension motions.

The results of our proposed method are comparable with those presented in several previous studies [3] and [19]. Pau et al. developed a model to estimate the elbow joint angle using the Hill-based method and a genetic algorithm in two muscle groups (biceps and triceps) [3]. In their study, they achieved an RMSE value of $18.6 \pm 6.5°$ for five continuous cycles.

Tang et al. studied the elbow joint angle estimation problem using artificial neural networks as a classifier. In their research, they utilized four muscle groups (biceps brachii, brachioradialis, triceps brachii and anconeus). The RMSE values of their model were $10.7°$, $9.67°$, $12.42°$ for motion period of 2 seconds, 4 seconds, and 8 seconds, respectively [19].

4.2. Response of Estimated Angle to Time

Figure 4(a), Fig. 4(b), Fig. 4(c), Fig. 4(d), Fig. 4(e) and Fig. 4(f) show a typical response of the estimated angle to time for a motion period of 10 seconds (red line indicate the measured angle). Ideally, the estimated angle should be comparable to the measured angle. To test this proximity, the Euclidean Distance (ED) was calculated to present the closeness between the pattern of the estimated angle and the measured angle as shown in Eq. (14):

$$ED = \sqrt{\sum_{i=1}^{N} (EMG_L - Angle_i)^2}, \qquad (14)$$

Tab. 1: The Euclidean Distance between elbow joint angle and features for flexion and extension motions (period of motion: 10 seconds, 8 seconds, 4 seconds, and 2 seconds). The bold text indicates the lowest value of the Euclidean Distance.

Features	Flexion motion (NORM)				Extension motion (NORM)			
	$T = 10{\sim}$s	$T = 8{\sim}$s	$T = 4{\sim}$s	$T = 2{\sim}$s	$T = 10{\sim}$s	$T = 8{\sim}$s	$T = 6{\sim}$s	$T = 2{\sim}$s
RMS	0.671	0.921	0.615	0.729	0.741	0.606	0.52	0.547
IEMG	0.625	0.898	0.61	0.761	0.682	0.597	0.519	0.563
VAR	1.34	1.598	1.023	1.088	1.493	1.202	0.686	0.742
MAV	0.625	0.898	0.61	0.761	0.682	0.597	0.519	0.563
LOG	0.654	0.884	0.622	0.786	0.66	0.602	0.545	0.587
WL	0.76	1.091	0.645	0.715	0.952	0.699	0.548	0.595
AAC	0.76	1.091	0.645	0.715	0.952	0.699	0.548	0.595
DASDV	0.696	1.039	0.661	0.718	0.827	0.632	0.551	0.575
ZC	0.445	**0.313**	0.289	0.244	0.805	0.766	0.602	0.642
SSC	1.449	1.547	1.08	0.974	1.709	1.003	0.763	0.618
WAMP	**0.312**	0.325	**0.242**	**0.341**	0.452	0.445	0.549	0.565
MYOP	0.581	0.624	0.532	0.678	**0.329**	**0.191**	**0.43**	**0.496**

Tab. 2: The Linear Regression R^2 values between the estimated and measured angles. The R^2 values were calculated for all periods of motion (10 second, 8 seconds, 4 seconds, and 2 seconds) for the flexion and extension movements.

Features	Flexion				Extension			
	$T = 10{\sim}$s	$T = 8{\sim}$s	$T = 4{\sim}$s	$T = 2{\sim}$s	$T = 10{\sim}$s	$T = 8{\sim}$s	$T = 6{\sim}$s	$T = 2{\sim}$s
RMS	0.952	0.949	0.965	0.969	0.974	0.995	0.993	0.981
IEMG	0.956	0.947	0.967	0.968	0.981	0.993	0.995	0.985
VAR	0.824	0.812	0.86	0.813	0.916	0.949	0.993	0.997
MAV	0.956	0.947	0.967	0.968	0.981	0.993	0.995	0.985
LOG	0.957	0.942	0.973	0.967	0.988	0.99	0.994	0.988
WL	0.962	0.939	0.971	0.978	0.977	0.993	0.995	0.988
AAC	0.962	0.939	0.971	0.978	0.977	0.993	0.995	0.988
DASDV	0.955	0.943	0.966	0.974	0.974	0.996	0.993	0.986
ZC	0.946	0.99	0.977	0.964	0.856	0.889	0.914	0.91
SSC	0.852	0.831	0.847	0.863	0.968	0.964	0.987	0.993
WAMP	0.995	0.998	0.995	0.993	0.976	0.996	0.942	0.93
8 MYOP	0.979	0.987	0.997	0.993	0.989	0.991	0.983	0.971

Tab. 3: The Linear Regression Slope values between the estimated and the measured angle. The slopes were calculated for all periods of motion (10 seconds, 8 seconds, 4 seconds, and 2 seconds) for the flexion and extension movements.

Features	Flexion				Extension			
	$T = 10{\sim}$s	$T = 8{\sim}$s	$T = 4{\sim}$s	$T = 2{\sim}$s	$T = 10{\sim}$s	$T = 8{\sim}$s	$T = 6{\sim}$s	$T = 2{\sim}$s
RMS	0.752	0.693	0.608	0.466	-0.735	-0.619	-0.535	-0.434
IEMG	0.761	0.703	0.609	0.446	-0.752	-0.622	-0.534	-0.418
VAR	0.68	0.569	0.499	0.299	-0.806	-0.51	-0.533	-0.394
MAV	0.761	0.703	0.609	0.446	-0.752	-0.622	-0.534	-0.418
LOG	0.733	0.712	0.597	0.432	-0.757	-0.616	-0.507	-0.399
WL	0.707	0.639	0.593	0.484	-0.652	-0.585	-0.505	-0.394
AAC	0.707	0.639	0.593	0.484	-0.652	-0.585	-0.505	-0.394
DASDV	0.743	0.656	0.583	0.483	-0.701	-0.613	-0.502	-0.406
ZC	0.766	0.852	0.804	0.825	-0.454	-0.811	-0.548	-0.452
SSC	0.637	0.624	0.488	0.426	-0.726	-0.627	-0.514	-0.47
WAMP	0.822	0.889	0.831	0.726	-0.702	-0.88	-0.666	-0.511
MYOP	0.892	0.851	0.727	0.554	-0.917	-0.909	-0.621	-0.497

where N indicates the number of samples, $Angle_i$ stands for the-ith measured angle and the EMG_L shows the filtered features (estimated angle). In general, a small value of ED indicates a close relationship between the estimated angle and the measured angle. ED was measured for all periods of motion (10 seconds, 8 seconds, 4 seconds and 2 seconds) and for all of the TD features.

Table 1 shows the summary of the ED values for all motion periods and features. The estimated angle based on the WAMP feature tended to show smaller ED values in the elbow flexion trajectory (for all periods of motion) compared to those based on the other features. For the elbow extension trajectory, the estimated angles based on the MYOP feature showed the lowest value (0.329, 0.191, 0.430, and 0.496 for motion period of 10 seconds, 8 seconds, 4 seconds and 2 seconds, respectively). The estimated angle based on the VAR feature tended to have higher Euclidean Distance values compared to other features for all periods of motion.

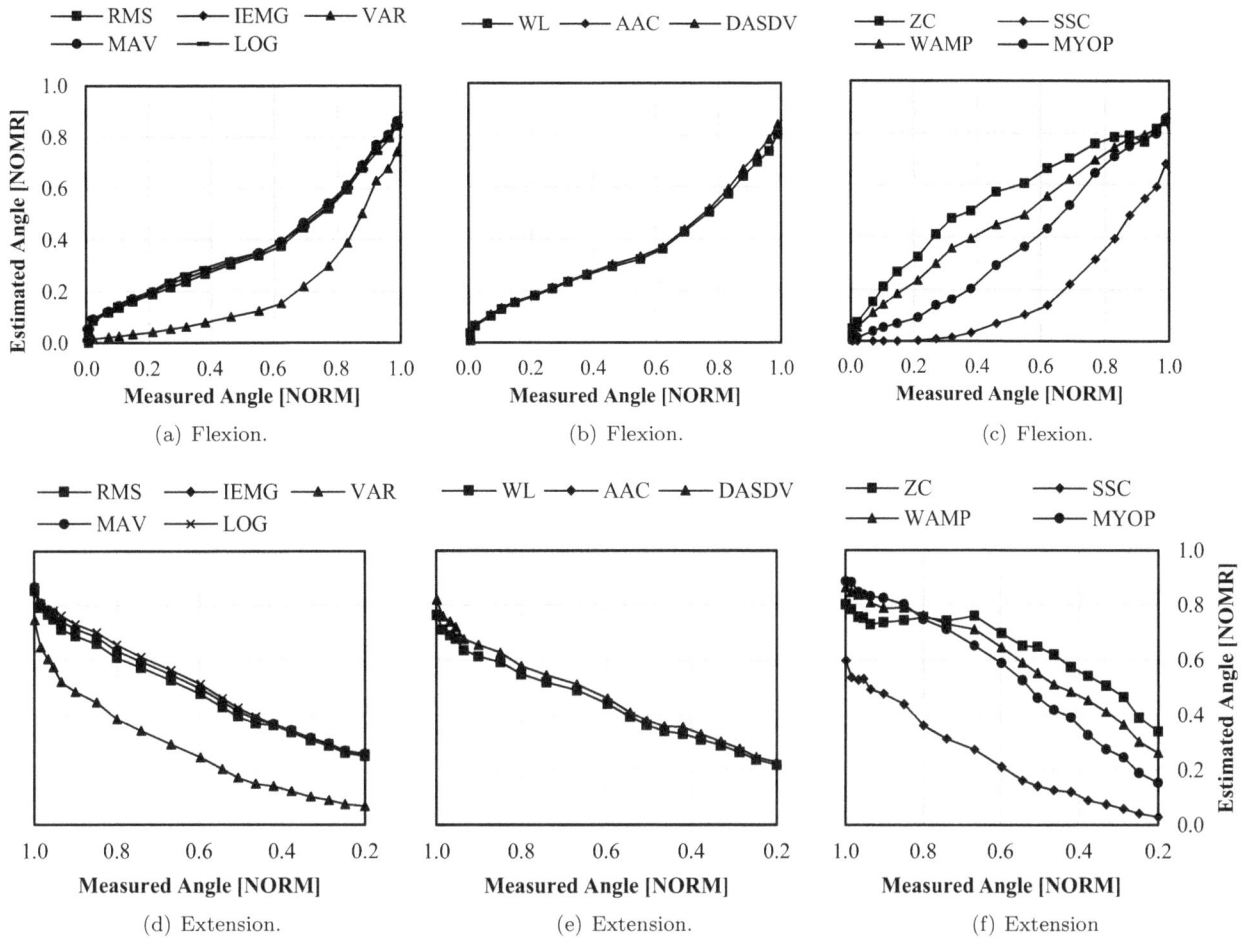

Fig. 5: The relationship between the normalized measured angle to the normalized EMG features during flexion (a), (b), and (c), and extension (d), (e), and (f) for period of motion = 10 seconds.

4.3. The Effects of the Elbow Joint Angle on EMG Signal Features

Figure 5 shows a typical relationship between the estimated and measured angles for both flexion and extension movements (period of motion = 10 seconds). The R^2 and Slope values were calculated to evaluate the linear regression between the estimated and measured angle as shown in Tab. 2 and Tab. 3. Table 2 shows that the R^2 of the estimated angle from the MYOP and WAMP features are higher and more consistent for all periods of motion (ranged between 0.938 and 0.998) compared to those calculated from other features. This means that the estimated angles were fitted closely to the measured angles.

The R^2 values of the model developed by Tang et al. were 0.83, 0.87 and 0.79 for the periods of motion 2 seconds, 4 seconds and 8 seconds, respectively [19]. Table 3 shows several of the Slope values for various EMG features. In the flexion trajectory, the estimated angle based on the WAMP and MYOP features showed the best Slope value (ranged from 0.727 to

0.889). In the extension trajectory, the estimated angle based on the MYOP feature had the best Slope value (ranged from −0.917 to −0.909 for periods of motion 10 seconds and 8 seconds, respectively). These values indicated that the estimated angle was almost linearly related to the measured angle. The negative value indicated a negative response between the measured angle and the estimated angle. From the ANOVA test, unfortunately, we found that there was a significant difference (p-value < 0.05) between the Slopes during extension and flexion movements. The Slope values decreased for the periods of motions from 10 seconds, 8 seconds, 4 seconds and 2 seconds, respectively. Ideally, the slope of the features should be constant so that the model can be used for any periods of motion. In the future, a model that can compensate the decrement of the slope is needed.

5. Conclusion

This study presents an investigation of TD features to estimate the elbow joint angle using EMG features based on a non-pattern recognition method. Some parameters such as the: Pearson's Correlation Coefficient, Root Mean Squared Error, Euclidean Distance, R^2, Linear Regression Slope were evaluated to obtain the best EMG features in estimating the elbow joint angle. The EMG signals for this study were collected from biceps alone enabled us to estimate the elbow joint angle. Our findings show that for a 10 second long recording period, the MyoPulse Percentage (MYOP) Rate produced the best accuracy: with PCC of 0.97 ± 0.02 (Mean±SD) and RMSE of $11.37 \pm 3.04°$ (Mean±SD), respectively. The MYOP feature also showed the highest R^2 and Slope value 0.986 ± 0.0083 (Mean±SD) and 0.746 ± 0.17 (Mean±SD), respectively for flexion and extension motions during all recorded periods.

References

[1] OSKOEI, M. A. and H. HU. Myoelectric control systems-A survey. *Biomedical Signal Processing and Control*. 2007, vol. 2, iss. 4, pp. 275–294. ISSN 1746-8094. DOI: 10.1016/j.bspc.2007.07.009.

[2] TRIWIYANTO, O. WAHYUNGGORO, H. A. NUGROHO and HERIANTO. Quantitative relationship between feature extraction of sEMG and upper limb elbow joint angle. In: *International Seminar on Application for Technology of Information and Communication (ISemantic)*. Semarang: IEEE, 2016, pp. 44–50. ISBN 978-1-5090-2326-4. DOI: 10.1109/ISEMANTIC.2016.7873808.

[3] PAU, J. W. L., S. S. Q. XIE and A. J. PULLAN. Neuromuscular Interfacing: Establishing an EMG-Driven Model for the Human Elbow Joint. *IEEE Transactions on Biomedical Engineering*. 2012, vol. 59, iss. 9, pp. 2586–2593. ISSN 1558-2531. DOI: 10.1109/TBME.2012.2206389.

[4] TANG, Z., H. YU and S. CANG. Impact of Load Variation on Joint Angle Estimation From Surface EMG Signals. *IEEE Transactions on Neural Systems and Rehabilitation Engineering*. 2016, vol. 24, iss. 12, pp. 1342–1350. ISSN 1558-0210. DOI: 10.1109/TNSRE.2015.2502663.

[5] BENITEZ, L. M. V., M. TABIE, N. WILL, S. SCHMIDT, M. JORDAN and E. A. KIRCHNER. Exoskeleton Technology in Rehabilitation: Towards an EMG-Based Orthosis System for Upper Limb Neuromotor Rehabilitation. *Journal of Robotics*. 2013, vol. 2013, iss. 1, pp. 1–13. ISSN 1687-9600. DOI: 10.1155/2013/610589.

[6] JANG, G., J. KIM, Y. CHOI and J. YIM. Human shoulder motion extraction using EMG signals. *International Journal of Precision Engineering and Manufacturing*. 2014, vol. 15, iss. 10, pp. 2185–2192. ISSN 2005-4602. DOI: 10.1007/s12541-014-0580-x.

[7] LI, Q., Y. SONG, Z. HOU and B. ZHU. sEMG Based Joint Angle Estimation of Lower Limbs Using LS-SVM. In: *20th International Conference on Neural Information Processing (ICONIP)*. Daegu: Springer, 2013, pp. 292–300. ISBN 978-364242053-5. DOI: 10.1007/978-3-642-42054-2_37.

[8] LOCONSOLE, C., S. DETTORI, A. FRISOLI, C. A. AVIZZANO and M. BERGAMASCO. An EMG-based approach for on-line predicted torque control in robotic-assisted rehabilitation. In: *2014 IEEE Haptics Symposium (HAPTICS)*. Houston: IEEE, 2014, pp. 181–186. ISBN 978-1-4799-3131-6. DOI: 10.1109/HAPTICS.2014.6775452.

[9] REZA, S. M. T., N. AHMAD, I. A. CHOUDHURY and R. A. R. GHAZILLA. A Fuzzy Controller for Lower Limb Exoskeletons during Sit-to-Stand and Stand-to-Sit Movement Using Wearable Sensors. *Sensors*. 2014, vol. 14, iss. 3, pp. 4342–4363. ISSN 1424-8220. DOI: 10.3390/s140304342.

[10] TSAI, A.-C., T.-H. HSIEH, J.-J. LUH and T.-T. LIN. A comparison of upper-limb motion pattern recognition using EMG signals during dynamic and isometric muscle contractions. *Biomedical Signal Processing and Control*. 2014, vol. 11, iss. 1, pp. 17–26. ISSN 1746-8094. DOI: 10.1016/j.bspc.2014.02.005.

[11] LENZI, T., S. M. M. D. ROSSI, N. VITIELLO and M. C. CARROZZA. Intention-Based EMG Control for Powered Exoskeletons. *IEEE Transactions on Biomedical Engineering*. 2012, vol. 59, iss. 8, pp. 2180–2190. ISSN 1558-2531. DOI: 10.1109/TBME.2012.2198821.

[12] LEE, S., H. KIM, H. JEONG and J. KIM. Analysis of musculoskeletal system of human during lifting task with arm using electromyography. *International Journal of Precision Engineering and Manufacturing*. 2015, vol. 16, iss. 2, pp. 393–398. ISSN 2005-4602. DOI: 10.1007/s12541-015-0052-y.

[13] PHINYOMARK, A., P. PHUKPATTARANONT and C. LIMSAKUL. Feature reduction and selection for EMG signal classification. *Expert Systems with Applications*. 2012, vol. 39, iss. 8, pp. 7420–7431. ISSN 0957-4174. DOI: 10.1016/j.eswa.2012.01.102.

[14] CHOWDHURY, R. H., M. B. I. REAZ, M. A. B. M. ALI, A. A. A. BAKAR, K. CHELLAPPAN and T. G. CHANG. Surface Electromyography Signal Processing and Classification Techniques. *Sensors*. 2013, vol. 13, iss. 9, pp. 12431–12466. ISSN 1424-8220. DOI: 10.3390/s130912431.

[15] FOUGNER, A. L. *Proportional Myoelectric Control of a Multifunction Upper-limb Prosthesis*. Trondheim, 2007. Thesis. Norwegian University of Science and Technology. Supervisor Tor Engebret Onshus.

[16] LUCA, C. J. D. Surface Electromyography: Detection and Recording. In: *DelSys* [online]. 2002. Available at: http://www.ti.com/lit/an/slva372c/slva372c.pdf.

[17] JANG, G., J. KIM, Y. CHOI and J. YIM. Human shoulder motion extraction using EMG signals. *International Journal of Precision Engineering and Manufacturing*. 2014, vol. 15, iss. 10, pp. 2185–2192. ISSN 2234-7593. DOI: 10.1007/s12541-014-0580-x.

[18] TAN, L. *Digital Signal Processing Fundamentals and Applications*. 1st ed. San Diego: Elsevier, 2008. ISBN 978-0-0805-5057-2.

[19] TANG, Z., K. ZHANG, S. SUN, Z. GAO, L. ZHANG and Z. YANG. An upper-limb power-assist exoskeleton using proportional myoelectric control. *Sensors*. 2014, vol. 14, iss. 10, pp. 6677–6694. ISSN 1424-8220. DOI: 10.3390/s140406677.

About Authors

TRIWIYANTO was born in Surabaya, Indonesia. He received his M.Sc. degree in Electronic Engineering in Institute of Technology Sepuluh Nopember in 2004, Surabaya, Indonesia. He is currently a Ph.D. candidate in Electrical Engineering at Gadjah Mada University, Yogyakarta, Indonesia. His research interests include biomedical signal analysis, embedded system, electronic instrumentation, assistive and rehabilitation devices.

Oyas WAHYUNGGORO was born in Jogjakarta, Indonesia. He received his Ph.D. degree in Electrical and Electronic Engineering from the Universiti Teknologi Petronas, Malaysia in 2011. His research interests include biomedical signal processing, intelligent system, and control system.

Hanung Adi NUGROHO was born in Jogjakarta, Indonesia. He received his Ph.D. degree in Electrical and Electronic Engineering from the Universiti Teknologi Petronas, Malaysia in 2012. His research interests include biomedical signal processing and image processing.

HERIANTO was born in Jogjakarta, Indonesia. He received his D.Eng. degree in the Department of Mechanical and Control Engineering, Tokyo Institute of Technology, Japan, in 2009. His research interests include robotics and manufacture. His current research is product design and development especially in rehabilitation robot.

19

COMPARATIVE ANALYSIS OF THE DISCRIMINATIVE CAPACITY OF EEG, TWO ECG-DERIVED AND RESPIRATORY SIGNALS IN AUTOMATIC SLEEP STAGING

Farideh EBRAHIMI[1,2]*, Seyed Kamaledin SETAREHDAN*[2]*,*
Radek MARTINEK[3]*, Homer NAZERAN*[4]

[1]Department of Computer, Faculty of Electrical and Computer Engineering,
Babol Noshirvani University of Technology, Shariati Avenue, Babol, Mazandaran, Iran
[2]Control and Intelligent Processing Center of Excellence, School of Electrical and Computer
Engineering, College of Engineering, University of Tehran, North Kargar Street, Tehran, Iran
[3]Department of Cybernetics and Biomedical Engineering, Faculty of Electrical Engineering and Computer
Science, VSB–Technical University of Ostrava, 17. listopadu 15, 708 33 Ostrava, Czech Republic
[4]Department of Electrical and Computer Engineering, College of Engineering, University of Texas El Paso,
500 W University Ave, El Paso, TX 79968, United States of America

f_ebrahimi88@alumni.ut.ac.ir, ksetareh@ut.ac.ir, radek.martinek@vsb.cz, hnazeran@utep.edu

Abstract. *Highly accurate classification of sleep stages is possible based on EEG signals alone. However, reliable and high quality acquisition of these signals in the home environment is difficult. Instead, electrocardiogram (ECG) and Respiratory (Res) signals are easier to record and may offer a practical alternative for home monitoring of sleep. Therefore, automatic sleep staging was performed using ECG, Res (thoracic excursion) and EEG signals from 31 nocturnal recordings of the Sleep Heart Health Study (SHHS) polysomnography Database. Feature vectors were extracted from 0.5 min (standard) epochs of sleep data by time-domain, frequency domain, time-frequency and nonlinear methods and optimized by using the Support Vector Machine -Recursive Feature Elimination (SVM-RFE) method. These features were then classified by using a SVM. Classification based upon EEG features produced a Correct Classification Ratio CCR = 0.92. In comparison, features derived from ECG signals alone, that is the combination of Heart Rate Variability (HRV), and ECG-Derived Respiration (EDR) signals produced a CCR = 0.54, while those features based on the combination of HRV and (thoracic) Res signals resulted in a CCR = 0.57. Overall comparison of the results based on standard epochs of EEG signals with those obtained from 5-minute (long) epochs of cardiorespiratory signals, revealed that acceptable CCR = 0.81 and discriminative capacity (Accuracy = 89.32 %, Specificity = 92.88 % and Sensitivity = 78.64 %) were also achievable when using optimal feature sets derived from long epochs of the latter signals in sleep staging. In addition, it was observed that the presence of some artifacts (like bigeminy) in the cardiorespiratory signals reduced the accuracy of automatic sleep staging more than the artifacts that contaminated the EEG signals.*

Keywords

Automatic sleep staging, ECG-derived respiration signals, Electroencephalogram (EEG) signal, Heart Rate Variability (HRV) signal, Res (thoracic excursion) signal.

1. Introduction

Generally speaking we can categorize 2 types of sleep: Non-Rapid Eye Movement (NREM) sleep, and Rapid Eye Movement (REM) sleep. The NREM sleep can be in turn sub-categorized as Stages 1 through 4, with Stage 1 being the lightest and Stage 4 being the deepest sleep state [1]. Similarly, in the AASM sleep standards, the NREM stage is sub-grouped into three Stages of N1, N2 and N3 [2]. Polysomnography (PSG) or *"multiple recording of physiological signals during sleep"* is widely used as the "gold standard" clinical technique for the evaluation of sleep and diagnosis of its disorders. In PSG signals such as EEG, ECG, EMG, EOG,

Respiration (Res) and others are recorded simultaneously during sleep. Among these signals, EEG is the most commonly used for sleep staging [3].

Because of the significant and pivotal roles that EEG signals play in sleep studies, a wide variety of approaches and techniques have been proposed for Automatic Sleep Staging based on these signals [4], [5], [6], [7] and [8]. In reference [4], the authors used single-channel EEG data to perform sleep stage scoring by leveraging a method called Complete Ensemble Empirical Mode Decomposition (EMD) with Adaptive Noise (CEEMDAN). They used bagging to classify the different sleep states. This work achieved an accuracy of 90.69 % in classifying 5 sleep stages. In another study [5], the investigators extracted many spectral features based on the Fast Fourier Transform (FFT) of multichannel PSG data to classify sleep stages by using a rule-based Decision Tree (DT) classifier and achieved an accuracy of 84 %.

A wide range of time- and frequency-domain features have been explored by the authors in reference [6] based on PSG signals that included two EEG channels, two EOG channels and one EMG channel for automatic sleep stage scoring. Their method based on a Dendrogram-SVM (DSVM), resulted in 92 % accuracy, 94 % specificity, and 82 % sensitivity. Kayikcioglu et al. [7] extracted Auto-Regressive (AR) coefficient features from a single channel EEG signal to classify both sleep and wake stages with an accuracy of 91 % using a Partial Least Squares Regression (PLSR) classifier.

Despite being highly accurate in classifying sleep stages automatically, EEG signals do not easily lend themselves to reliable acquisition in the home environment. This is in stark contrast to ECG and Res signals that have proved to be far easier to record in such environments and may suggest a viable alternative to home monitoring of sleep. Moreover, it has been indicated that the Heart Rate Variability (HRV) signal spectral components produce quantitative markers of sympathetic and parasympathetic activities of the Autonomic Nervous System (ANS), which differ significantly during wake and different sleep stages. As such, using HRV signals to extract information for automatic sleep staging is a promising exploration [9] and [10].

Recently, there has been a surge in the number of approaches to sleep scoring based on ECG and Res signals. For instance, Penzel et al. indicated that the dynamics of HRV signals are different in wake and sleep stages by deploying Detrended Fluctuation Analysis (DFA), [10]. In addition, they utilized spectral analysis and DFA so as to extract information from HRV signals separately for automatic sleep staging and discovered that in comparison with spectral analysis, DFA is a more verifiable approach [12]. Likewise, Redmond

et al. [13], [14] and [15] extracted a variety of useful spectral parameters and time-domain features from 0.5-minute epochs of HRV and Res signals. They managed to attain an accuracy of 79 % in an effort to distinguish among Wake, NREM and REM Sleep Stages by a subject-independent classifier. The accuracy in their investigation plummeted to 67 % in a subject-specific system [13]. Additionally, they could achieve an accuracy of 76.1 % for a 3-class (Wake, NREM Sleep and REM) system employing cardiorespiratory signals [14]. Andane et al. [16], used HRV signals for sleep analysis, extracted features by using spectral, time-domain, and DFA methods and were able to separate Wake and Sleep Stages (2 classes) attaining an accuracy of 79.99 %. Similarly, Mendez et al. [17], deployed a time-varying autoregressive model to extract features from HRV signals and used a Hidden Markov Model (HMM) for the purpose of classification. They managed to separate REM and NREM sleep (2 classes) with 79 % accuracy. Kesper et al. [18], also utilized spectral parameters of HRV signals and were able to correctly classify (wake, light sleep, deep sleep and REM sleep) with an accuracy of 57.7 % using 0.5-minute epochs separated from 18 overnight PSG recordings. Carskadon et al. [19], explored the application of the Res signal variability in Wake and Sleep Stages in children. They noticed that the respiration rate and its regularity dropped in NREM Sleep compared to Wake State. Along the same line of work, Miyata et al. initially created surrogate data from raw Res signals as a linear stochastic time series so that the Fourier transform of the surrogate data would be the same as that of the raw Res signal. In their work, they computed the correlation dimensions of the original and the surrogate Res signals and discovered that there was a considerable difference between these values. This further proved that Res signals were generated from a nonlinear underlying system [20]. Motivated by these findings, a number of nonlinear techniques have been used in sleep analysis based on cardiorespiratory signals alone. For example, the significance of using approximate entropy of Res signals in Wake and Sleep Stages has been researched [21]. Having nonlinear characteristics does not generally lead to the conclusion that the signal exhibits fractal characteristics. Some studies, however, have maintained that Res signals could manifest fractal characteristics [21], [22], [23] and [24]. Although a growing body of research has demonstrated that cardiorespiratory signals play a pivotal role in automatic sleep staging, more studies and extensive research are required to establish that cardiorespiratory signals could be reliably used to perform automatic sleep staging. The literature shows that a large number of the recent investigations conducted on automatic sleep staging based on cardiorespiratory signals just separate two stages of sleep [13], [14], [15], [16] and [17]. Moreover, as no previous studies have compared the results of automatic sleep staging based on EEG and cardiorespiratory signals, we are highly motivated to perform such an exploration and attempt to fill this gap.

In a previous work [25], we evaluated the utility of ECG as well as the combination of ECG and Res signals in automatic sleep staging based on 5-min (long) epochs. It was observed that acceptable discriminative capacity could be achieved when features were extracted from these long segments. As the standard epoch length for sleep scoring is 0.5-minute, it seemed natural to extend the previous study and use the available long segments to investigate the efficacy of standard epochs in automatic sleep staging. Since standard epochs from Stage 1 were also available in the database (continuous long segments for this Sleep Stage were not available), we were able to include Stage 1 in automatic sleep staging in our current study.

In this investigation, automatic sleep staging was performed as follows:

- by using EEG signal features,

- by combining HRV and EDR signal features,

- by combining HRV and Res signals features.

The feature vectors for each approach were extracted from standard epochs of sleep data by time-domain, frequency-domain, time-frequency, and non-linear methods and then optimized by using the SVM-RFE method. A SVM classifier was then used to perform classification. The results of automatic sleep staging in this study (standard epochs) were compared with those of our previous research, in which features were extracted from long epochs of cardiorespiratory signals [25]. Moreover, automatic sleep staging based on cardiorespiratory signals in the presence of some artifacts (like bigeminy in ECG signals) was explored.

2. Materials and Methods

Figure 1 shows the block diagram of the algorithm used in this study. First, HRV and EDR signals were extracted from ECG signals. Then, feature vectors were extracted from EEG, ECG-derived, and Res signals. Subsequently, the extracted feature vectors were optimized by the SVM-RFE method. Finally, the classification of Wake and Sleep Stages was performed by using:

- EEG,

- combined HRV and RS-EDR (ECG) as well,

- combined HRV and Res (Cardiorespiratory) signals.

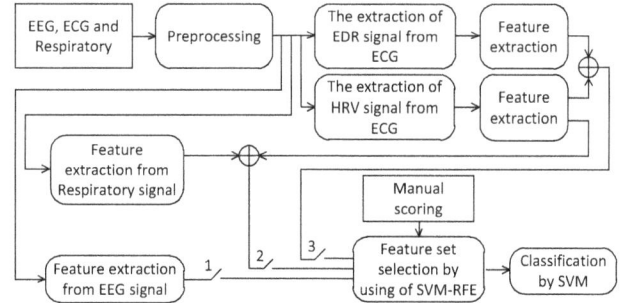

Fig. 1: Automatic sleep staging algorithm using EEG, ECG and Res (thoracic excursion) signals.

These 3 states are represented by three switches.

2.1. Polysomnographic Data

The Sleep Heart Health Study (SHHS) database was used to provide the sleep data for this investigation [26]. Subjects who participated in the acquisition of data for this database did not use beta-blockers, alpha-blockers or inhibitors. In this database, EEG signals are sampled at a sampling rate of $Fs = 125$ Hz, ECG signals at a rate of 250 Hz and thoracic Res signals are recorded by inductive plethysmography bands and have a sampling rate of 10 Hz, all extracted from 31 overnight (nocturnal) polysomnographic recordings from men and women (age $\geq$ 40). Sleep architecture for these data was determined in each subject according to the Rechtschaffen and Kales (R&K) criteria on standard epochs [1]. Generally, atrial fibrillation frequently happens in patients with sleep apnea [27], [28], [29], [30], [31] and [32]. In our study, the recordings were selected by considering a Respiratory Disturbance Index 3 Percent (RDI3P) < 5 to have near-normal characteristics. Therefore, we assumed there were no atrial fibrillations in our database. Regarding the analysis of HRV, it has been suggested to use 5-minute ECG signals to perform a "short-term" HRV analysis [33].

Table 2 shows the number of long epochs of Wake and different Sleep Stages except Stage 1 and standard epochs from Wake and all Sleep Stages for all recordings. First, for each recording, continuous parts of EEG, ECG and Res signals in each sleep cycle were separated and then, long segments of different Sleep Stages were selected manually. Some parts of the data (from the end segments of each cycle) were not continuous 5-minute segments.

Typically about 50–60 % of the total duration of sleep is spent in Light Sleep, 15–20 % in Deep Sleep, 20–25 % in REM Sleep, and 5 % or less in Wake [34]. In the database, Wake times records are much more than 5 % but we considered just relaxed Wake or the total time between sleep onset and final wake-up. Therefore, we included a small part of the Wake data (8.11 % of the total data for this stage), which happens during sleep times. As there were no continuous long (5-min) epochs available for Stage 1 in the database, we could not include this sleep stage in our previous study [25].

However, when we carried out the current investigation, we were able to make use of the 0.5-minute segments of the cardiorespiratory signals during Stage 1, for further analysis. All of the existing continuous long segments in the database were used in this study. Overall, 113.49 hours (59.45 %) out of the 190.88 hours of sleep data available were used, as the remaining data were not continuous long segments during different Sleep Stages. In summary, 84.73 % of Stage 1, 55.9 % of Stage 2, 47.33 % of SWS (Slow Wave Sleep) and 71.4 % of REM Sleep data were used in this investigation.

Tab. 1: The number of 5-minute (long) and 0.5-minute (standard) epochs of EEG, ECG and Res signals in the Wake and Sleep Stages for 31 subjects.

	The number of clean 5-min (long) Epochs	Remaining number of 5-min (long) Epochs	Total number of 0.5-min (standard) Epochs
Wake	79	19	790+190
Sleep Stage1	-	-	953
Sleep Stage2	434	239	4340+2390
SWS	185	16	1850+160
REM Sleep	259	131	2590+1310
Total	957	405	14573

As the main goal of this study was to perform the comparative utility and analysis of EEG and cardiorespiratory signals in automatic sleep staging, we excluded continuous long segments polluted with artifacts in the first part of the study so that the results would be independent of artifacts. For example, continuous long segments with bigeminy were not used. The algorithm was tested on the remaining continuous long segments. It is important to note that in the selection of continuous long segments, there were still transition points between Sleep Stages, as some of these segments belonged to the beginning and the end portions of each sleep cycle.

2.2. Feature Extraction

As mentioned above, we investigated automatic sleep staging by applying a variety of advanced digital processing methods to EEG, ECG, and Res signals to preprocess and analyze these signals. Then we extracted different sets of features from EEG, ECG (HRV, EDR) and Res (thoracic excursion) signals, the details of which are presented in the following sections.

1) Feature Extraction from EEG Signals

The 0.5-minute (standard) epochs of electroencephalographic signals were acquired from EEG channels C3-A2 and were used for feature extraction. These signals were first normalized based on their means and standard deviations for each subject. They were then filtered by an 8th order elliptic band pass filter with cutoff frequencies of 0.5 and 40 Hz, and subsequently 5-feature sets (a total of 34 features) were extracted from these data.

Time-Frequency Features

The Daubechies10 (db10) mother wavelet was used for analyzing the EEG signals. A Wavelet Packet Transform (WPT) with 7 levels was applied and the frequency domain information of the following 6 bands was selected.

- $\{0.45 - 3.9\}$, Delta, Wavelet coefficient $= [C_{22}, C_{20}, C_{18}]$.
- $\{3.9 - 7.8\}$, Theta, Wavelet coefficient $= C_{14}$.
- $\{7.8 - 11.7\}$, Alpha, Wavelet coefficient $= C_{15}$.
- $\{11.7 - 15.6\}$, Spindle, Wavelet coefficient $= C_{16}$.
- $\{15.6 - 23.4\}$, Beta1, Wavelet coefficient $= C_9$.
- $\{23.4 - 39.05\}$, Beta2, Wavelet coefficient $= [C_{10}, C_{11}]$.

Figure 2 show the decomposition of the EEG signals by using Wavelet Packet Transform (WPT) at 7 levels.

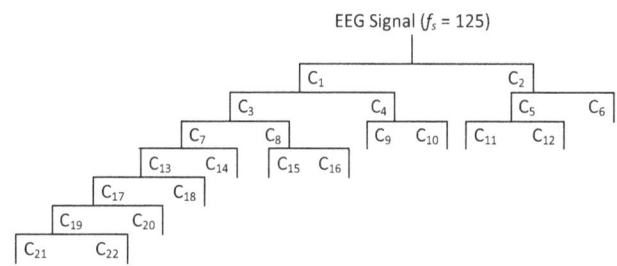

Fig. 2: Decomposition of EEG signals by using Wavelet Packet Transform (WPT) of 7 levels.

In general, the wavelet decomposition of a given signal $x(t)$ is presented by:

$$x(t) = \sum_{k=-\infty}^{\infty} C_{N,k}\varphi_{N,k}(t) + \sum_{j=1}^{N}\sum_{k=-\infty}^{\infty} d_{j,k}\psi_{j,k}(t). \quad (1)$$

In wavelet decomposition, Parseval's theorem relates the energy of the signal $x(t)$ to the energy in each one of the components and their wavelet coefficients, provided that the scaling functions ($\varphi(t)$) and the wavelets

$(\psi(t))$ form an orthonormal basis [35]. Parseval's theorem for discrete wavelet transform is given by Eq. (2):

$$E = \sum_{k=-\infty}^{\infty} c_{N,k}^2 + \sum_{j=1}^{N} \sum_{k=-\infty}^{\infty} d_{j,k}^2. \qquad (2)$$

Subsequently, Shannon entropy in these frequency bands was calculated from wavelet coefficients in standard epochs of EEG signals using Eq. (3) below where $C(i)s$ represent wavelet coefficients in each frequency band:

$$en = \sum_i p(i)\log_2 p(i),$$
$$p(i) = \frac{C^2(i)}{\sum_i C^2(j)}. \qquad (3)$$

Next, the following features were calculated for the frequency bands 1 to 6, which were then used to represent the time-frequency distribution of the EEG signals:

- Mean quadratic value or Energy of Wavelet Packet (WP) coefficients for each of the 6 bands $(E_1, E_2, \ldots, E_6)$,

- Total Energy (E_7),

- Ratio of different Energy values (E_8, E_9, E_{10}),

- Shannon Entropy of wavelet packet (WP) coefficients for each of the 6 bands (Entropy$_{11}$, Entropy$_{12}, \ldots,$ Entropy$_{16}$)

  $E_7 = \sum_{i=1}^{6} E_i \rightarrow$ The total Energy of all 6 bands,
  $E_8 = E_3/(E_1 + E_2) \rightarrow$ Alpha / (Delta + Theta),
  $E_9 = E_1/(E_2 + E_3) \rightarrow$ Delta / (Theta + Alpha),
  $E_{10} = E_2/(E_1 + E_3) \rightarrow$ Theta / (Delta + Alpha).

Frequency-Domain and Time-Domain Features

Frequency-domain and time-domain features included: the relative spectral energy in 6 frequency bands, the power value and frequency related to the peak point of the power spectrum, the harmonic parameters and Hjorth parameters, as well as the mean absolute value of EEG amplitude, which will be defined in the following section. In order to extract all these features (except the mean absolute value of EEG amplitude), it was necessary to calculate the power spectrum of the EEG signals. Among the many different methods for the calculation of power spectrum, the AutoRegressive (AR) modeling-based method was used as it offers better accuracy, smoother spectra, and higher spectral resolution compared to other methods. For the selection of an appropriate order for the AR model, the Minimum Description Length (MDL) and Akaike methods were used. Here we implemented a 10th-order AR model for 10-second long EEG segments [36], [37]

and [38]. The AR coefficients were estimated by the Burg method. These features were extracted in three 10-second segments in each standard epoch and the average of three feature vectors formed the final one.

- **Relative Spectral Energies**

The power spectrum of EEG signals was first estimated and the total power was then calculated in the frequency interval 0.5–40 Hz. Then power in 6 frequency bands, as described in section a (above), were calculated and rounded up to the upper integer (Delta: 0.5–4Hz, Theta: 4–8 Hz, Alpha: 8–12 Hz, Spindle: 12–16 Hz, Beta1: 16–24 Hz, Beta2: 24–40 Hz). Finally, these power values were normalized with respect to the total power in order to obtain the relative spectral energies.

- **Power Value and Frequency at Power Spectrum Peak**

There were several peaks and valleys in the power spectrum of the EEG signals. First, a peak with the maximum power value was selected. Then, the peak power value and the related frequency were selected as 2 features.

- **Harmonic Parameters**

Harmonic parameters are defined as follows:

$$f_c = \frac{\int_{f_L}^{f_H} f \cdot p(f)\mathrm{d}f}{\int_{f_L}^{f_H} p(f)\mathrm{d}f}, \qquad (4)$$

$$f_\sigma = \sqrt{\frac{\int_{f_L}^{f_H} (f - f_c)^2 p(f)\mathrm{d}f}{\int_{f_L}^{f_H} p(f)\mathrm{d}f}}, \qquad (5)$$

$$p_{f_c} = p(f_c). \qquad (6)$$

In Eq. (4), Eq. (5), and Eq. (6), $p(f)$ is signal power spectrum, and f_L and f_H are minimum and maximum frequencies, respectively. The $f(c)$ and $f(\sigma)$ in Eq. (4) and Eq. (5) are similar to normalized values of mean frequency and standard deviation of frequency.

- **Hjorth Parameters**

The n-order spectral moment is defined in Eq. (7):

$$a_n = \int_{-\infty}^{\infty} (2\phi f)^n p(f)\mathrm{d}f, \qquad (7)$$

where $p(f)$ is the power spectrum. The power spectrum is the Fourier transform of the autocorrelation function. In the following, $R(\tau)$ is the autocorrelation function, and $R(0)$ is the variance of the signal:

$$
\begin{aligned}
R(\tau) &= E\left[x(t)x(t+\tau)\right], \\
R(0) &= E\left[x(t)^2\right] = \sigma_0^2, \\
R(\tau) &= \int_{-\infty}^{\infty} p(f)e^{j2\phi f\tau}\mathrm{d}f, \\
R(\tau) &= \int_{-\infty}^{\infty} p(f)\mathrm{d}f = a_0.
\end{aligned} \tag{8}
$$

Therefore, a_0 is the signal's variance, a_2 is the variance of the derivative of the signal, and a_4 is the variance of the second derivative of the signal. In addition, there is a relationship between the signal and its spectral moment: $a_{2n} = \sigma_n^2$. Based on these spectral moments, Hjorth parameters were calculated as follows:

$$
\text{Activity} = a_0 = \sigma_0^2, \tag{9}
$$

$$
\text{Mobility} = \left[\frac{a_2}{a_0}\right]^{1/2} = \frac{\sigma_1}{\sigma_0}, \tag{10}
$$

$$
\begin{aligned}
\text{Complexity} &= \left[\left(\frac{a_4}{a_2}\right) - \left(\frac{a_2}{a_0}\right)\right]^{1/2} = \\
&= \left[\left(\frac{\sigma_2}{\sigma_1}\right)^2 - \left(\frac{\sigma_1}{\sigma_0}\right)^2\right]^{1/2}.
\end{aligned} \tag{11}
$$

In these equations, all a parameters were first estimated by Eq. (12), and then Hjorth parameters were calculated:

$$
a_n = \sum_{f_H}^{f_i = f_L} (2\phi f_i)^2 p(f_i)\Delta f. \tag{12}
$$

As sleep EEG is a nonstationary signal, its spectral moments are variable with time. Therefore, these features can be useful in sleep EEG analysis [37], [38] and [39].

- Mean Absolute Value of EEG Signal Amplitude

The amplitudes of EEG signals are different in Wake and various Sleep Stages. For example, the amplitude of these signals increases in Deep Sleep and decreases in REM Sleep. Therefore, the mean absolute value of the EEG signal amplitude was considered as a feature for further analysis.

Nonlinear Features

The nonlinear dynamic features included DFA-based features and entropy measures. The DFA-based feature extraction consisted of five steps. First, the profile of a zero-mean normalized time series of length N was determined:

$$
Y(i) = \sum_{k=1}^{i} x_k. \tag{13}
$$

Then, the profile was divided into $N_n = N/n$ non-overlapping segments. Subsequently, a local trend for each segment of the data was calculated (by using a first-degree polynomial) and then subtracted from the profile:

$$
\begin{aligned}
z_n(j) &= Y(j) - P_n(j, s), \\
s &= 1, 2, \ldots, N_n, \\
j &= \left[(s-1)n + 1 : sn\right].
\end{aligned} \tag{14}
$$

Afterwards, the variance of the detrended time series $Z_n(j)$ was calculated for each segment:

$$
F_n^2(s) = \frac{1}{n} \sum_{k=(s-1)n+1}^{sn} Z_n^2(k). \tag{15}
$$

Finally, the square root of the average over all N_n segments was calculated to obtain the DFA fluctuation function as follows:

$$
F_n = \left[\frac{1}{N_n} \sum_{s=1}^{N_n} F_2^n(s)\right]^{1/2}. \tag{16}
$$

For calculation of α feature of the DFA in standard epochs of EEG signals, 30 values for n in the interval $\{4.30 \times fs(3750)\}$ were considered [40]. The value of α was then calculated as the slope of the $(\log F(n))$ versus $\log(n)$ for different scale values n:

$$
\begin{aligned}
ApEn(m, r, N) &= \varphi^m(r) - \varphi^{m+1}(r), \\
\varphi^m(r) &= [N - (m-1)\tau]^{-1} \cdot \\
&\quad \cdot \sum_{i=1}^{N-(m-1)\tau} \ln C_i^m(r),
\end{aligned} \tag{17}
$$

where:

$$
C_i^m(r) = \frac{B_i}{N - (m-1)\tau}, \tag{18}
$$

$$
B_i = \text{number of } j \text{ such that } d\left|X_i, X_j\right| \le r.
$$

In the above equations (X_i, X_j) are m-dimensional pattern vectors, whose components are time-delayed versions of the elements in the original time series with delay τ, a multiple of the sampling period, as follows:

$$
\begin{aligned}
X_i &= (X_i, X_{i+\tau}, X_{i+2\tau}, \ldots, X_{i+(m-1)\tau}), \\
X_j &= (X_j, X_{j+\tau}, X_{j+2\tau}, \ldots, X_{j+(m-1)\tau}), \\
&\quad X_i \in R^m \; X_j \in R^m,
\end{aligned} \tag{19}
$$

and $d\left|X_i, X_j\right|$ is a measure of the distance between X_i and X_j. For large values of N, the $ApEn$ is given by:

$$
ApEn(m, r, N) = [N - m\tau]^{-1} \sum_{i=1}^{N-m\tau} -\ln\left(\frac{A_i}{B_i}\right), \tag{20}
$$

where A_i is the number of $X_i's$ within tolerance r of $X_j's$ for the $(m+1)$ - dimensional pattern vector and

B_i is the number of $X'_i s$ with tolerance r of $X'_i s$ in the m-dimensional pattern vector.

Sample Entropy ($SampEn$) is another measure of complexity [41], which is very similar to $ApEn$. The main difference between these two measures is how self-counting is handled in their computation. In $ApEn$ calculation, self-counting is included at each iteration to prevent computing the natural logarithm of zero. However, in the calculation of $SampEn$, the natural logarithm is computed once and self-counting is excluded by requiring that $i \neq j$ in Eq. (20). SampEn is computed by modifying the ApEn formula given in Eq. (20) to:

$$SampEn(m, r, N) = -\ln\frac{A}{B} = -\ln\frac{\sum_{i=1}^{N-m\tau} A_i}{\sum_{i=1}^{N-m\tau} B_i}. \quad (21)$$

The calculation of $ApEn$ and $SampEn$ of EEG signals requires a priori specification of some unknown parameters as: m, the Embedding Dimension (ED)$-r$, a tolerance value and τ, the time delay. For our investigation the following values were selected: $m = 2$, $r = 0.5$ times the standard deviation of the data [21], [41] and $\tau = 11$ samples (0.09 second), [21]. The time delay was determined as the lag at the point for which the autocorrelation function of the signal was near zero for the first time.

2)   Feature Extraction from HRV, ECG-Derived Respiration (EDR) and Res Signals

In this section we describe the processing of the ECG and Res signals listed in Tab. 2. First, HRV and RS-EDR signals were derived from ECG signals and then features were extracted.

HRV and ECG-Derived Respiration (EDR) Signals

The selected ECG segments were filtered by using a FIR band pass filter with low and high cut-off frequencies of 8 and 20 Hz, respectively [42]. In order to extract HRV signals, QRS complexes were first detected by using an Enhanced Hilbert Transform (EHT) algorithm [43] and then were assessed manually for correcting the missing beats. The details of our HRV signal derivation algorithm from ECG segments are reported elsewhere [44].

For the extraction of the EDR signal, some studies in the literature have used 2 leads of ECG signals [45], [46], [47] and [48], while others took advantage of more than 2 leads [46]. In our investigation, however, we made use of ECG lead II alone as this lead seems to be very popular in the most recent literature [50], [51],

[52], [53], [54], [55], [56], [57] and [58]. Therefore, the EDR signals were extracted from lead II ECG signals by using the RSampl or RS-EDR method [56]. The procedure for the extraction of RS-EDR signal from ECG segments is explained in detail in our previous work [25].

Feature Extraction from HRV, Res and RS-EDR Signals

• HRV Signals

A 4-feature set was extracted from the HRV signals using standard epochs as follows:

- The time-domain features including: the median, the Inter-Quartile Range (IQR), the Mean Absolute Difference (MAD), the mean, the standard deviation and the range.

- The nonlinear dynamics features including: DFA-based feature α, which represents the slope of the $\log F(n)$ versus $\log(n)$ in the range $10 \leq n \leq 30$ (before DFA feature extraction, a 1 Hz re-sampling was applied to HRV data) and entropy measures (Shannon entropy, $ApEn$ and $SampEn$).

- DWT-based features including: the normalized values of energy in the VLF, LF, and HF bands (Waves 1:3), the energies in LF/HF (Wave 4), Shannon entropy in VLF, LF, and HF bands (Waves 5:7), the ratio of entropies in the LF and HF bands (Wave 8).

- The Empirical Mode Decomposition-based (EMD-based) features consisting of: normalized values of energy in the VLF, LF, and HF bands computed by the Hilbert energy spectrum (EMD 1:3), the ratio of energies in the LF and HF bands (EMD 4), harmonic parameters such as the central frequency, the deviation of central frequency and the energy in central frequency were extracted from the Hilbert amplitude spectrum (EMD 5-7), $ApEn$ (EMD 8) and SampEn (EMD 9) were calculated from the most significant IMF of the standard epochs. To calculate ApEn and $SampEn$ from HRV signals, $m = 2$, $r = 0.2$ times the standard deviation of the data and $\tau = 1$ sample were selected [59].

When exploring the utility of the DFA and EMD methods in feature extraction from HRV signals using long and standard epochs, we had to make some provisions in our approach. In long epochs of HRV we extracted 10 Intrinsic Mode Functions (IMFs) by the EMD method and the ApEn and SampEn from the 4 most important IMFs were calculated, while in standard epochs we could only extract 2 IMFs and the most important IMF was used in calculation of entropies.

When using the DFA method here, as compared to our previous work [44], we only extracted α from standard epochs of HRV signals, while for long epochs, both α_1 and α_2 were extracted. The n values in Eq. (13), Eq. (14), Eq. (15) and Eq. (16), were considered as $n = 10 : 30$ for α_1 and $n = 60 : 300$ for α_2.

In summary, 34 features were extracted from long epochs, and 27 features from standard epochs of HRV signals. Also, 11 features were extracted from Res signals, which will be explained in the following sections. Automatic sleep staging was then performed using features extracted from a combination of HRV and Res signals in standard and long epoch lengths.

• Res (Thoracic Respiratory) Signals

For this investigation, the Res signals (thoracic excursion) recorded by inductive plethysmography bands and sampled at 10 Hz were used. First, Res signals were filtered by a 10th order Butterworth low pass filter with a 0.8 Hz cut-off frequency in order to remove high frequency noise and variations above the respiratory frequency. Then, a 6-point moving average filtering method was used to smooth out the Res signals and to remove additional peaks. Finally, since the signal was not calibrated in terms of absolute Tidal Volume (TV), we normalized it for each subject and considered only relative differences. The thoracic Res signals were normalized by first detecting the turning points and then calculating the differences between sequential peaks and troughs. The median peak-to-trough amplitude over the entire record was subsequently determined and the signal was normalized by dividing into this value, resulting in median peak-to-trough amplitude equal to unity [39]. The median was more robust to outliers and did not move (change) unless more than half of the signal was contaminated with noise, which in the plethysmogram can be extreme and create very large peak-to-trough values, and can skew the mean.

The steps involved for feature extraction from 0.5-minute epochs of Res signals in Wake and different Sleep Stages were as follows. First, peaks and valleys of thoracic Res signals were detected. Then the missed peaks were corrected by the procedure explained in our previous work [25]. Finally, the TV and the Respiration Rate (ResR) were calculated as the amplitude difference between a successive peak and valley, and the number of breaths per minute, respectively. Typically, the Tidal Volume is the volume of air exhaled during a breath. Therefore, the TV by definition is the difference between the maximum and minimum volume or the integral of the area under the expiratory flow. ResR is the breathing frequency per minute. In this study, the mean, standard deviation and coefficient of variation (standard deviation $\cdot$ 100/mean) of the TV and ResR were computed as features. Other features,

such as $ApEn$, $SampEn$ and Shannon Entropy, were extracted from Res signals. We chose $m = 2$, $r = 0.2$ times the standard deviation of the data, and $\tau = 11$ sample (1.1 second), [18] for this purpose. Finally, the peak frequency of Res signals and the power value in peak frequency were extracted. The Power Spectral Densities (PSDs) of Res signals were estimated by using the nonparametric method and a 1024-point FFT (Fast Fourier Transform). Totally, 11 features were extracted from the 0.5-minute epochs of thoracic Res signals.

• RS-EDR Signals

The procedure for the feature extraction from RS-EDR signals was the same as that used for Res signals. First, peaks and valleys were detected by the method explained in our previous work [25]. Then, the mean, standard deviation and coefficient of the variation of the TV and ResR were computed. The Shannon Entropy, $ApEn$, $SampEn$, peak frequency and power value at peak frequency were calculated from RS-EDR signals too. Therefore, the feature extraction procedure in 0.5-minute epochs of Res and RS-EDR signals was the same as the feature extraction procedure in 5-minute epochs.

Classification

A Support Vector Machine (SVM) classifier was used for automatic sleep staging. The SVM separating hyperplane was calculated by solving the quadratic optimization problem. The Radial Basis Function (RBF) kernel $\left(k(x, y) = e^{-\gamma \|x - y\|^d} \right)$, and the one-against-one method were used for the SVM multi-class classification. The details of classifier training and test procedure are reported in our previous work [25].

The dataset was divided into the training and test sets. It can be seen in Tab. 2 that the Wake Stage has the minimum number of epochs. In staging 0.5-minute epochs, 30 % of the clean epochs in the Wake Stage $(0.30 \cdot 790 = 237)$ were randomly selected for training. In order to get comparable results for all Sleep Stages, equal number of epochs in other Stages was used for training. In staging 5, 4, 3, 2, 1-minute epochs, 80 % of clean epochs in each Stage, were randomly used for training. The performance of the classifier was tested on the remaining epochs in each Stage.

The SVM Recursive Feature Elimination (RFE) method was used for feature selection. This method was developed by Guyon, et al. and has been used in gene selection for cancer classification [61]. In the SVM-RFE method, the effect of removing a feature on an objective function is used as a ranking criterion. Guyon and co-workers used the margin [61] or Total Error Rate (TER), [16] as objective function. In our study here, Correct Classification Ratio (CCR) was used as ranking criterion. We optimized 3 sets of features, which were extracted from 0.5-minute epochs of:

- EEG (34 features),

- HRV and Res signals (38 features with 27 features from HRV and 11 from Res signals),

- combined HRV and RS-EDR signals (38 features with 27 features from HRV and 11 from RS-EDR).

In order to investigate automatic sleep staging by the combination of HRV and Res signals in different epoch lengths, different feature sets were optimized.

3.   Results

In the first phase of this study, we automatically classified the Wake, Stage 1, Stage 2, SWS and REM Sleep by using EEG, combined HRV and RS-EDR, as well as combined HRV and Res features. Feature vectors were extracted from 0.5-minute (standard) epochs. We used a SVM classifier and the optimal parameters for this classifier were found by the procedure described in the Classification section. The SVM-RFE ranking method was applied to 34 features extracted from EEG signals. The best results were obtained when using 25 features.

The discriminative capacity (classification results) for the best 25 features derived from the EEG signals is (are) presented in Tab. 3.

Tab. 2: Automatic Sleep Staging using EEG signals. Results are reported for optimum $C = 8$ and $\sigma = 8$ ($\gamma = 0.0156$). In addition CCR1 (on training data) = 0.945 and CCR2 (on test data) = 0.9228 (25 best features selected), features were extracted from 0.5-minute (standard) epochs.

Sleep Stages	Accuracy	Specificity	Sensitivity
Wake	98.99	99.52	90.59
Stage 1	99.97	100	99.72
Stage 2	90.88	94.32	86.49
SWS	97.42	97.49	97.14
REM	92.17	93.77	87.42
Total on test data	**95.87**	**97.41**	**89.83**
Total on training data	97.79	98.62	94.51

Similarly, the SVM-RFE ranking method was applied to the 38 features (extracted from HRV and RS-EDR signals). The best results were obtained when using 24 features.

The classification results for the best 24 features derived from the HRV and RS-EDR signals are presented in Tab. 4.

Finally, the SVM-RFE ranking method was applied to the 38 features (extracted from HRV and Res sig-

Tab. 3: Automatic Sleep Staging using the combination of HRV and RS-EDR signals. Results are reported for optimum $C = 8192$ and $\sigma = 64$ ($\gamma = 0.00024$). In addition CCR1 (on training data) = 0.63 and CCR2 (on test data) = 0.539 (24 best features selected), features were extracted in 0.5-minute (standard) epochs.

Sleep Stages	Accuracy	Specificity	Sensitivity
Wake	85.9	87.51	60.39
Stage 1	95.15	95.87	86.59
Stage 2	60.45	83.34	31.24
SWS	67.30	74.13	43.14
REM	71.31	79.02	48.04
Total on test data	**76.42**	**85.09**	**43.59**
Total on training data	84.79	90.42	63.03

nals). The best results were obtained when using 22 features.

The classification results for the best 22 features derived from the HRV and Res signals are presented in Tab. 5.

Tab. 4: Automatic Sleep Staging using the combination of HRV and Res signals. Results are reported for optimum $C = 32$ and $\sigma = 16$ ($\gamma = 0.0039$). In addition CCR1 (on training data) = 0.6464 and CCR2 (on test data) = 0.5723. (22 best features selected), features were extracted in 0.5-minute (standard) epochs.

Sleep Stages	Accuracy	Specificity	Sensitivity
Wake	85.63	87.74	52.26
Stage 1	95.32	96.31	83.37
Stage 2	63.92	85.48	36.41
SWS	71.98	76.46	55.42
REM	77.53	83.89	58.64
Total on test data	**79.14**	**86.83**	**49.83**
Total on training data	85.41	90.84	64.64

In our previous work, we performed automatic sleep staging to distinguish between Wake, Stage 2, SWS and REM Sleep by using combined HRV and RS-EDR, as well as combined HRV and Res features based on feature vectors extracted from 5-minute (long) epochs. The results are presented here for comparison. The classification results for the best 35 features derived from the combination of HRV and RS-EDR signals are presented in Tab. 6, [25].

Similarly, the classification results for the best 27 features derived from the combined HRV and thoracic Res signals are presented in Tab. 7, [25].

Figure 3 shows the CCR values versus the number of best features in a decreasing fashion (from n features to 1 feature) in classification by EEG signals (in 0.5-minute), HRV+EDR signals (in 0.5-minute), HRV+RES signals (in 0.5-minute), HRV+EDR signals (in 5-minute) and HRV+RES signals (in 5-minute). In

Tab. 5: Automatic Sleep Staging by using the combination of HRV and RS-EDR signals. Results are reported for optimum $C = 2048$ and $\sigma = 64$ ($\gamma = 0.00024$) when using the 35 best features. In addition, CCR1 (on training data) = 0.73 and CCR2 (on test data) = 0.7, features were extracted in 5-minute (long) epochs [25].

Sleep Stages	Accuracy	Specificity	Sensitivity
Wake	92.70	93.18	87.50
Stage 2	75.00	82.85	65.51
SWS	83.85	89.67	59.45
REM	83.85	88.57	71.15
Total on test data	**83.85**	**89.23**	**67.7**
Total on training data	84.11	89.41	68.23

Tab. 6: Automatic Sleep Staging by using the combination of HRV and Res signals. Results are reported for optimum $C = 8$ and $\sigma = 8$ ($\gamma = 0.0156$) when using the 27 best features. In addition, CCR1 (on training data) =0.83 and CCR2 (on test data) = 0.81, features were extracted in 5-minute (long) epochs [25].

Sleep Stages	Accuracy	Specificity	Sensitivity
Wake	95.83	96.59	87.50
Stage 2	81.25	87.61	73.56
SWS	89.06	91.61	78.37
REM	91.14	93.57	84.61
Total on test data	**89.32**	**92.88**	**78.64**
Total on training data	90.58	93.72	81.17

classification by different signals, the best outcomes (using the best combination of features) are inserted in the diagram.

Finally, the results of sleep staging based on 0.5-minute epochs including both clean test epochs and epochs polluted with artifacts of EEG signals are shown in Tab. 8.

Tab. 7: Automatic Sleep Staging using EEG signals on all 0.5-minute epochs (all data in Tab. 2 except training part of artifact free data). Results are reported for optimum $C = 8$ and $\sigma = 8$ ($\gamma = 0.0156$), and the best combination of 25 features, CCR1 (on training data) = 0.94 and CCR2 (on test data) = 0.85.

Sleep Stages	Accuracy	Specificity	Sensitivity
Wake	95.93	99.47	69.58
Stage 1	99.32	100	99.72
Stage 2	86.28	97.57	80.07
SWS	94.88	99.32	96.5
REM	88.2	98.73	79.68
Total on test data	**92.8**	**95.5**	**82.6**
Total on training data	97.7	98.61	94.51

Comparing the data in Tab. 8 and Tab. 3 shows that the Accuracy value decreased from 95.87 % to 92.8 % (CCR value decreased from 0.92 to

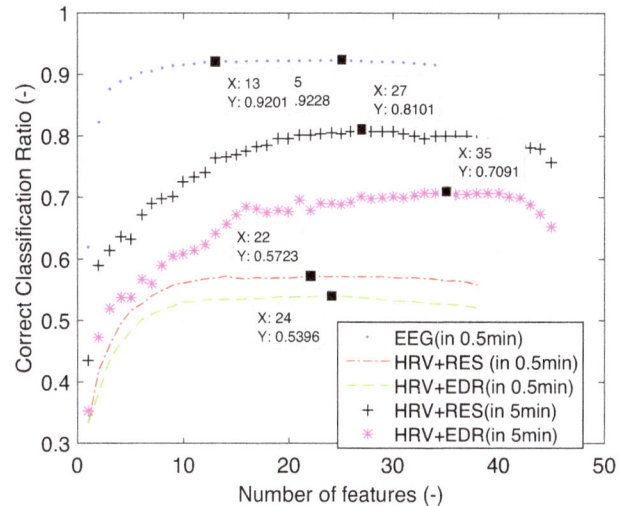

Fig. 3: From right to left, each point in this diagram is the best possible subset (which maximizes CCR) of m features (m is changed from n to 1). When a feature is removed, all possible *n-1* combinations of n remaining features is checked, it means that the removed features in previous steps cannot be tested in the subsequent combinations. The best result in 0.5-min was obtained by the combination of 25 features for EEG signal, 24 features for HRV + EDR signals, and 22 features for HRV + RES signals. For comparison, in 5-min epochs, the best results was obtained by the combination of 35 features for HRV + EDR signals, and 27 features for HRV + RES signals.

0.85). Our previous observations revealed that in automatic sleep staging by cardiorespiratory signals the best result was obtained when feature vectors were extracted from 5-minute epochs. The results based on the HRV + Res signals using 5-minute epoch lengths (clean test epochs and all artefactual epochs) are presented in Tab. 8. In comparison with the results in Tab. 7, we observe that the Accuracy decreased from 89.32 % to 81.1 % (CCR decreased from 0.81 to 0.63).

Tab. 8: Automatic Sleep Staging by using the combination of HRV and Res signals on all 5-minute epochs (all data in Tab. 2 except training part of artifact free data). Results are reported for optimum $C = 8$ and $\sigma = 8$ ($\gamma = 0.0156$) and the best combination of 27 features, CCR1 (on training data) = 0.83 and CCR2 (on test data) = 0.63.

Sleep Stages	Accuracy	Specificity	Sensitivity
Wake	87.7	90.2	48.5
Stage 2	68.3	81.5	57.3
SWS	84.9	86.7	66
REM	83.5	88.4	72.6
Total on test data	**81.1**	**87.4**	**62.3**
Total on training data	90.5	93.72	81.17

4.   Discussion

The main objective of this study was to perform a comparative analysis of the discriminative capacity of EEG and cardiorespiratory signals in automatic sleep staging. Therefore, we investigated a number of related research questions. Firstly, we probed how closely the sleep staging results based on EEG signals compared with the results when discriminative features were solely derived from cardiorespiratory signals in standard epochs. These findings were compared with those achieved by using cardiorespiratory signals in long epochs. Secondly, we investigated automatic sleep staging by cardiorespiratory signals in the presence of some artifacts (like bigeminy) in ECG signals.

The feature vectors were extracted intelligently to appropriately satisfy the requirements of a research study focused on performing automatic sleep staging. Our careful review of the sleep physiology literature revealed that heart and respiration rates varied considerably during Wake and different Sleep Stages. The time-domain and time-frequency methods were used for measuring the variations of the heart rate in Wake and different Sleep Stages [9], [33] and [62] as well as their manifestations in the Autonomic Nervous System (ANS) activity. In addition, some of the features were extracted by nonlinear dynamics system analysis methods. This approach seemed plausible, as it has been shown in the literature that both the linear and nonlinear characteristics of physiologic systems like the one underlying the generation of HRV signals should be considered simultaneously [63]. It should be mentioned that the EMD method was used for the extraction of some new features in our investigation. This method is more compatible with nonlinear characteristics of physiological systems than the DWT (and it does not have the limitation of having to choose the mother wavelet). Ordinarily, the respiration rate and depth are different in Wake and Sleep Stages. These differences were measured by the calculation of the mean, standard deviation, and the Coefficient of Variation (COV) of Respiration Rate and Tidal Volume. Calculating entropy and deploying the DFA method enabled us to evaluate the regularity and fractal characteristics of Respiratory Signals. For performing automatic sleep staging based on EEG signals, we extracted different features to analyze the amplitude and frequency variability and nonlinear characteristics of these signals.

Different sets of features were extracted from HRV, RS-EDR, Thoracic Respiratory and EEG signals. The significance of most of these features was evaluated by applying popular statistical methods (ANOVA and t-test) and was reported in our previous works [25] and [44]. The SVM-RFE method was used for the selection of sub- optimal feature sets and the classification of Sleep Stages was achieved by using a SVM classi-

fier. Automatic sleep staging was performed based on features extracted from 0.5-minute (standard) epochs of EEG signals and by combining features from HRV and Respiratory (reference or RS-EDR) signals derived from both 0.5-minute (standard) and 5-minute (long) epochs of these cardiorespiratory signals. In automatic sleep staging of Wake, Stage 1, Stage 2, SWS and REM Sleep we were able to generate the following results:

- an accuracy of 95.89 % with a CCR = 0.92 when we used 0.5-minute epochs of EEG signals,

- an accuracy of 76.02 % with a CCR = 0.54, when features were derived from standard epoch lengths of a HRV + RS-EDR signals,

- an accuracy of 78.87 % with a CCR = 0.57 when features were extracted from standard epoch lengths of a combination of HRV + Res signals.

It should be pointed out that we used the same labels in training for classifications based on the EEG, HRV + RS-EDR, as well as HRV + Res signals. We also tested our algorithm by using the same label of data samples so that the results would be comparable. We achieved excellent results by performing sleep staging using the standard epoch length of EEG signals and showed that cardiorespiratory signals would not produce acceptable outcomes based on this short epoch length.

As HRV and Respiratory signals are signals with slow dynamic and as "short-term" HRV analysis requires a recommended epoch length of 5 minutes [33] to perform automatic sleep staging by cardiorespiratory signals, we also extracted feature vectors based on 5-minute epochs of these signals. With these feature vectors we were able to distinguish among Wake, Stage 2, SWS and REM Sleep with the following results:

- an accuracy of 89.32 % with a CCR = 0.81 based on HRV + Res signals,

- an accuracy of 83.85 % with a CCR = 0.7 by using HRV + RS-EDR signals [25].

In summary, we observed that automatic sleep staging results when using 0.5-minute (standard) epochs of EEG signals were comparable with results obtained based on HRV + Res signals in 5-minute epochs. In addition, sleep staging results based on HRV + RS-EDR signals (or ECG signal alone) were also acceptable when classification was done in 5-minute epochs.

It is important to emphasize that the above-mentioned results were obtained when features were extracted from artifact-free (clean) signals listed in Tab. 2. However, when we applied our algorithm to the entire 0.5-minute epochs of EEG data (both clean and with artifact), to discriminate among Wake, Stage 1,

Stage 2, SWS and REM Sleep Stages, the accuracy decreased from 95.8 % with a CCR = 0.92 to 92.8 % with a CCR = 0.85. In distinguishing between Wake, Stage 2, SWS and REM Sleep, using all 5-minute epochs (both clean and with artifact) based on HRV +Res signals, the accuracy decreased from 89.3 % with a CCR = 0.81 to 81.1 % with a CCR = 0.63. Therefore, when the algorithm was applied to all epochs (clean and artefactual), the reduction of the accuracy of automatic sleep staging by cardiorespiratory signals was more pronounced than the case when noisy EEG signals were used. This happens due to the presence of some artifacts (like bigeminy) in ECG signals.

5. Conclusion

In this study, we successfully applied a variety of feature extraction methods to derive discriminative and informative features from EEG and cardiorespiratory signals. Sub-optimal feature sets were found by the SVM-RFE method and classified by using a SVM classifier. We observed that the EEG signals could produce excellent outcomes in automatic sleep staging when feature vectors were extracted from 0.5-minute (standard) epochs. We also made the general observation that reasonably good results could be achieved when sleep staging is performed based on features derived from 5-minute epochs of the combination of HRV and Res signals. Moreover, ECG signals alone could produce acceptable results (when feature vectors were extracted in 5-minute epochs). Therefore, the closer the RS-EDR signals resembled the reference respiratory signals, the better the results by ECG signals alone became.

Here we strived to perform a comprehensive investigation into automatic sleep staging based on simultaneous analysis of EEG and cardiorespiratory signals. We demonstrated that automatic classification of Wake and different Sleep Stages is possible by extracting feature vectors from long epochs of cardiorespiratory signals alone. In addition, we observed that the presence of some artifacts (like bigeminy) decreases the classification results based on cardiorespiratory signals more than those achieved from noisy EEG signals. We extracted features by using a combination of linear and nonlinear methods. More features could be added to these extracted features to improve the classification results based on cardiorespiratory signals alone. For example, recent research has shown that current algorithms used to perform spectral analysis of HRV signals are not able to completely separate the sympathetic and parasympathetic components of the HRV signals as manifested in the LF and HF bands [10], [33], [62] and [64]. Recently a method called Principal Dynamic Mode (PDM) analysis of HRV signals that allows more precise decomposition of the HRV signal spectral components and hence complete separation of the sympathetic and parasympathetic influences of the ANS activity was presented in the literature [65]. In our future work, we envision using the PDM approach in achieving more precise spectral analysis of the HRV signals and investigating the variations of features extracted from them in the Wake and Sleep Stages. Moreover, it has been shown that cardiac and respiratory rhythms could synchronize with different ratios. A pronounced sleep-stage dependency has been observed with the degree of synchronization: low during REM and Wake, higher during Light Sleep, and most pronounced during Deep Sleep [66] and [67]. These synchronization ratios could serve as useful features and would inform our future algorithm enhancements in performing automatic sleep staging based on cardiorespiratory signals alone.

References

[1] RECHTSCHAFFEN, A. and A. KALES. A Manual of standardized terminology, techniques and scoring system for sleep stages of human subjects. *Electroencephalography and Clinical Neurophysiology.* 1968, vol. 26, iss. 6, pp. 644. ISSN 0013-4694. DOI: 10.1016/0013-4694(69)90021-2.

[2] IBER, C., S. ANCOLI-ISRAEL, A. CHESSON and S. F. QUAN. The AASM Manual for the scoring of sleep and associated events: Rules, Terminology and Technical Specifications. In: *American Academy of Sleep Medicine* [online]. 2007. https://aasm.org/.

[3] SHEPARD, J. W. *Atlas of sleep medicine.* Mount Kisco, NY: Futura Publishing Company, 1991. ISBN 978-0879935092.

[4] HASSAN, A. R. and M. I. H. BHUIYAN. Computer-aided sleep staging using complete ensemble empirical mode decomposition with adaptive noise and bootstrap aggregating. *Biomedical Signal Processing and Control.* 2016, vol. 24, iss. 1, pp. 1–10. ISSN 1746-8094. DOI: 10.1016/j.bspc.2015.09.002.

[5] LAN, K. C., D. W. CHANG, C. E. KUO, M. Z. WEI, Y. H. LI, F. Z. SHAW and S. F. LIANG. Using off-the-shelf lossy compression for wireless home sleep staging, *Journal of Neuroscience Methods.* 2015, vol. 246, iss. 1, pp. 142–152. ISSN 0165-0270. DOI: 10.1016/j.jneumeth.2015.03.013.

[6] LAJNEF, T., S. CHAIBI, P. RUBY, P. E. AGUERA, J. B. EICHENLAUB, M. SAMET, A. KACHOURI and K. JERBI. Learning machines

and sleeping brains: Automatic sleep stage classification using decision-tree multi-class support vector machines. *Journal of Neuroscience Methods*. 2015, vol. 250, iss. 1, pp. 94–105. ISSN 0165-0270. DOI: 10.1016/j.jneumeth.2015.01.022.

[7] KAYIKCIOGLU, T., M. MALEKI and K. EROGLU. Fast and accurate PLS-based classification of EEG sleep using single channel data. *Expert Systems with Applications*. 2015, vol. 42, iss. 21, pp. 7825–7830. ISSN 0957-4174. DOI: 10.1016/j.eswa.2015.06.010.

[8] BRIGNOL, A., T. AL-ANI and X. DROUOT. Phase space and power spectral approaches for EEG-based automatic sleep–wake classification in humans: A comparative study using short and standard epoch lengths. *Computer Methods and Programs in Biomedicine*. 2013, vol. 109, iss. 1, pp. 227–238. ISSN 0169-2607. DOI: 10.1016/j.cmpb.2012.10.002.

[9] NAZERAN, H., Y. PAMULA and K. BEHBEHANI. Heart rate variability (HRV): Sleep disorder breathing. *Wiley Encyclopedia of Biomedical Engineering*. Hoboken: John Wiley & Sons, 2006. ISBN 0471740365. DOI: 10.1002/9780471740360.ebs1387.

[10] MOORCROFT, W. and P. BELCHER. Understanding Sleep and Dreaming. *The Body During Sleep, Part II*. Boston: Springer, 2005. ISBN 978-1-4614-6466-2.

[11] PENZEL, T., J. W. KANTELHARDT, L. C. CHANG, K. VOIGT and C. VOGELMEIER. Dynamics of Heart Rate and Sleep Stages in Normals and Patients with Sleep Apnea. *Neuropsychopharmacology*. 2003, vol. 28, iss. 1, pp. 48–53. ISSN 0893-133x. DOI: 10.1038/sj.npp.1300146.

[12] PENZEL, T., J. W. KANTELHARDT, L. GROTE, J. H. PETER and A. BUNDE. Comparison of Detrended Fluctuation Analysis and Spectral Analysis for Heart Rate Variability in Sleep and Sleep Apnea. *IEEE Transaction on Biomedical Engineering*. 2003, vol. 50, no. 10, pp. 1143–1151. ISSN 1558-2531. DOI: 10.1109/TBME.2003.817636.

[13] REDMOND, S. J. and C. HENEGHAN. Cardiorespiratory-based Sleep Staging in Subject with Obstructive Sleep Apnea. *IEEE Transaction on Biomedical Engineering*. 2006, vol. 53, iss. 3, pp. 485–496. ISSN 1558-2531. DOI: 10.1109/TBME.2005.869773.

[14] REDMOND, S. J., P. CHAZAL, C. BRIEN, S. RYAN, W. MCNICHOLAS and C. HENEGHAN. Sleep Staging Using Cardiorespiratory Signals. *Somnologie - Schlafforschung und Schlafmedizin*. 2007, vol. 11, iss. 4, pp. 245–256. ISSN 1432-9123. DOI: 10.1007/s11818-007-0314-8.

[15] REDMOND, S. J. and C. HENEGHAN. Electrocardiogram-Based Automatic Sleep Staging in Sleep Disordered Breathing. In: *Computers in Cardiology*. Thessaloniki Chalkidiki: IEEE, 2003, pp. 609–612. ISBN 0-7803-8170-X. DOI: 10.1109/CIC.2003.1291229.

[16] ADNANE, M., Z. JIANG and Z. YAN. Sleep–wake stages classification and sleep efficiency estimation using single-lead electrocardiogram. *Expert Systems with Applications*. 2012, vol. 39, iss. 1, pp. 1401–1413. ISSN 0957-4174. DOI: 10.1016/j.eswa.2011.08.022.

[17] MENDEZ, M. O., M. MATTEUCCI, V. CASTRONOVO, L. FERINI-STRAMBI, S. CERUTTI and A. M. BIANCHI. Sleep staging from Heart Rate Variability: time-varying spectral features and Hidden Markov Models. *International Journal of Biomedical Engineering and Technology*. 2010, vol. 3, no. 3/4, pp. 246–263. ISSN 1752-6418. DOI: 10.1504/IJBET.2010.032695.

[18] KESPER, K., S. CANISIUS, T. PENZEL, T. PLOCH and W. CASSEL. ECG signal analysis for the assessment of sleep-disordered breathing and sleep pattern. *Medical & Biological Engineering & Computing*. 2012, vol. 50, iss. 2, pp. 135–144. ISSN 0140-0118. DOI: 10.1007/s11517-011-0853-9.

[19] CARSKADON, M. A., K. ELARVEY, W. DEMENT, F. B. SIMMONS and T. F. ANDERS. Respiration During Sleep in Children. *The Western journal of medicine*. 1978, vol. 128, no. 6, pp. 477–481. ISSN 0093-0415.

[20] MIYATA, M., N. BURIOKA, H. SUYAMA, T. SAKO, T. NOMURA, T. TAKESHIMA, S. HIGAMI and E. SHIMIZU. Non-linear behaviour of respiratory movement in obstructive sleep apnoea syndrome. *Clinical Physiology and Functional Imaging*. 2002, vol. 22, iss. 5, pp. 320–327. ISSN 1475-0961. DOI: 10.1046/j.1475-097X.2002.00438.x.

[21] BURIOKA, N., G. CORNELISSEN, F. HALBERG, D. T. KAPLAN, H. SUYAMA, T. SAKO and E. SHIMIZU. Approximate Entropy of Human Respiratory Movement During Eye-Closed Waking and Different Sleep Stages. *Chest*. 2003, vol. 123, iss. 1, pp. 80–86. ISSN 0012-3692. DOI: 10.1378/chest.123.1.80.

[22] FADEL, P. J., S. M. BARMAN, S. W. PHILLIPS and G. L. GEBBER. Fractal Fluctuations in Human Respiration. *Journal of Applied Physiology*. 2004, vol. 97, no. 6, pp. 2056–2064. ISSN 8750-7587. DOI: 10.1152/japplphysiol.00657.2004.

[23] LARSEN, P. D., D. E. ELDER, Y. C. TZENG, A. J. CAMPBELL and D. C. GALLETLY. Fractal Characteristics of Breath to Breath Timing in Sleeping Infants. *Respiratory Physiology & Neurobiology*. 2004, vol. 139, iss. 3, pp. 263–270. ISSN 1569-9048. DOI: 10.1016/j.resp.2003.11.001.

[24] WYSOCKI, M., M. N. FIAMMA, C. STRAUS, C. S. POON and T. SIMILOWSKI. Chaotic Dynamics of Resting Ventilatory Flow in Humans Assessed through Noise Titration. *Respiratory Physiology & Neurobiology*. 2006, vol. 153, iss. 1, pp. 54–65. ISSN 1569-9048. DOI: 10.1016/j.resp.2005.09.008.

[25] EBRAHIMI, F., S. K. SETAREHDAN and H. NAZERAN. Automatic sleep staging by simultaneous Analysis of ECG and Respiratory Signals in long epochs. *Biomedical Signal Processing and Control*. 2015, vol. 18, iss. 1, pp. 69–79. ISSN 1746-8094. DOI: 10.1016/j.bspc.2014.12.003.

[26] QUAN, S. F., B. V. HOWARD, C. IBER, F. J. NIETO and J. P. KILEY. The Sleep Heart Health Study: Design, Rationale, and Methods. *Sleep*. 1997, vol. 20, iss. 12, pp. 1077–1085. ISSN 1550-9109. DOI: 10.1093/sleep/20.12.1077.

[27] JAVAHERI, S., T. PARKE, J. LIMING, W. S. CORBETT and H. NISHIYAMA. Sleep apnea in 81 ambulatory male patients with stable heart failure. Types and their prevalences, consequences, and presentations. *Circulation*. 1998, vol. 99, iss. 12, pp. 2709–2712. ISSN 0009-7322. DOI: 10.1161/01.cir.97.21.2154.

[28] MEHRA, R., E. BENJAMIN, E. SHAHAR, D. J. GOTTLIEB, R. NAWABIT, H. L. KIRCHNER and S. REDLINE. Association of nocturnal arrhythmias with sleep-disordered breathing: the sleep heart health study. *American Journal of Respiratory and Critical Care Medicine*. 2006, vol. 173, no. 8, pp. 910–916. ISSN 1073-449x. DOI: 10.1164/rccm.200509-1442OC.

[29] BARANCHUK, A., C. S. SIMPSON, D. P. REDFEARN, K. MICHAEL and M. FITZPATRICK. Understanding the association between sleep apnea & cardiac arrhythmias. *Revista Electrofisiologia y Arritmias*. 2008, vol. 1, iss. 5, pp. 5–6. ISSN 1851-7595.

[30] HERSI, A. S. Obstructive sleep apnea and cardiac arrhythmias. *Annals of Thoracic Medicine*. 2010, vol. 5, iss. 1, pp. 10–17. ISSN 1817-1737. DOI: 10.4103/1817-1737.58954.

[31] MONAHAN, K., A. STORFER-ISSER, R. MEHRA, E. SHAHAR, M. MITTLEMAN, J. ROTTMAN, N. PUNJABI, M. SANDERS S. QUAN, H. RESNICK and S. REDLINE. Triggering of Nocturnal Arrhythmias by Sleep-Disordered Breathing Events. *Journal of the American College of Cardiology*. 2009, vol. 54, iss. 19, pp. 1797–1804. ISSN 0735-1097. DOI: 10.1016/j.jacc.2009.06.038.

[32] BARANCHUK, A. Sleep apnea, cardiac arrhythmias, and conduction disorders. *Journal of Electrocardiology*. 2012, vol. 45, iss. 5, pp. 508–512. ISSN 0022-0736. DOI: 10.1016/j.jelectrocard.2012.03.003.

[33] Task Force of the European Society of Cardiology and North American Society of Pacing and Electrophysiology Heart rate variability, standards of measurement, physiological interpretation and clinical use. *European Heart Journal*. 1996, vol. 17, iss. 1, pp. 354–381. ISSN 0195-668X.

[34] HIRSHKOWITZ, M. Normal human sleep: an overview. *Medical Clinics of North America*. 2004, vol. 88, iss. 3, pp. 551–655. ISSN 0025-7125. DOI: 10.1016/j.mcna.2004.01.001.

[35] BURROS, C. S., R. A. GOPINATH and H. GUO. *Introduction to wavelets and wavelet transforms: a primer*. Upper Saddle River, N.J.: Prentice Hall, 1998. ISBN 978-0134896007.

[36] DURKA, P., U. MALINOWSKA, U. SYELENBERGER, A. WAKAROW and K. BLINOWSKA. High resolution parametric description of slow wave sleep. *Journal of Neuroscience Methods*. 2005, vol. 147, iss. 1, pp. 15–21. ISSN 0165-0270. DOI: 10.1016/j.jneumeth.2005.02.010.

[37] ESTRADA, E., H. NAZERAN, P. NAVA, K. BEHBEHANI, J. BURK and E. LUCAS. EEG Feature extraction for classification of Sleep Stages. In: *The 26th Annual International Conference of the IEEE Engineering in Medicine and Biology Society*. San Francisco: IEEE, 2004, pp. 196–199. ISBN 0-7803-8439-3. DOI: 10.1109/IEMBS.2004.1403125.

[38] VAN HESE, P., W. PHILIPS, J. KONINCK, R. VAN DE WALLE and I. LEMAHIEU. Automatic Detection of Sleep Stages using the EEG. In: *2001 Conference Proceedings of the 23rd Annual International Conference of the IEEE Engineering in Medicine and Biology Society*. Istanbul: IEEE, 2001, pp. 1944–1947. ISBN 0-7803-7211-5. DOI: 10.1109/IEMBS.2001.1020608.

[39] NIEDERMEYER, E., F. LOPES DA SILVA. *Electroencephalography: basic principles, clinical applications, and related fields*. 3th ed. Baltimore: Williams & Wilkins, 1993. ISBN 978-0683302844.

[40] FARAG, A. F., S. M. EL-METWALLY and A. ABDEL AAL MORSY. *Automated Sleep Staging Using Detrended Fluctuation Analysis of Sleep EEG*. Soft Computing Applications. Berlin: Springer, 2013, pp. 501–510. ISBN 978-3-642-33941-7. DOI: 10.1007/978-3-642-33941-7_44.

[41] RICHMAN J. S. and J. R. MOORMAN. Physiological Time-series Analysis Using Approximate Entropy and Sample Entropy. *American Journal of Physiology - Heart and Circulatory Physiology*. 2000, vol. 278, no. 6, pp. 2039–2049. ISSN 0363-6135.

[42] BENITEZ, D., P. A. GAYDECKI, A. ZAIDI and A. P. FITZPATRICK. The Use of the Hilbert Transform in ECG Signal Analysis. *Computers in Biology and Medicine*. 2001, vol. 31, iss. 5, pp. 399–406. ISSN 0010-4825. DOI: 10.1016/S0010-4825(01)00009-9.

[43] CHATLAPALLI, S., H. NAZERAN, V. MELARKOD, R. KRISHNAM, E. ESTRADA, Y. PAMULA and S. CABRERA. Accurate Derivation of Heart Rate Variability Signal for Detection of Sleep Disordered Breathing in Children. In: *The 26th Annual International Conference of the IEEE Engineering in Medicine and Biology Society*. San Francisco: IEEE, 2004, pp. 538–541. ISBN 0-7803-8439-3. DOI: 10.1109/IEMBS.2004.1403213.

[44] EBRAHIMI, F., S. K. SETAREHDAN, J. A. MOYEDA and H. NAZERAN. Automatic sleep staging using empirical mode decomposition, discrete wavelet transform, time-domain, and nonlinear dynamics features of heart rate variability signals. *Computer Methods and Programs in Biomedicine*. 2013, vol. 112, iss. 1, pp. 47–57. ISSN 0169-2607. DOI: 10.1016/j.cmpb.2013.06.007.

[45] MOODY, G. B., R. G. MARK, A. ZOCCOLA and S. MANTERO. Derivation of respiratory signals from multi lead ecgs. *Computers in Cardiology*. 1985, vol. 12, iss. 1, pp. 113–116. ISSN 0276-6574.

[46] BEHBEHANI, K., S. VIJENDRA, J. R. BURKE and E. A. LUCAS. An investigation of the mean electrical axis angle and respiration during sleep. In: *Proceedings of the Second Joint 24th Annual Conference and the Annual Fall Meeting of the Biomedical Engineering Society*. Houston: IEEE, 2002, pp. 1550–1551. ISBN 0-7803-7612-9. DOI: 10.1109/IEMBS.2002.1106531.

[47] ZHAO, L., S. REISMAN and T. FINDLEY. Derivation of respiration from electrocardiogram during heart rate variability studies. In: *Computers in Cardiology*. Bethesda: IEEE, 1994, pp. 53–56. ISBN 0-8186-6570-X. DOI: 10.1109/CIC.1994.470251.

[48] CAGGIANO, D. and S. REISMAN. Respiration derived from the electrocardiogram: A quantitative comparison of three different methods. In: *Proceedings of the IEEE 22nd Annual Northeast Bioengineering Conference*. New Brunswick: IEEE, 1996, pp. 103–104. ISBN 0-7803-3204-0. DOI: 10.1109/CIC.1998.731718.

[49] TRAVAGLINI, A., C. LAMBERTI, J. DEBIE and M. FERRI. Respiratory Signal Derived from Eight-lead ECG. In: *Computers in Cardiology*. Cleveland: IEEE, 1998, pp. 65–68. ISBN 0-7803-3204-0. DOI: 10.1109/NEBC.1996.503238.

[50] BRIEN, C. O. and C. HENEGHAN. A comparison of algorithms for estimation of a respiratory signal from the surface electrocardiogram. *Computers in Biology and Medicine*. 2007, vol. 37, iss. 3, pp. 305–314. ISSN 0010-4825. DOI: 10.1016/j.compbiomed.2006.02.002.

[51] MASON, C. and L. TARASSENKO. Quantitative Assessment of Respiratory Derivation Algorithm. In: *2001 Conference Proceedings of the 23rd Annual International Conference of the IEEE Engineering in Medicine and Biology Society*. Istanbul: IEEE, 2001, pp. 1998–2001. ISBN 0-7803-7211-5. DOI: 10.1109/IEMBS.2001.1020622.

[52] CHAYAL, P., T. PENZEL and C. HENEGHAN. Automated detection of obstructive sleep apnea at different time scales using the electrocardiogram. *Physiological Measurement*. 2004, vol. 25, no. 4, pp. 967–983. ISSN 0967-3334. DOI: 10.1088/0967-3334/25/4/015.

[53] WIDJAJA, D., J. TAELMAN, S. VANDEPUT, M. BRAEKEN, R. OTTE, B. BERGH and S. HUFFE. ECG-Derived Respiration: Comparison and New Measures for Respiratory variability. In: *Computing in Cardiology*. Belfast: IEEE, 2010, pp. 149–152. ISBN 978-1-4244-7319-9.

[54] YI, W. J. and S. PARK. Derivation of Respiration from ECG Measured without Subject's Awareness Using Wavelet Transform. In: *Proceedings of the Second Joint 24th Annual Conference and the Annual Fall Meeting of the Biomedical Engineering Society*. Houston: IEEE, 2002, pp. 130–131. ISBN 0-7803-7612-9. DOI: 10.1109/IEMBS.2002.1134420.

[55] BOYLE, J., N. BIDARGADDI, A. SARELA and M. KARUNANITHI. Automatic Detection of Respiration Rate From Ambulatory Single-Lead ECG. *IEEE Transaction on Information Technology in Biomedicine.* 2009, vol. 13, iss. 6, pp. 890–896. ISSN 1089-7771. DOI: 10.1109/TITB.2009.2031239.

[56] BABAEIZADEH, S., S. H. ZHOU, S. D. PITTMAN and D. P. WHITE. Electrocardiogram-derived respiration in screening of sleep-disordered breathing. *Journal of Electrocardiography.* 2011, vol. 44, iss. 6, pp. 700–706. ISSN 00220736. DOI: 10.1016/j.jelectrocard.2011.08.004.

[57] DING, S., X. ZHU, W. CHEN and D. WEI. Derivation of Respiratory Signal from Single-Channel ECGs Based on Source Statistics. *International Journal of Bioelectromagnetism.* 2004, vol. 6, no. 2, pp. 43–49. ISSN 1456-7865.

[58] BALOCCHI, R., D. MENICUCCI, E. SANTARCANGELO, L. SEBASTIANI, A. GEMIGNANI, B. GHELARDUCCI and M. VARANINI. Deriving the respiratory sinus arrhythmia from the heartbeat time series using empirical mode decomposition. *Chaos, Solitons & Fractals.* 2004, vol. 20, iss. 1, pp. 171–177. ISSN 0960-0779. DOI: 10.1016/S0960-0779(03)00441-7.

[59] CHON, K., C. G. SCULLY and S. LU. Approximate Entropy for All Signals. *IEEE Engineering in Medicine and Biology Magazine.* 2009, vol. 28, iss. 6, pp. 18–23. ISSN 0739-5175. DOI: 10.1109/MEMB.2009.934629.

[60] HSU, C. W., C. C. CHANG and C. J. LIN. A Practical Guide to Support Vector Classification. In: *CSIE NTU* [Online]. Available at: https://www.csie.ntu.edu.tw/~cjlin/libsvm/index.html.

[61] GUYON, I., J. WESTON, S. BARNHILL and V. VAPNICK. Gene selection for cancer classification using support vector machines. *Machine Learning.* 2002, vol. 46, iss. 1–3, pp. 389–422. ISSN 0885-6125. DOI: 10.1023/A:1012487302797.

[62] MANCIA, G. Autonomic modulation of the cardiovascular system during sleep. *New England Journal of Medicine.* 1993, vol. 328, iss. 1, pp. 347–349. ISSN 0028-4793. DOI: 10.1056/NEJM199302043280511.

[63] VIGO, D., J. DOMINGUEZ, S. GUINJOAN, M. SCARAMAL, E. RUFFA, J. SOLERNO, L. N. SIRI and D. CARDINALI. Nonlinear analysis of heart rate variability within independent frequency components during the sleep-wake cycle. *Autonomic Neuroscience.* 2010, vol. 154, iss. 1–2, pp. 84–88. ISSN 1566-0702. DOI: 10.1016/j.autneu.2009.10.007.

[64] AKO, M., T. KAWARA, S. UCHIDA, S. MIYAZAKI, K. NISHIHARA, J. MUKAI, K. HIRAO, J. AKO and Y. OKUBO. Correlation between electroencephalography and heart rate variability during sleep. *Psychiatry and Clinical Neurosciences.* 2003, vol. 57, iss. 1, pp. 59–65. ISSN 1323-1316. DOI: 10.1046/j.1440-1819.2003.01080.x.

[65] ZHONG, Y., H. WANG, K. H. JU, K.-M. JAN and K. H. CHON. Nonlinear Analysis of the Separate Contributions of Autonomic Nervous Systems to Heart Rate Variability Using Principal Dynamic Modes. *IEEE Transactions on Biomedical Engineering.* 2004, vol. 51, no. 2, pp. 255–262. ISSN 0018-9294. DOI: 10.1109/TBME.2003.820401.

[66] BARTSCH, R., J. KANTELHARDT, T. PENZEL and S. HAVLIN. Experimental evidence for phase synchronization transitions in the human cardiorespiratory system. *Physical Review Letters.* 2007, vol. 98, iss. 5, pp. 54–102. ISSN 0031-9007. DOI: 10.1103/PhysRevLett.98.054102.

[67] BARTSCH, R., A. SCHUMANN, J. KANTELHARDT, T. PENZEL and P. IVANOV. Phase transitions in physiologic coupling. *Proceedings of the National Academy of Sciences.* 2012, vol. 109, no. 26, pp. 10181–10186. ISSN 0027-8424. DOI: 10.1073/pnas.1204568109.

About Authors

Farideh EBRAHIMI Farideh Ebrahimi received the B.Sc. degree in Biomedical Engineering from Amirkabir University of Technology (Tehran Polytechnic), Tehran, Iran, in 2004, the M.Sc. degree in Biomedical Engineering from Shahed University, Tehran, Iran, in 2008, and the Ph.D. degree in Biomedical Engineering from University of Tehran, Tehran, Iran, in 2014 under supervision of Professor Setaredan. She has published over 15 papers in international journals & conferences. Her teaching and research areas include Digital Signal Processing, Biomedical Signal Processing, Artificial Neural Networks and Pattern recognition.

Seyed Kamaledin SETAREHDAN Seyed Kamaledin Setarehdan received the B.Sc. degree in Electronic Engineering from the University of Tehran, Tehran, Iran, in 1989 and the M.Sc. degree in Biomedical Engineering from Sharif University of Technology, Tehran, Iran in 1992. He received the Ph.D. degree from the Electrical and Electronic

Engineering Department of the University of Strath-clyde in Glasgow, UK in 1998 in the field of medical image and signal processing. From March 1998 to December 2000 he was a postdoctoral research fellow in the Signal Processing Division of the Electrical and Electronic Engineering Department, University of Strathclyde in Glasgow, UK. In January 2000, he joined the School of Electrical and Computer Engineering, College of Engineering, University of Tehran, Tehran, Iran, as an assistant professor where he is currently a full professor. Dr. Setarehdan's main research interests are medical signal and image processing in general, medical ultrasound, EEG based brain computer interfacing and medical applications of the Near-Infrared Spectroscopy.

Radek MARTINEK Radek Martinek was born in 1984 in Czech Republic. In 2009 he received Master's degree in Information and Communication Technology from VSB–Technical University of Ostrava. Since 2012 he worked here as a Research Fellow. In 2014 he successfully defended his dissertation thesis titled "The Use of Complex Adaptive Methods of Signal Processing for Refining the Diagnostic Quality of the Abdominal Fetal Electrocardiogram". He became an Associate Professor in Technical Cybernetics in 2017 after defending the habilitation thesis titled "Design and Optimization of Adaptive Systems for Applications of Technical Cybernetics and Biomedical Engineering Based on Virtual Instrumentation". He works as an Associate Professor at VSB–Technical University of Ostrava since 2017. His current re-search interests include: Digital Signal Processing (Linear and Adaptive Filtering, Soft Computing - Artificial Intelligence and Adaptive Fuzzy Sys-tems, Non-Adaptive Methods, Biological Signal Processing, Digital Processing of Speech Signals); Wireless Communications (Software-Defined Radio); Power Quality Improvement. He has more than 70 journal and conference articles in his research areas.

Homer NAZERAN Nazeran holds B.Sc., M.Sc. and Ph.D. degrees in Electrical (Honors), Clinical and Biomedical Engineering from UT Austin, Case Western Reserve and University of Texas South-western Medical Center (UTSWM) at Dallas/UTA, respectively. He has close to 3 decades of experience in industry and academia and has practiced and taught biomedical engineering in the Middle East, Europe, Australia and USA. In Australia, with Professor Andrew Downing he co-founded the School of Engi-neering at the Flinders University of South Australia, introduced and established the electrical and electron-ics and biomedical engineering degree programs (1991 to 2001). He returned to the University of Texas at Arlington as a visiting professor in 1997 and 2001. He joined UTEP in 2002 to create and establish biomed-ical engineering degree programs at the Department of Electrical and Computer Engineering. His research interests are in the areas of computer modeling of physiological systems, intelligent biomedical instru-mentation and biomedical signal processing as applied to chronic health conditions and telemedicine. He has more than 150 journal and conference articles in his research areas published in IEEE Engineering in Medicine and Biology Society (EMBS) and other flagship international conference proceedings. He is a reviewer for several national and international jour-nals in his related fields including IEEE Transactions on Biomedical Engineering, Medical and Biological Engineering and Computing, Biomedical Engineering Online and others. His teaching interests are in electronics, biomedical instrumentation, physiological systems, and biomedical signal processing. He is also interested in development of novel teaching methods, lifelong learning and critical thinking habits in the classroom and interdisciplinary education based on application of nonlinear dynamics systems (complex-ity) theory. His research, teaching and professional activities have been supported by NIH, NSF, and DOE among others.

20

VDCC Based Dual-Mode Quadrature Sinusoidal Oscillator with Outputs at Appropriate Impedance Levels

Mayank SRIVASTAVA[1], Dinesh PRASAD[2]

[1]Department of Electrical, Electronics and Communication Engineering, The Northcap University,
HUDA Sector-23 A, 122017 Gurgaon, India
[2]Department of Electronics and Communication Engineering, Faculty of Engineering and Technology,
Jamia Millia Islamia, 110025 New Delhi, India

mayank2780@gmail.com, dprasad@jmi.ac.in

Abstract. *This article presents a new dual-mode (i.e. both current-mode and voltage-mode) quadrature sinusoidal oscillator using two Voltage Differencing Current Conveyors (VDCCs), two resistors and two capacitors. The proposed configuration use only grounded passive elements and enjoys independent resistor/electronic tuning of both Condition of Oscillation (CO) as well as Frequency of Oscillation (FO). The quadrature current and voltage mode outputs of this circuit are available at appropriate impedance terminals. The behavior of presented oscillator is also examined under non ideal/parasitic conditions. The validity of the proposed configuration has been confirmed by SPICE simulations with TSMC 0.18 µm process parameters.*

Keywords

Dual-mode oscillator, electronic control, high impedance CM outputs, low impedance VM output, VDCC.

1. Introduction

Oscillator is an important circuit configuration, which finds several applications in measurement, signal processing, communication, instrumentation and control systems [1], [2], [3], [4], [5]. A quadrature sinusoidal oscillator which provides two 900 phase shifted sinusoidal outputs simultaneously is widely used in telecommunication engineering applications such as in quadrature mixers, single sideband modulators and direct-conversion receivers or for measurement purpose in se-lective voltmeters and vector generators [6], [7], [8]. Several VM and CM quadrature sinusoidal oscillators employing different Active Building Blocks (ABBs) have been reported in open literature. Careful investigation of literature available on quadrature oscillators reveals that most of the oscillators provide either voltage mode quadrature outputs or current mode quadrature outputs but only very few circuits are those which provide voltage mode and current mode quadrature outputs simultaneously. This simultaneous availability of quadrature outputs is very useful in mixed mode applications where current and voltage signals are required together.

The earlier work on dual-mode quadrature sinusoidal Oscillators employing different active element(s) has been reported in [9], [10], [11], [12], [13], [14], [15], [16], [17], [18], [19], [20], [21]. Unfortunately, these reported circuits suffer from one or more of following drawbacks: (i) presence of floating passive component(s), which is not desirable from the viewpoint of monolithic integration (ii) lack of independent electronic control of both FO and CO (iii) lack of independent resistive control of both FO and CO (iv) non-availability of quadrature VM outputs at low impedance terminals (which need additional voltage followers for cascading) and (v) non-availability of high impedance explicit quadrature CM outputs (which need additional current followers for sensing and taking out the currents) (vi) use of ABB with current copy terminal(s) to get explicit CM outputs (which need additional MOSFETs/BJTs to realize current copy terminal(s).

Therefore, the purpose of this article is to present a new CM/VM sinusoidal quadrature oscillator employing two VDCCs, two resistors and two capacitors which enjoys following advantageous features simultaneously

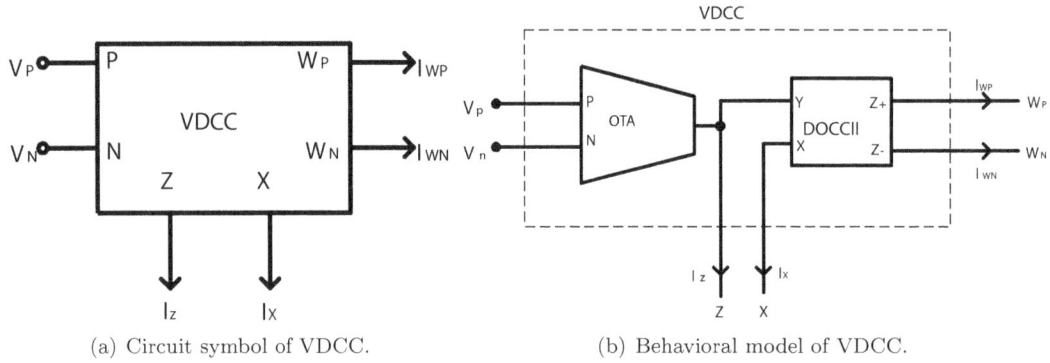

(a) Circuit symbol of VDCC.   (b) Behavioral model of VDCC.

Fig. 1: VDCC circuit symbol and behavioral model.

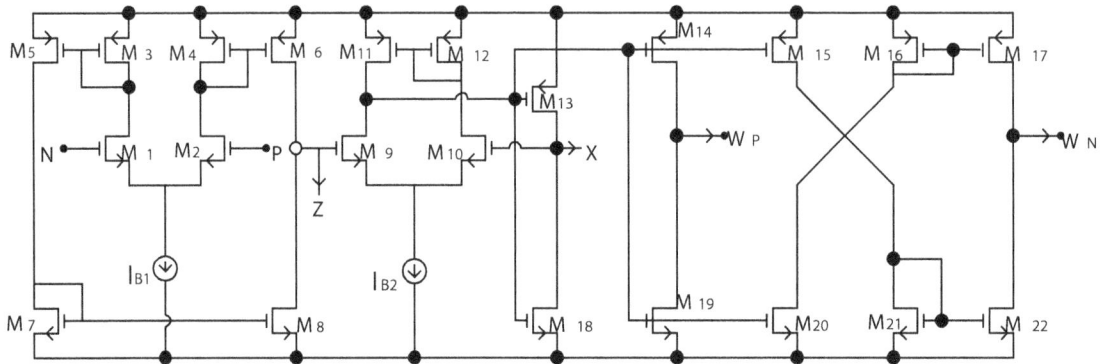

Fig. 2: CMOS implementation of VDCC [23].

which are not available with any of the previously proposed quadrature oscillator configurations;

- Use of all grounded passive components.

- Independent electronic control of both FO and CO.

- Independent resister control of both FO and CO.

- Independent tuning of FO and CO under non-ideal conditions.

- Independent tuning of FO even under the influence of parasitics.

- Availability of high impedance explicit CM quadrature outputs.

- Availability of low impedance quadrature VM outputs.

- Low active/passive sensitivities.

- Good frequency stability.

VDCC is a versatile ABB proposed by Biolek in 2008, which provides electronically tunable transconductance gain in addition to transferring both current and voltage in its relevant terminals [22]. Some applications of VDCC as grounded/floating inductor simulators [23], [24], Single Resistance Controlled Oscillator

(SRCO) [25] and voltage mode biquad filter [26] have been reported in available literature.

The circuit symbol and behavioral model of VDCC are shown in Fig. 1, where P and N are input terminals and Z, X, WP and WN are output terminals. The terminals Z, X, WP and WN exhibit high impedances while X is a low impedance terminal. The CMOS implementation of VDCC [23] has been shown in Fig. 2. The ideal terminal characteristics of VDCC can be defined by the hybrid matrix as given by Eq. (1):

$$
\begin{bmatrix} I_N \\ I_P \\ I_Z \\ V_X \\ I_{W_P} \\ I_{W_N} \end{bmatrix} = \begin{bmatrix} 0 & 0 & 0 & 0 \\ 0 & 0 & 0 & 0 \\ g_m & -g_m & 0 & 0 \\ 0 & 0 & 1 & 0 \\ 0 & 0 & 0 & 1 \\ 0 & 0 & 0 & -1 \end{bmatrix} \cdot \begin{bmatrix} V_P \\ V_N \\ V_Z \\ I_X \end{bmatrix}. \tag{1}
$$

2. Proposed Quadrature Oscillator

The proposed dual-mode quadrature oscillator is shown in Fig. 3.

The routine circuit analysis of proposed dual mode quadrature oscillator configuration as shown in Fig. 3,

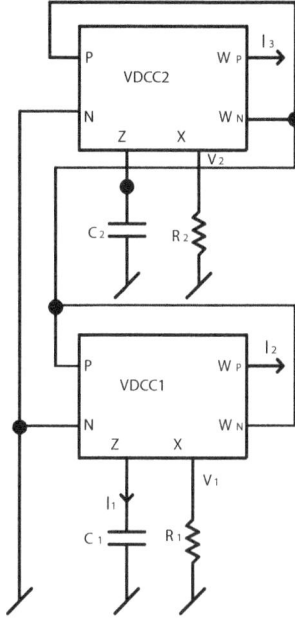

Fig. 3: The proposed dual-mode quadrature oscillator configuration.

yields the following characteristic equation:

$$s^2 + s\frac{1}{C_2}\left(\frac{1}{R_2} - g_{m_2}\right) + \frac{g_{m_1}}{R_1 C_1 C_2} = 0. \tag{2}$$

Thus, from Eq. (2), are the CO Eq. (3) and FO Eq. (4).

$$\left(\frac{1}{R_2} - g_{m_2}\right) \leq 0. \tag{3}$$

$$\omega_0 = \sqrt{\frac{g_{m_1}}{R_1 C_1 C_2}}. \tag{4}$$

From Eq. (3) and Eq. (4) it is clear that CO can be set by R_2 or g_{m_2} and FO is tuned by R_1 or g_{m_1}. Hence, CO and FO both enjoy the independent electronic as well as resistive tuning.

The expressions of CO and FO do not have any common term. Thus, the proposed circuit has the feature of completely non-interactive control of CO and FO. The current relationships derived from Fig. 3 are:

$$\frac{I_2(s)}{I_1(s)} = \frac{1}{sR_1 C_1}. \tag{5}$$

$$\frac{I_2(s)}{I_3(s)} = \frac{g_{m_1} R_2}{sC_1 R_1}. \tag{6}$$

For sinusoidal steady state, Eq. (5) and Eq. (6) become:

$$\frac{I_2(j\omega)}{I_1(j\omega)} = \frac{1}{\omega R_1 C_1} e^{-j90^\circ}. \tag{7}$$

$$\frac{I_2(j\omega)}{I_3(j\omega)} = \frac{g_{m_1} R_2}{\omega C_1 R_1} e^{-j90^\circ}. \tag{8}$$

It is evident from Eq. (7) and Eq. (8) that the phase difference between currents (I_{o2} and I_{o1}) and (I_{o2} and I_{o3}) is -90°. Hence, the currents (I_{o1} and I_{o2}), and (I_{o2} and I_{o3}) are in phase quadrature. The currents I_{o2} and I_{o3} are explicit quadrature current outputs at high impedance terminals.

The voltage transfer function from V_1 (voltage at "X" terminal of VDCC1) to V_2 (voltage at "X" terminal of VDCC2) is:

$$\frac{V_1(s)}{V_2(s)} = \frac{g_{m_1}}{sC_1}. \tag{9}$$

In sinusoidal steady state:

$$\frac{V_1(j\omega)}{V_2(j\omega)} = \frac{g_{m_1}}{\omega C_1} e^{j-90^\circ}. \tag{10}$$

Hence, the phase difference between V_1 to V_2 is -90° i.e. V_2 and V_1 are in quadrature form and are available at low impedance terminals. Thus, the proposed circuit configuration can provide both VM and CM quadrature signals simultaneously at appropriate impedance levels.

3.   Effects of VDCC Parasitics

In this section, the proposed quadrature oscillator is investigated under the influence of VDCC terminal parasitics. In CMOS VDCC (shown in Fig. 2) parasitic resistance R_x appears in series with X terminal, parasitic resistance R_P and parasitic capacitance C_P appears in parallel between W_P terminal and ground, parasitic resistance R_N and parasitic capacitance C_N appears in parallel between W_N terminal and ground and a grounded parasitic resistance R_Z appears at Z terminal. The proposed oscillator configuration including VDCC terminal parasitics has been shown in Fig. 4.

The effect of low resistance parasitic resistors R_{X1} and R_{X2} can be eliminated by merging them with external resistors R_1 and R_2 respectively. It is further noted the low parasitic capacitances C_{N1} and C_{N2} are in parallel with external capacitor C_2. So, the effects of C_{N1} and C_{N2} can be alleviated by merging them with C_2. Therefore, the influence of parasitic capacitances can be completely removed from the proposed oscillator configuration but grounded parallel parasitic resistances R_{N1}, R_{N2}, R_{Z1} and R_{Z2} cannot be balanced and will affect the circuit. To minimize the influence of theses parasitic resistances, the operating frequency can be chosen as Eq. (11).

$$\omega_0 \gg \max \left\{ \cfrac{1}{(C_{N1} + C_{N2} + C_2)\left(\cfrac{1}{\cfrac{1}{R_{N1}} + \cfrac{1}{R_{N2}} + \cfrac{1}{R_{Z2}}}\right)}, \; \frac{1}{C_1 R_{Z1}} \right\}. \tag{11}$$

$$\max \left\{ \cfrac{1}{(C_{N1} + C_{N2} + C_2+)\left(\cfrac{1}{\cfrac{1}{R_{N1}} + \cfrac{1}{R_{N2}} + \cfrac{1}{R_{Z2}}}\right)}, \; \frac{1}{C_1 R_{Z1}} \right\} \ll \omega_0 \ll \min \frac{1}{R_{P1} C_{P1}}, \; \frac{1}{R_{P2} C_{P2}}. \tag{12}$$

$$\frac{1}{C_1 R_{Z1}} + \frac{\left(\cfrac{1}{R_{N1}} + \cfrac{1}{R_{N2}} + \cfrac{1}{R_{Z2}}\right)}{(C_{N1} + C_{N2} + C_2)} + \frac{\cfrac{1}{R_2 + R_{X2}}}{C_{N1} + C_{N2} + C_2} - \frac{g_{m2}}{C_{N1} + C_{N2} + C_2} \leq 0. \tag{13}$$

$$\omega_0 = \sqrt{\frac{\cfrac{g_{m1}}{(R_1 + R_{X1})} - \cfrac{g_{m2}}{R_{Z1}} + \cfrac{1}{R_{Z1}}\left(\cfrac{1}{R_{N1}} + \cfrac{1}{R_{N2}} + \cfrac{1}{R_{Z2}}\right) + \cfrac{1}{R_{Z1}}\left(\cfrac{1}{R_2 + R_{X2}}\right)}{(C_{N1} + C_{N2} + C_2)\,C_1}}. \tag{14}$$

The parasitic resistance R_{P1} (or R_{P2}) and parasitic capacitance C_{P1} (or C_{P2}) appear between W_p terminal of VDCC1 (or W_p terminal of VDCC2) and ground. The phase relationship between current I_2 and I_3 will be changed due to these parasitic components. So, to reduce the effect of these parasitics, the operating frequency (ω_0) has to be chosen as:

$$\omega_0 \ll \min \left\{ \frac{1}{R_{P1} C_{P1}}, \; \frac{1}{R_{P2} C_{P2}} \right\}. \tag{15}$$

Therefore, the useful operating frequency range of presented quadrature oscillator circuit can be described by combining conditions given in Eq. (11) and Eq. (15) we obtain Eq. (12)

The CO and FO of oscillator shown in Fig. 4 are given for CO Eq. (13) and for FO Eq. (14).

From Eq. (13) and Eq. (14) it is clear that even under the influence of parasitics the FO can be independently tunable by g_{m1}. So, proposed configuration exhibits low parasitic effects.

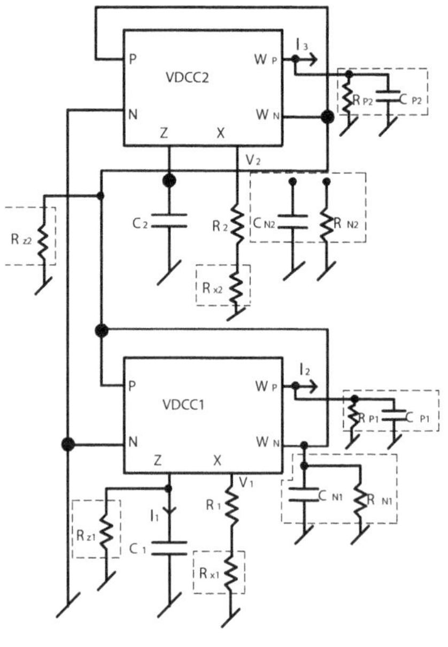

Fig. 4: The proposed dual-mode quadrature oscillator configuration with VDCC terminal parasitics.

4. Non-Ideal Analysis and Sensitivity Calculations

In the non-ideal case, the VDCC can be characterized by the following equations:

$$I_Z = \alpha g_{m_1} (V_P - V_N), \tag{16}$$

$$V_Z = \beta V_X, \tag{17}$$

$$I_{WP} = \gamma_{WP} I_X, \tag{18}$$

$$I_{WN} = -\gamma_{WN} I_X, \tag{19}$$

where α, γ_{WP}, γ_{WN} are current tracking errors and β is voltage tracking error.

Considering the non-idealities of VDCC-1 and VDCC-2, the characteristic equation of circuit shown in Fig. 4 becomes:

$$s^2 + s\frac{1}{C_2\beta_1\beta_2}\left(\frac{\beta_1\gamma_{WN_2}}{R_2} - \beta_1\beta_2\alpha_2 g_{m_2}\right) + \dots$$
$$\dots + \frac{g_{m_1}\alpha_1\gamma_{WN_1}}{R_1 C_1 C_2 \beta_1} = 0, \tag{20}$$

where $(\alpha_1, \beta_1, \gamma_{WP_1}, \gamma_{WN_1})$ and $(\alpha_2, \beta_2, \gamma_{WP_2}, \gamma_{WN_2})$ are tracking errors of VDCC-1 and VDCC-2 respectively.

The CO and FO of proposed circuit under non ideal conditions can be found from Eq. (20), for CO: Eq. (21) and for FO: Eq. (22):

$$\left(\frac{\beta_1\gamma_{WN_2}}{R_2} - \beta_1\beta_2\alpha_2 g_{m_2}\right) \le 0. \tag{21}$$

$$\omega_0 = \sqrt{\frac{g_{m_1}\alpha_1\gamma_{WN_1}}{R_1 C_1 C_2 \beta_1}}. \tag{22}$$

It is noted from Eq. (21) and Eq. (22) that CO and FO are independently tunable under non-ideal conditions as well, CO by g_{m_2} or β_2 or α_2 or γ_{WN_2} and FO by g_{m_1} or α_1 or γ_{WN_2}, which confirm the good non-ideal behavior of presented oscillator. The sensitivities of ω_0 with respect to R_1, R_2, C_1, C_2, g_{m_1}, g_{m_2} under the influence of terminal parasitics of VDCCs are given by Eq. (23).

Also under non-ideal conditions, the sensitivities of ω_0 with respect to various active and passive elements are obtained in Eq. (24).

So, it can be seen from Eq. (24), that all the sensitivities under non-ideal conditions are low and not more than half in magnitude.

The frequency Stability Factor S^F of proposed oscillator is found to be $2\sqrt{n}$ for $C_1 = C_2$, $1/R_2 = g_{m_2}$, and $1/R_1 = ng_{m_1}$. Therefore, a high value of frequency stability can be achieved by selecting large value of n.

5. Simulation Results

The performance of proposed oscillator has been verified by SPICE simulations using TSMC CMOS 0.18 μm processes parameters. Simulations have been performed using CMOS VDCC (shown in Fig. 3) with supply voltages ±0.9 VDC and transconductances gains $g_{m_1} = g_{m_2} = 277$ μA·V^{-1}. The passive component values were chosen as: $R_1 = 5$ kΩ, $R_2 = 3.65$ kΩ, $C_1 = C_2 = 0.05$ nF. The dimensions of MOS transistors used in simulation have been given in Tab. 1.

Tab. 1: Dimensions of MOS transistors.

Transistor	W/L (μm)
M_1, M_2, M_3, M_4	3.6/1.8
M_5, M_6	7.2/1.8
M_7, M_8	2.4/1.8
M_9, M_{10}	3.06/1.72
M_{11}, M_{12}	9.0/1.72
M_{13}, M_{14}, M_{15}, M_{16}	14.4/1.72
M_{17}	13.85/1.72
M_{18}, M_{19}, M_{20}, M_{21}, M_{22}	0.72/0.72

Simulated current and voltage responses have been shown in Fig. 5 and Fig. 6 respectively. The lissajous figures shown in Fig. 7 and Fig. 8 are ellipses with no tilt in axis which confirm the quadrature relationship of current I_2 with current I_3 and voltage V_1 with V_2. Output spectrum of current I_2 is shown in Fig. 9, where the frequency of oscillation equals to 737.8 kHz and the Total Harmonic Distortion (THD) is found to be 2.66 %. Figure 10 shows the variation of amplitude and frequency of current output I_2 on varying resistance R_1. The electronic control of FO (of I_2) with the bias current I_{B1} was shown in Fig. 11. The THD values of current I_2 for different values of bias current I_{B1} havebeen shown in Fig. 12 and it is clear here that except $I_{B1} = 55$ μA (for which THD is 3.55 %), for all other values of I_{B1} THD is less than 3 % which is under expectable range.

The robustness of proposed configuration has been checked by Monte-Carlo simulations on ±10 % variation of R_2 and the sample results are shown in Fig. 13.

It can be illustrated from Fig. 13 that, the frequency of oscillation varies from 721.176 kHz (minimum value) to 739.103 kHz (maximum value) with mean value of 732.578 kHz. So, the variation from mean frequency is 1.5 % (lower side) and 0.088 % (upper side), which confirms the good frequency stability.

These results thus validate the feasibility of proposed configuration.

A comparison of proposed oscillator configuration with previously reported second order dual mode quadrature sinusoidal oscillators has been summarized in Tab. 2.

$$S_{R_1}^{\omega_0} = \cfrac{g_{m_1} R_1}{2\left[\cfrac{g_{m_1}}{R_1 + R_{X1}} - \cfrac{g_{m_2}}{R_{Z1}} + \cfrac{1}{R_{Z1}}\left(\cfrac{1}{R_{N1}} + \cfrac{1}{R_{N2}} + \cfrac{1}{R_{Z2}}\right) + \cfrac{1}{R_{Z1}}\left(\cfrac{1}{R_2 + R_{X2}}\right)\right]} (R_1 + R_{X1})^2.$$

$$S_{R_2}^{\omega_0} = \cfrac{R_2}{2\left[\cfrac{g_{m_1}}{R_1 + R_{X1}} - \cfrac{g_{m_2}}{R_{Z1}} + \cfrac{1}{R_{Z1}}\left(\cfrac{1}{R_{N1}} + \cfrac{1}{R_{N2}} + \cfrac{1}{R_{Z2}}\right) + \cfrac{1}{R_{Z1}}\left(\cfrac{1}{R_2 + R_{X2}}\right)\right]} (R_2 + R_{X2})^2 R_{Z1}.$$

$$S_{C_1}^{\omega_0} = -\frac{1}{2}.$$

$$\qquad\qquad\qquad\qquad\qquad\qquad\qquad\qquad\qquad\qquad\qquad\qquad\qquad (23)$$

$$S_{C_1}^{\omega_0} = -\frac{C_2}{2\left(C_{N1} + C_{N2} + C_2\right)}.$$

$$S_{g_{m_1}}^{\omega_0} = \cfrac{g_{m_1}}{2\left[\cfrac{g_{m_1}}{R_1 + R_{X1}} - \cfrac{g_{m_2}}{R_{Z1}} + \cfrac{1}{R_{Z1}}\left(\cfrac{1}{R_{N1}} + \cfrac{1}{R_{N2}} + \cfrac{1}{R_{Z2}}\right) + \cfrac{1}{R_{Z1}}\left(\cfrac{1}{R_2 + R_{X2}}\right)\right]} (R_1 + R_{X1}).$$

$$S_{g_{m_2}}^{\omega_0} = \cfrac{g_{m_2}}{2\left[\cfrac{g_{m_1}}{R_1 + R_{X1}} - \cfrac{g_{m_2}}{R_{Z1}} + \cfrac{1}{R_{Z1}}\left(\cfrac{1}{R_{N1}} + \cfrac{1}{R_{N2}} + \cfrac{1}{R_{Z2}}\right) + \cfrac{1}{R_{Z1}}\left(\cfrac{1}{R_2 + R_{X2}}\right)\right]} R_{Z1}.$$

$$S_{R_1}^{\omega_0} = -\frac{1}{2}, \;\; S_{R_2}^{\omega_0} = 0, \;\; S_{g_{m_1}}^{\omega_0} = \frac{1}{2}, \;\; S_{g_{m_2}}^{\omega_0} = 0, \;\; S_{C_1}^{\omega_0} = -\frac{1}{2}, \;\; S_{C_2}^{\omega_0} = -\frac{1}{2}, \;\; S_{\alpha_1}^{\omega_0} = \frac{1}{2},$$

$$\qquad\qquad\qquad\qquad\qquad\qquad\qquad\qquad\qquad\qquad\qquad\qquad (24)$$

$$S_{\alpha_2}^{\omega_0} = 0, \;\; S_{\beta_1}^{\omega_0} = -\frac{1}{2}, \;\; S_{\beta_2}^{\omega_0} = 0, \;\; S_{\gamma_{WP_1}}^{\omega_0} = 0, \;\; S_{\gamma_{WP_2}}^{\omega_0} = 0, \;\; S_{\gamma_{WN_1}}^{\omega_0} = 0, \;\; S_{\gamma_{WN_2}}^{\omega_0} = 0.$$

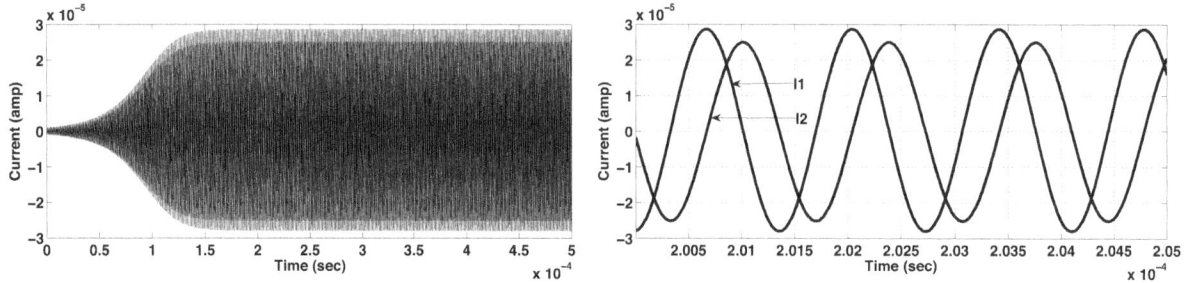

(a) Transient response of I_1, I_2 and I_3.

(b) Steady state response of I_1 and I_2.

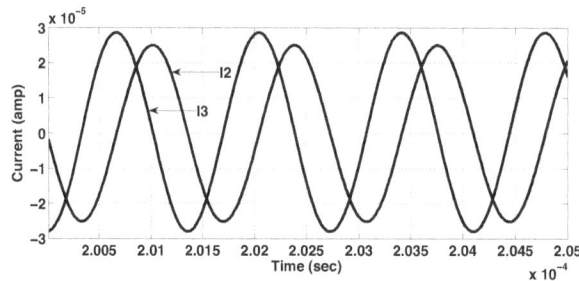

(c) Steady state response of I_2 and I_3.

Fig. 5: Simulated current responses.

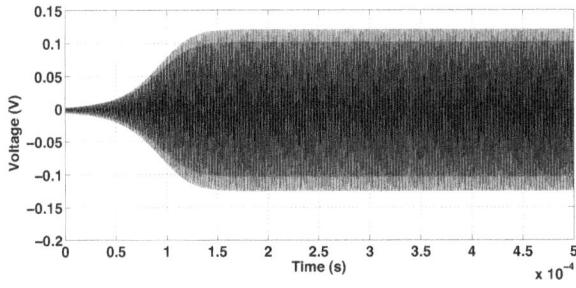

(a) Transient response of V_1 and V_2.

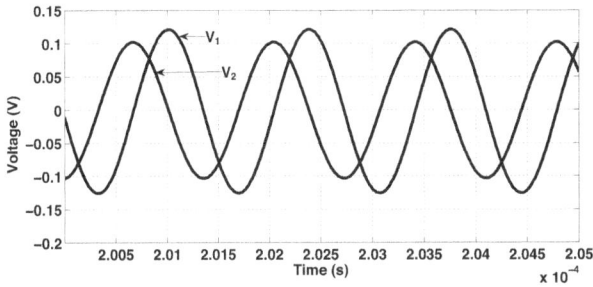

(b) Steady state response of V_1 and V_2.

Fig. 6: Simulated voltage responses.

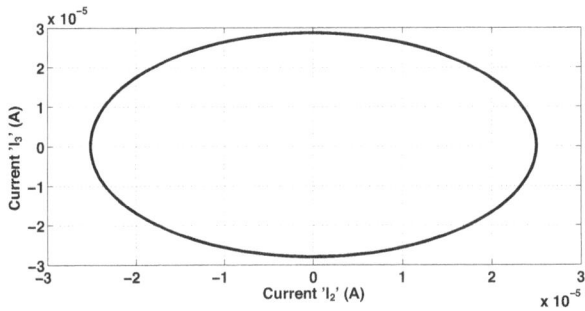

Fig. 7: Lissajous figure showing quadrature relationship between current I_2 and I_3.

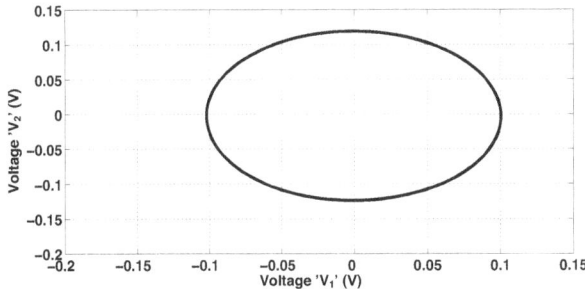

Fig. 8: Lissajous figure showing quadrature relationship between voltage V_1 and V_2.

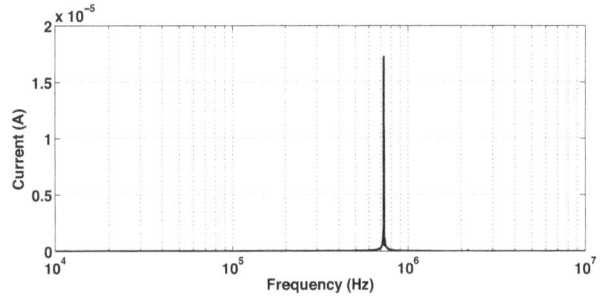

Fig. 9: Output spectrums of current I_2.

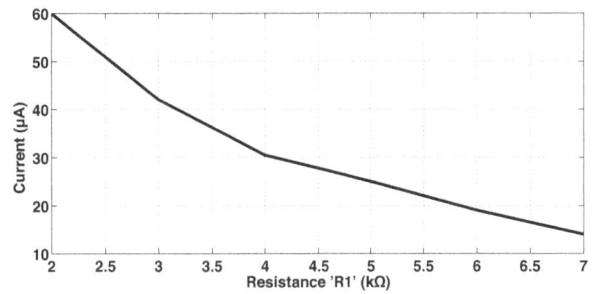

Fig. 10: Variation of of output current I_2 amplitude with resistance R_2.

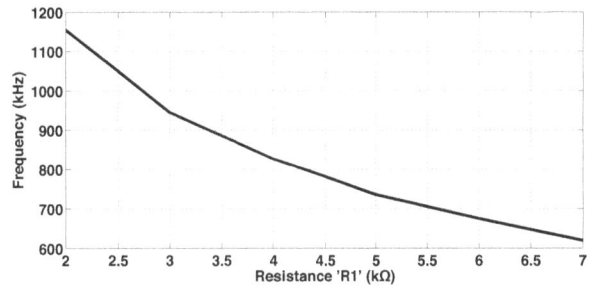

Fig. 11: Electronic tuning of operational frequency with bias current I_{B1}.

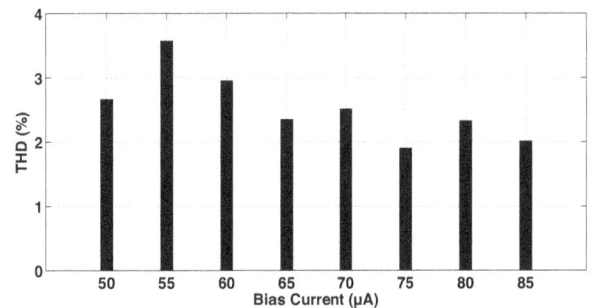

Fig. 12: THD at different values of bias current I_{B1}.

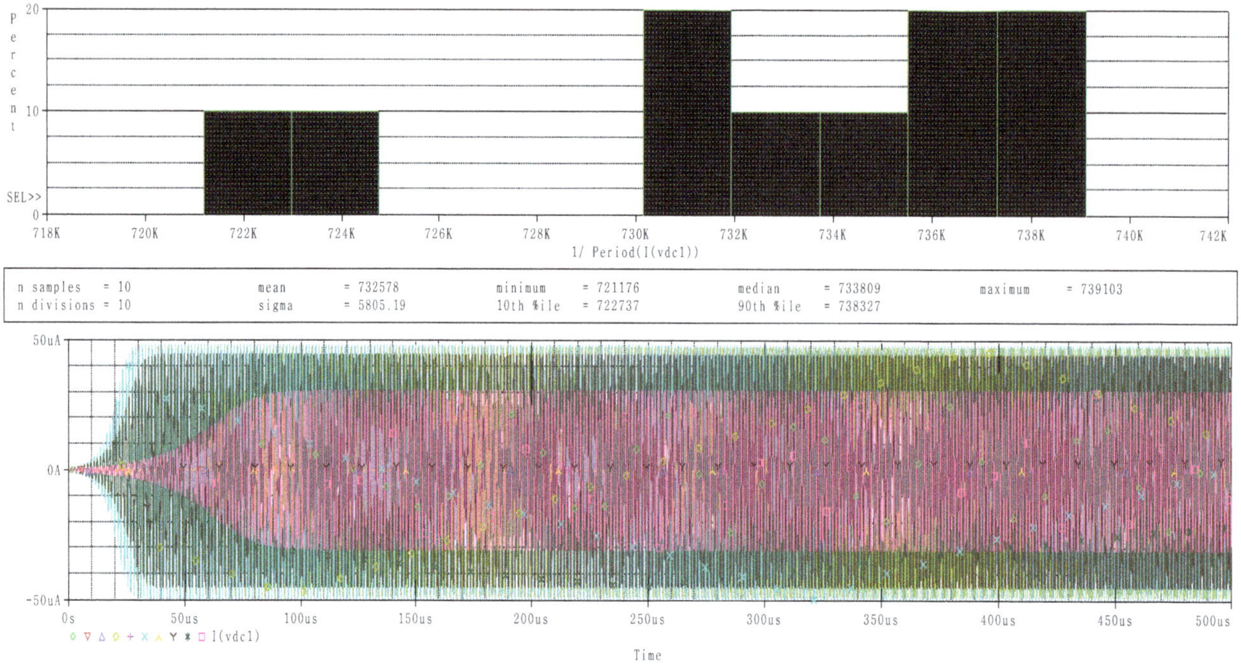

Fig. 13: Results of Monte-Carlo simulation of proposed oscillator.

Tab. 2: Comparison with other previously reported second order dual-mode quadrature oscillators.

Ref.	Number and name of active building blocks (ABBs)	Number of passive elements	All the passive elements grounded	Independent electro. tunability in both FO and CO	Independent resistor tunability in both FO and CO	Availability of VM quadrature outputs at low impedance terminals	Availability of explicit CM quadrature outputs at high impedance terminals	Use of ABB with additional copy terminal(s) to get explicit CM outputs
[9]	2-CDTA	3	Yes	No	No	No	Yes	Yes
[10]	1-MO-CCCDTA	3	Yes	No	No	No	Yes	Yes
[11]	3-CCII	5	No	No	Yes	No	Yes	Yes
[12]	1-FDCCII	5	Yes	No	Yes	No	Yes	Yes
[13]	1-FDCCII	4	Yes	No	No	No	No	Yes
[14]	3-DVCC	5	Yes	No	Yes	No	Yes	Yes
[15]	3-CCII	5	Yes	No	Yes	Yes	Yes	Yes
[16]	2-ZC-CG-CDBA	5	No	No	Yes	Yes	Yes	Yes
[17]	1- DO-CCCDTA	2	Yes	Yes	No	No	No	Yes
[18]	1-MCDTA + I-CFTA	2	Yes	Yes	No	No	Yes	Yes
[19]	2-CCCDTA+ 1-Voltage Buffer	2	Yes	Yes	No	No	No	No
[20]	3-CCCII	2	Yes	Yes	No	No	Yes	Yes
[21]	1-DVCCCTA	2	Yes	Yes	No	No	Yes	Yes
Prop.	**2-VDCC**	**4**	**Yes**	**Yes**	**Yes**	**Yes**	**Yes**	**No**

6. Conclusions

A new CM/VM quadrature sinusoidal oscillator employing two VDCCs, two grounded resistors and two grounded capacitors has been proposed. Use of grounded resistors and capacitors make the proposed configuration suitable for monolithic integration. The presented circuit configuration enjoys several advantageous features expected from a dual mode quadrature oscillator such as: independent electronic tuning of FO and CO, independent resistive control of FO and CO, explicit quadrature current outputs at high impedance terminals, quadrature voltage outputs at low impedance terminals, independent control of FO even under the influence of VDCC port parasitics, low active/passive sensitivities, good frequency stability and low non ideal effects. The SPICE simulation results confirm the validity of theoretical predictions.

References

[1] SENANI, R. New types of sine wave oscillators. *IEEE Transactions on Instrumentation and Measurement*. 1985, vol. 34, iss. 3, pp. 461–463. ISSN 0018-9456. DOI: 10.1109/TIM.1985.4315370.

[2] SENANI, R. and D. R. BHASKAR. Single op-amp sinusoidal oscillators suitable for generation of very low frequencies. *IEEE Transactions on Instrumentation and Measurement*. 1991, vol. 40, iss. 4, pp. 777–779. ISSN 0018-9456. DOI: 10.1109/19.85353.

[3] CHEN, J. J., C. C. CHEN, H. W. TSAO and S. I. LIU. Current-mode oscillators using single current follower. *Electronics Letters*. 1991, vol. 27, iss. 22, pp. 2056–2059. ISSN 0013-5194. DOI: 10.1049/el:19911276.

[4] ABUELMATTI, M. T. Grounded-capacitor current-mode oscillator using single current follower. *IEEE Transactions on Circuits and Systems*. 1992, vol. 39, iss. 12, pp. 1018–1020. ISSN 1057-7122. DOI: 10.1109/81.207726.

[5] BHASKAR, D. R. and R. SENANI. New CFOA-based single-element-controlled sinusoidal oscillators. *IEEE Transactions on Instrumentation and Measurement*. 2006, vol. 55, iss. 6, pp. 2014–2021. ISSN 0018-9456. DOI: 10.1109/TIM.2006.884139.

[6] GIBSON, J. D. *Communications handbook*. New York: CRC Press, 2002. ISBN 0-8493-0967-0.

[7] KHAN, I. A. and S. KHAWAJA. An integrable gm-C quadrature oscillator. *International Journal of Electronics*. 2001, vol. 87, iss. 11, pp. 1353–1357. ISSN 0020-7217. DOI: 10.1080/002072100750000150.

[8] BOLTON, W. *Measurement and Instrumentation Systems*. Oxford: Newnes, 1996. ISBN 978-0750631143.

[9] LAHIRI, A. Novel voltage/current-mode quadrature oscillator using current differencing transconductance amplifier. *Analog Integrated Circuits and Signal Processing*. 2009, vol. 61, iss. 2, pp. 1–8. ISSN 0925-1030. DOI: 10.1007/s10470-009-9291-0.

[10] CHEN, H. P. Electronically tunable quadrature oscillator using grounded components with current and voltage outputs. *The Scientific World Journal*. 2014, vol. 2014, ID 572165, pp. 1–8. ISSN 1537-744X. DOI: 10.1155/2014/572165.

[11] HORNG, J. W., H. P. CHOU and J. C. SHIU. Current-mode and voltage-mode quadrature oscillator employing multiple outputs CCIIs and grounded capacitors. In: *Proceeding of ISCAS-2006*. Island of Kos: IEEE, 2006, pp. 441–444. ISBN 0-7803-9389-9. DOI: 10.1109/ISCAS.2006.1692617.

[12] HORNG, J. W., C. L. HOU, C. M. CHANG, H. P. CHOU, C. T. LIN and Y. H. WEN. Quadrature oscillators with grounded capacitors and resistors using FDCCIIs. *ETRI Journal*. 2006, vol. 28, no. 4, pp. 486–494. ISSN 1225-6463. DOI: 10.4218/etrij.06.0105.0181.

[13] HORNG, J. W., C. L. HOU and C. M. CHANG. Current or/and voltage-mode quadrature oscillators with grounded capacitors and resistors using FDCCIIs. *WSEAS Transactions on circuits and Systems*. 2008, vol. 7, iss. 3, pp. 129–138. ISSN 1109-2734. DOI: 10.1.1.588.9029.

[14] MAHESHWARI, S. and B. CHATURVEDI. Versatile quadrature oscillator with grounded components. In: *International Conference IMPACT-2009*. Aligarh: IEEE, 2009, pp. 209–212. ISBN 978-1-4244-3602-6. DOI: 10.1109/MSPCT.2009.5164212.

[15] ABDALLA, K. K., D. R. BHASKAR and R. SENANI. Configuration for realising a current-mode universalfilter and dual-mode quadrature single resistor controlled oscillator. *IET Circuits, Devices and Systems*. 2012, vol. 6, iss. 3, pp. 159–167. ISSN 1751-858X. DOI: 10.1049/iet-cds.2011.0160.

[16] BIOLEK, D., A. LAHIRI, W. JAIKLA, M. SIRIPRUCHYANUN and J. BAJER. Realization of electronically tunable voltage-mode/current-mode quadrature sinusoidal oscillator using

ZC-CG-CDBA. *Microelectronics Journal.* 2011, vol. 42, iss. 10, pp. 1116–1123. ISSN 0026-2692. DOI: 10.1016/j.mejo.2011.07.004.

[17] JAIKLA, W. and M. SIRIPRUCHYANUN. A versetile quadrature oscillator and universal biquad filter using dual-output current controlled current differencing transconductance amplifier. In: *International Symposium on Communications and Information Technologies.* Bangkok: IEEE, 2006, pp. 1072–1075. ISBN 0-7803-9740-1. DOI: 10.1109/ISCIT.2006.339942.

[18] LAHIRI, A. Resistor-less mixed-mode quadrature sinusoidal oscillator. *International Journal of Computer and Electrical Engineering.* 2010, vol. 2, no. 1, pp. 63–66. ISSN 1793-8163. DOI: 10.7763/IJCEE.2010.V2.114.

[19] JAIKLA, W. and M. SIRIPRUCHYANUN. CCCDTAs-based versatile quadrature oscillator and universal biquad filter. In: *Proceedings of ECTI conference.* Chiang Rai: ETCI association, 2007, pp. 1065–1068. ISBN 978-1-4244-2101-5.

[20] MAHESHWARI, S. and I. A. KHAN. Mixed-mode quadrature oscillator using translinear conveyors and grounded capacitors. In: *International Conference on Multimedia, Signal Processing and Communication Technologies (IMPACT).* Aligarh: IEEE, 2011, pp. 153–156. ISBN 978-1-4577-1105-3. DOI: 10.1109/MSPCT.2011.6150462.

[21] JAIKLA, W., M. SIRIPRUCHYANUN and A. LAHIRI. Resistor-less dual mode quadrature sinusoidal oscillator using a single active building block. *Microelectronics Journal.* 2011, vol. 42, iss. 1, pp. 135–140. ISSN 0026-2692. DOI: 10.1016/j.mejo.2010.08.017.

[22] BIOLEK, D., R. SENANI, V. BIOLKOVA and Z. KOLKA. Active elements for analog signal processing; classification, review and new proposals. *Radioengineering.* 2008, vol. 17, no. 4, pp. 15–32. ISSN 1210-2512. DOI: 10.1.1.476.9997.

[23] KACAR, F., A. YESIL, S. MINAEI and H. KUNTMAN. Positive/ negative lossy/ lossless grounded inductance simulators employing single VDCC and only two passive elements. *International Journal of Electronics and Communication (AEU).* 2014, vol. 68, iss. 1, pp. 73–78. ISSN 1434-8411. DOI: 10.1016/j.aeue.2013.08.020.

[24] PRASAD, D. and J. AHMAD. New electronically-controllable lossless synthetic floating inductance circuit using single VDCC. *Circuits and Systems.* 2014, vol. 5, no. 1, pp. 13–17. ISSN 1549-8328.

DOI: 10.4236/cs.2014.51003.

[25] PRASAD, D., D. R. BHASKAR and M. SRIVASTAVA. New single VDCC-based explicit current-mode SRCO employing all grounded passive components. *Electronics Journal.* 2014, vol. 18, no. 2, pp. 81–88. ISSN 2079-9292. DOI: 10.7251/ELS1418081P.

[26] KACAR, F., A. YESIL and K. GURKAN. Design and experiment of VDDC-based voltage mode universal filter. *Indian Journal of Pure and Applied physics.* 2015, vol. 53, no. 5, pp. 341–349. ISSN 0218-1266.

About Authors

Mayank SRIVASTAVA was born on in Agra, Uttar Pradesh, India. He obtained B. E. degree in Electronics and Communication Engineering from Dr. B. R. A. University, Agra, M. Tech. with specialization in Engineering Systems from Dayalbagh Educational Institute (Deemed University) Agra and Ph.D. in Analog integrated circuits and signal processing from Jamia Millia Islamia, New Delhi, India. Currently he is working as an Associate Professor with Department of Electrical Electronics and Communication Engineering, The Northcap University, Gurgaon (Haryana), India. His research interests are in the areas of Bipolar and CMOS Analog Integrated Circuits and Current Mode Signal Processing. Dr. Srivastava has authored or co-authored 19 research papers in International Journals and Conferences.

Dinesh PRASAD was born on 5. July 1977 in Gorakhpur, Uttar Pradesh, India. He obtained B. Tech. degree from the Institute of Engineering and Technology (IET), Lucknow, M. Tech. with specialization in Electronic Circuits and Systems Design from Aligarh Muslim University, Aligarh, India, and Ph.D. in the area of Analog Integrated Circuits and Signal Processing, from Jamia Millia Islamia (a Central University). Dr. Prasad joined the Electronics and Communication Engineering Department of the Faculty of Engineering and Technology, Jamia Millia Islamia, in February 2002 as a Lecturer. He became Sr. Lecturer on February 2007. His teaching and current research interests are in the areas of Bipolar and CMOS Analog Integrated Circuits, Current Mode Signal Processing and Circuits and Systems. Dr. Prasad has authored or co-authored 30 research papers, all in international journals. Dr. Prasad acted as reviewer of various international journal of repute, on invitation of editor.

21

Hopfield Lagrange Network Based Method for Economic Emission Dispatch Problem of Fixed-Head Hydro Thermal Systems

Thang NGUYEN TRUNG[1], Dieu VO NGOC[2]

[1]Department of Electrical Engineering, Faculty of Electrical and Electronics Engineering, Ton Duc Thang University, 19 Nguyen Huu Tho Street, Ho Chi Minh City, Vietnam
[2]Department of Power Systems, Faculty of Electrical and Electronics Engineering, Ho Chi Minh City University of Technology, 268 Ly Thuong Kiet Street, Ho Chi Minh City, Vietnam

trungthangttt@tdt.edu.vn, vndieu@gmail.com

Abstract. *This paper proposes a Hopfield Lagrange Network (HLN) based method (HLNM) for economic emission dispatch of fixed head hydrothermal systems. HLN is a combination of Lagrange function and continuous Hopfield neural network where the Lagrange function is directly used as the energy function for the continuous Hopfield neural network. In the proposed method, HLN is used to find a set of non-dominated solutions and a fuzzy based mechanism is then exploited to determine the best compromise solution among the obtained ones. The proposed method has been tested on four hydrothermal systems and the obtained results in terms of total fuel cost, emission, and computational time have been compared to those other methods in the literature. The result comparisons have indicated that the proposed method is favorable for solving the economic emission dispatch problem of fixed-head hydrothermal systems.*

Keywords

Economic emission dispatch, fixed head, hopfield Lagrange network, hydrothermal systems.

1. Introduction

The short term hydro-thermal scheduling (HTS) problem is to determine the power generation among the available thermal and hydro power plants so that the total fuel cost of thermal units is minimized over a scheduled time of a single day or a week while satisfying both equality and inequality constraints including power balance, available water, and generation lim-its of both thermal and hydro plants [1]. In practical systems, thermal power generating stations are the sources of carbon dioxide (CO_2), sulfur dioxide (SO_2), and nitrogen oxides (NO_x) causing atmospheric pollution [2]. Therefore, the optimal scheduling of generation in a hydrothermal system involves the allocation of generation among the hydro and thermal plants to simultaneously minimize the fuel cost and emission level of thermal plants satisfying the various constraints on the hydraulic and system network becomes a practical requirement. In the past decades, several conventional methods have been used to solve the HTS problem neglecting environmental aspects such as lambda-gamma iteration method (LGM) [1], an effective conventional method (ECM) based on Lagrange multiplier theory [3], dynamic programming (DP) [4], Lagrange relaxation (LR) method [5], and decomposition and coordination method [6]. Among these methods, Lagrange multiplier theory based method does not find out optimal solution and it must be used together with other optimization techniques [7] whilethe DP and LR methods are more popular ones. However, the computational and dimensional requirements of the DP method increase drastically with large-scale system planning horizon, which is not appropriate for dealing with large-scale problems [8]. On the contrary, the LR method is more efficient and can deal with large-scale problems. However, the solution quality of the LR for optimization problems depends on its duality gap which is a result of the dual problem formulation and might oscillate, leading to divergence for some problems with operation limits and non-convexity of incremental heat rate curves of the generators. Besides, the other methods require simplifications to solve the original model which may yield sub-optimal solutions [2]. Several optimization techniques have been proposed to deal with the economic emission dispatch problems. A particle swarm optimization and gamma based method (γ-PSO) has been suggested in [1] to solve the problem. Similar to LGM [1], the coordination equations are used in the iterative algorithm to obtain optimal solution in the γ-PSO method. Unlike existing PSO [9], each particle in the method is represented with respect to gamma, leading to easier convergence. Two novel search methods have been presented in [10] for dealing with the problem. Those are hybrid algorithm and heuristic searches with genetic algorithm (GA). Both techniques can achieve convergence with a smallermaximum number of generations. However, the computational time of the heuristic searches with GA is slower than the one of the hybrid algorithm. An improved bacterial foraging algorithm (BFA) has been applied to solve the short-term HTS problem considering the environmental aspects given in [11]. A non-dominated sorting genetic algorithm-II (NSGA II) method [12] has been applied to economic environmental dispatch of fixed head hydrothermal scheduling problem with both convex and non-convex fuel cost and emission functions. Another method based on integration of predator-prey optimization and Powell search method (PPO-PS) [13] has been implemented for solving economic emission dispatch for fixed-head hydrothermal systems. The PPO-PS is a powerful method for solving the problem,however, there are many control parameters in this method and an appropriate selection of penalty parameters for a good performance is really a difficult work This paper proposes a Hopfield Lagrange network (HLN) based method (HLNM) for solving the economic emission dispatch of fixed-head hydrothermal systems. The proposedHLN method is a combination of Lagrange function and continuous Hopfield neural network where the Lagrange function is directly used as the energy function for the continuous Hopfield neural network. In addition, the HLN is developed by applying the augmented Hopfield terms;therefore, HLN can tackle oscillation of conventional Hopfield network and get faster convergence as well as obtain higher quality solutions. There is a fact that the proposed HLN is a family of deterministic algorithms, so it also copes with the limited applicability to objective function not to be differentiable. Consequently, the HLN cannot deal with systems where fuel cost and emission functions are represented as nonconvex curves.In the proposed method, HLN is used to find a set of non-dominated solutions and a fuzzy based mechanism is then exploited to determine the best compromise solution among the obtained ones. The proposed method has been tested on four hydrothermal systems and the obtained results in terms of total fuel cost, emission, and computational time have been compared to those other methods in the literature.

2. Problem Formulation

Consider an electric power system having N1 thermal plants and N2 hydro plants. The problem is to find the active power generation of each plant in the system so as the total generation cost and emission of thermal plants is minimized over an M-schedule period time satisfying power balance, water availability constraint, and generation limits.

2.1. Fuel Cost Objective

The fuel cost function F1 for all thermal units is approximated by a quadratic function as follows [12]:

$$F_1 = \sum_{k=1}^{M} \sum_{i=1}^{N_1} t_k \left\{ a_{fsi} + b_{fsi} P_{sik} + c_{fsi} P_{sik}^2 \right\}, \quad (1)$$

where a_{fsi}, b_{fsi}, cfsiarefuel cost coefficients of thermal plant i; P_{sik} is power output of thermal unit i at subinterval k; t_k is the duration of subinterval k.

2.2. Emission Objective

The atmospheric pollutants such as sulphur oxides (SO_x) and nitrogen oxides (NO_x) caused by fossil-fueled thermal generator can be modeled separately. Each gaseous emission is represented by quadratic function as follows [2]:

$$NO_{sik} = \alpha_{1si} + \beta_{1si} P_{sik} + \gamma_{1si} P_{sik}^2, \quad (2)$$

$$SO_{sik} = \alpha_{2si} + \beta_{2si} P_{sik} + \gamma_{2si} P_{sik}^2, \quad (3)$$

$$CO_{sik} = \alpha_{3si} + \beta_{3si} P_{sik} + \gamma_{3si} P_{sik}^2, \quad (4)$$

and then the total emission can be calculatedas follows [2]:

$$F_2 = w_1 NO_{sik} + w_2 SO_{sik} + w_3 CO_{sik}, \quad (5)$$

where w_1, w_2, and w_3 are positive weighting factors of the individualgaseous emission contribution to the emission objective; α_{1si}, β_{1si}, and γ_{1si} are emission coefficients for NO_x; α_{2si}, β_{2si}, and γ_{2si} are emission coefficients for SO_x; and α_{3si}, β_{3si}, and γ_{3si} are emission coefficients for CO_2.

1) Load Demand Equality Constraint

The total power generation from thermal and hydro plants satisfies the total power demand of the system and transmission losses:

$$\sum_{i=1}^{N_1} P_{sik} + \sum_{j=1}^{N_2} P_{hjk} - P_{LK} - P_{DK} = 0, \quad (6)$$
$$k = 1, 2, \ldots, M,$$

where the power losses in transmission lines are calculated as follows:

$$P_{LK} = \sum_{i=1}^{N_1+N_2} \sum_{j=1}^{N_1+N_2} P_{ik} B_{ij} P_{jk} + \\ + \sum_{i=1}^{N_1+N_2} B_{0i} P_{ik} + B_{00}, \tag{7}$$

where P_{Dk}, P_{Lk} are load demand, transmission loss during subinterval k, in MW; P_{hjk} is generation output of hydro unit j during subinterval k, in MW; B_{ij}, B_{0i}, and B_{00} are loss formula coefficients of transmission system.

2) Water Availability Constraints

The total water discharge for each hydro plant during the schedule time is fixed:

$$\sum_{k=1}^{M} t_k q_{jk} = W_j, \quad j = 1, 2, \ldots, N_2, \tag{8}$$

where W_j is volume of water available for generation by hydro plant j during the scheduled period, and the water discharge q_{jk} for hydro unit j at subinterval k is determined by:

$$q_{jk} = a_{hj} + b_{hj} P_{hjk} + c_j P_{hjk}^2, \tag{9}$$

where a_{hj}, b_{hj}, c_{hj} are water discharge coefficients of hydro unit j.

3) Generator Operating Limits

The power output of thermal and hydro plants should be limited between their upper and lower boundaries:

$$P_{si\,min} \leq P_{sik} \leq P_{si\,max}, \\ i = 1, 2, \ldots, N_1, \; k = 1, 2, \ldots, M. \tag{10}$$

$$P_{hj\,min} \leq P_{hjk} \leq P_{hj\,max}, \\ i = 1, 2, \ldots, N_2, \; k = 1, 2, \ldots, M. \tag{11}$$

where $P_{si\,max}$, $P_{si\,min}$ are maximum and minimum power output of thermal unit i, respectively; and $P_{hj\,max}$, $P_{hj\,min}$ are maximum and minimum power output of hydro plant j, respectively.

3. HLN for the Problem

The Lagrange function L of the problem is formulated as follows:

$$L = \sum_{k=1}^{M} \sum_{i=1}^{N_1} t_k \left(a_{si} + b_{si} P_{sik} + c_{si} P_{sik}^2 \right) \\ + \sum_{k=1}^{M} \lambda_k \left(P_{Lk} + P_{Dk} - \sum_{i=1}^{N_1} P_{sik} - \sum_{j=1}^{N_2} P_{hjk} \right) \\ + \sum_{j=1}^{N_2} \gamma_{hj} \sum_{k=1}^{M} \left(t_k q_{jk} - W_j \right). \tag{12}$$

In Eq. (12) λ_k, γ_{hj} are Lagrangian multipliers associated with power balance and water constraint, respectively. Further:

$$a_{si} = \psi a_{fsi} + (1-\psi)(w_1 \alpha_{1si} + w_2 \alpha_{2si} + w_3 \alpha_{3si}), \tag{13}$$

$$b_{si} = \psi b_{fsi} + (1-\psi)(w_1 \beta_{1si} + w_2 \beta_{2si} + w_3 \beta_{3si}), \tag{14}$$

$$c_{si} = \psi c_{fsi} + (1-\psi)(w_1 \gamma_{1si} + w_2 \gamma_{2si} + w_3 \gamma_{3si}), \tag{15}$$

$$0 \leq \psi \leq 1, \tag{16}$$

where ψ is weighting factor for combination of objectives [14].

The energy function E of the problem is described in terms of neurons is determined in Eq. (17),

$$E = \sum_{k=1}^{M} \sum_{i=1}^{N_1} t_k \left(a_{si} + b_{si} V_{sik} + c_{si} V_{sik}^2 \right) \\ + \sum_{k=1}^{M} V_{\lambda_k} \left(P_{Lk} + P_{Dk} - \sum_{i=1}^{N_1} V_{sik} - \sum_{j=1}^{N_2} V_{hjk} \right) \tag{17} \\ + \sum_{j=1}^{N_2} V_{\gamma_{hj}} \left(\sum_{k=1}^{M} t_k q_{jk} - W_j \right) \\ + \sum_{k=1}^{M} \left(\sum_{i=1}^{N_1} \int_0^{V_{sik}} g^{-1}(V) dV + \sum_{j=1}^{N_2} \int_0^{V_{hjk}} g^{-1}(V) dV \right),$$

where $V_{\lambda k}$ and $V_{\gamma hj}$ are outputs of the multiplier neurons associated with power balance and water constraint, respectively; V_{hjk}, Vsik are output of continuous neuron hjk, sik representing P_{hjk}, P_{sik}, respectively. The dynamics of the model for updating neuron inputs are defined as follows:

$$\frac{dU_{sik}}{dt} = \frac{\partial E}{\partial V_{sik}} = - \left\{ \begin{array}{c} t_k(b_{si} + 2c_{si}V_{sik}) \\ + V_{\lambda k} \left(\frac{\partial P_{Lk}}{\partial V_{sik}} - 1 \right) + U_{sik} \end{array} \right\} \tag{18}$$

$$\frac{dU_{hjk}}{dt} = \frac{\partial E}{\partial V_{hjk}} = - \left\{ \begin{array}{c} V_{\lambda k} \left(\frac{\partial P_{Lk}}{\partial V_{hjk}} - 1 \right) \\ + V_{\gamma hj} \left(t_k \frac{\partial q_{jk}}{\partial V_{hjk}} \right) + U_{hjk} \end{array} \right\} \tag{19}$$

$$\frac{dU_{\lambda k}}{dt} = +\frac{\partial E}{\partial V_{\lambda k}} = P_{Dk} + P_{Lk} - \sum_{i=1}^{N_1} V_{sik} - \sum_{j=1}^{N_2} V_{hjk} \tag{20}$$

$$\frac{dU_{\gamma hj}}{dt} = +\frac{\partial E}{\partial V_{\gamma hj}} = \sum_{k=1}^{M} t_k q_{jk} - W_j. \tag{21}$$

The inputs of neurons at step n are updated:

$$U_{sik}^{(n)} = U_{sik}^{(n-1)} - \alpha_{si} \frac{\partial E}{\partial V_{sik}}, \tag{22}$$

$$U_{hjk}^{(n)} = U_{hjk}^{(n-1)} - \alpha_{hj} \frac{\partial E}{\partial V_{hjk}}, \tag{23}$$

$$U_{\lambda k}^{(n)} = U_{\lambda k}^{(n-1)} + \alpha_{\lambda k} \frac{\partial E}{\partial V_{\lambda k}}, \tag{24}$$

$$U_{\gamma hj}^{(n)} = U_{\gamma hj}^{(n-1)} + \alpha_{\gamma hj} \frac{\partial E}{\partial V_{\gamma hj}}, \tag{25}$$

where $U_{\lambda k}$, $U_{\gamma h j}$ are inputs of the multiplier neurons; U_{sik} and U_{hjk} are inputs of the neurons sik and hjk respectively; $\alpha_{\lambda k}$, $\alpha_{\gamma h}$ are step sizes for updating of multiplier neurons; α_{si}, α_{hj} are step sizes for updating of continuous neurons.

The outputs of continuous neurons and multiplier neurons:

$$V_{sik} = g(U_{sik}) =$$
$$(P_{si\,\max} - P_{si\,\min})\left(\frac{1+\tanh(\sigma U_{sik})}{2}\right) + P_{si\,\min}, \quad (26)$$

$$V_{hjk} = g(U_{hjk}) =$$
$$(P_{hj\,\max} - P_{hj\,\min})\left(\frac{1+\tanh(\sigma U_{hjk})}{2}\right) + P_{hj\,\min}, \quad (27)$$

where σ is slope of the sigmoid functionwhich determines the shape of the sigmoid function. The outputs of multiplier neurons are determined using a transfer function:

$$V_{\lambda k} = U_{\lambda k}, \quad (28)$$

$$V_{\gamma h j} = U_{\gamma h j}, \quad (29)$$

3.1. Initialization

The initial outputs of continuous neurons are set at their middle limits and the multiplier neurons are set as follows:

$$V_{\lambda k}^{(0)} = \frac{1}{N_1}\sum_{i=1}^{N_1} t_k\left(b_{si} + 2c_{si}V_{sik}^{(0)}\right) / \left(1 - \frac{\partial P_{Lk}}{\partial V_{sik}}\right), \quad (30)$$

$$V_{\gamma h j}^{(0)} = \frac{1}{M}\sum_{k=1}^{M} V_{\lambda k}^{(0)}\left(1 - \frac{\partial P_{Lk}}{\partial V_{hjk}}\right) / \left(t_k\frac{\partial q_{jk}}{\partial V_{hjk}}\right). \quad (31)$$

3.2. Stopping Criteria

The algorithm will be terminated when either the maximum error $Err_{\max}$ is lower than a predefined threshold ϵ or maximum number of iterations $N_{\max}$ is reached.

4.   Best Compromise Solution by Fuzzy-Based Mechanism

The economic emission dispatch of hydrothermal system is a very complex problem due to many variables and objectives. Moreover, three cases of dispatch for each system consisting of economic dispatch, emission dispatch and economic emission dispatch are carried out. For economic dispatch, only fuel cost is minimized while emission is neglected and for emission dispatch, only emission is minimized whereas the fuel cost is neglected. On the contrary, for economic emission dispatch, both fuel cost and emission are considered and

the compromise solution for the economic emission dispatch must satisfy both fuel cost and emission objectives. However, the determination of the compromise is not simple since there is a conflict between the two objectives for an optimal solution. In fact, if a solution tends to have good fuel cost, its emission will become worse and vice versa. Consequently, the Fuzzy-Based Mechanism is carried out to determine the best compromise. In the technique, two weight factors associate with fuel objective and emission objective are employed to determine a set of non-dominated solutions and then the cardinal priority of each non-dominated solution is calculated. As a result, solution with the highest value of cardinal priority is chosen as a compromise solution. On the other hand, the set of non-dominated solutions has a significant impact on the determination of the compromise solution. If the number of non-dominated solutions is low, a good compromise can be skipped and if a large number of non-dominated solutions is calculated, the task for obtaining the solution is time consuming. Therefore, the determination of the best compromise is not simple and must be carefully carried out. In this paper, the best compromise solution for the problem is determined using the fuzzy satisfying method [14]. The fuzzy goal is represented in linear membership function as follows [14]:

$$\mu(F_j) = \begin{cases} 1 & \text{if} & F_j \le F_{j\,\min}, \\ \dfrac{F_{j\,\max} - F_j}{F_{j\,\max} - F_{j\,\min}} & \text{if} & F_{j\,\min} < F_j < F_{j\,\max}, \\ 0 & \text{if} & F_j \ge F_{j\,\max}, \end{cases}$$
$$(32)$$

where F_j is the value of objective j; $F_{j\,\max}$ and $F_{j\,\min}$ are maximum and minimum values of objective j, respectively. For each k non-dominated solution, the membership function is normalized as follows [15]:

$$\mu_D^k = \sum_{i=1}^{N_{obj}} \mu(F_i^k) / \sum_{k=1}^{N_p}\sum_{i=1}^{N_{obj}} \mu(F_i^k), \quad (33)$$

where μ_D^k is the cardinal priority of $k-th$ non-dominated solution, $\mu(F_j)$ is membership function of objective j, N_{obj} is number of objective functions, and N_p is number of Pareto-optimal solutions. The solution that attains the maximum membership μ_D^k in the fuzzy set is chosen as the 'best' solution based on cardinal priority ranking [16]:

$$\text{Max}\left\{\mu_D^k : k = 1, 2, \ldots, N_p\right\}. \quad (34)$$

5.   Numerical Results

The proposed method has been tested on four systems where the first system has one thermal and one hydro power plant, the second one consists of one thermal and two hydropower plants, the third and last ones

Tab. 1: Result comparison for the economic dispatch for the first three systems ($\psi = 1, w_1 = w_2 = w_3 = 0$).

System	Method	Fuel cost ($)	Emission (kg)			CPU time (s)
			NO$_x$	SO$_2$	CO$_2$	
1	LGM [2]	96 024.42	14 829.94	44 111.89	247 838.53	-
	EPSO [2]	96 024.61	14 830.00	44 111.98	247 839.50	-
	γ-PSO [2]	96 024.40	14 829.93	44 111.88	247 838.43	-
	HLN	96 024.37	14 834.48	44 112.91	247 696.31	0.92
2	LGM [2]	848.241	575.402	4 986.16	2 951.46	-
	EPSO [2]	848.204	575.513	4 986.00	2 952.00	-
	γ-PSO [2]	847.908	575.477	4 985.74	2 951.65	-
	HLN	848.349	575.261	4 986.42	2 950.19	0.4
3	LGM [2]	53 053.79	28 199.21	74 867.81	454 063.64	-
	EPSO [2]	53 053.79	28 199.21	74 867.80	454 063.56	-
	γ-PSO [2]	53 053.79	28 199.21	74 867.80	454 063.63	-
	HLN	53 051.61	28 556.53	74 954.09	458 621.31	0.32

Tab. 2: Result comparison for the emission dispatch for the first three problems ($\psi = 0, w_1 = w_2 = w_3 = 1/3$).

System	Method	Fuel cost ($)	Emission (kg)				CPU time (s)
			NO$_x$	SO$_2$	CO$_2$	NO$_x$+SO$_2$+CO$_2$	
1	LGM [2]	96488.08	14376.32	44202.36	242406.08	300984.76	-
	EPSO [2]	96488.38	14376.41	44202.51	242407.42	300986.33	-
	γ-PSO [2]	96488.08	14376.32	44202.36	242406.08	300984.76	-
	HLN	96809.80	14267.87	44312.40	241263.61	299843.87	0.49
2	LGM [2]	851.98	571.99	4993.75	2922.82	8488.56	-
	EPSO [2]	853.15	571.73	4995.19	2922.14	8489.06	-
	γ-PSO [2]	851.98	571.99	4993.75	2922.82	8488.56	-
	HLN	851.91	572.00	4993.66	2922.81	8488.47	1.8
3	LGM [2]	54359.64	21739.27	74131.82	373122.57	468993.66	-
	EPSO [2]	54359.66	21739.27	74131.82	373122.57	468993.66	-
	γ-PSO [2]	54359.53	21739.19	74131.68	373121.27	468992.14	-
	HLN	55392.75	19986.58	73824.88	350972.26	444783.71	0.12

have two thermal and two hydropower plants. The data for the thermal and hydro plants in the first three systems are from [3] whereas emission data are from [16]. The data for the last one are from [12]. The proposed method is coded in Matlab 7.2 programming language and run on an Intel 1.8 GHz with 4GB of RAM PC.

5.1. The First Three Systems

The objectives of the test systems in this section include one fuel cost and three emissions of NO$_x$, SO$_2$ and CO$_2$ scheduled in 24 subintervals with one hour for each. For each system, three cases of dispatches are considered including economic dispatch ($\psi = 1, w_1 = w_2 = w_3 = 0$), emission dispatch ($\psi = 0, w_1 = w_2 = w_3 = 1/3$), and economic emission dispatch ($\psi = 0.5, w_1 = w_2 = w_3 = 1/3$). The obtained results from the proposed method for three dispatch cases including economic dispatch, emission dispatch, and economic emission dispatch for the three test systems are compared to those from other methods including LGM, EPSO, and γ-PSO in [2] as given in Tab. 1, Tab. 2, and Tab. 3. For the economic dispatch, the proposed HLN can obtain better total costs than the others except for the System 2 where the cost is slightly higher than for the others. For the emission dispatch, the proposed HLN can obtain less total emission than the others for

all test systems. In the economic emission dispatch, there is a trade-off between total cost and emission objectives and the obtained solutions from the methods are non-dominated as in Tab. 3. The total computational time for each system for the three cases is given in Tab. 4. The study in [2] has not reported computer processor and we fail to compare the processor. However, as indicated in Tab. 4 in the paper, HLN is very fast compared to LGM [2], EPSO [2], γ-PSO [2] since HLN has gotten optimal solutions with 1.51 seconds for System 1, 3 seconds for System 2 and 0.740 second for System 3 whereas that time from LGM is 10 seconds higher, from EPSO is about 100 seconds and from γ-PSO is about 40 seconds. Clearly, these methods are time consuming and it is very slow for convergence as compared to HLN. Convergence characteristics obtained by HLN in terms of maximum error and number of iterations for economic dispatch of System 1, System 2 and System 3 are depicted in Fig. 1, Fig. 2 and Fig. 3. Clearly, HLN has obtained the optimal solution with the lowest number of iterations at System 3, 2594 iterations and with the highest number of iterations at System 1, 6263 iterations. Consequently, the convergence time for economic dispatch of the System 1 is the longest meanwhile this time for System 3 is the fastest and they are respectively 0.92 and 0.32 as reported in Tab. 1. The optimized control variables for test System 1 is given in table Tab. A in Appendix section.

Tab. 3: Result comparison for the economic emission dispatch of the first three system ($\psi = 0.5, w_1 = w_2 = w_3 = 1/3$).

System	Method	Fuel cost ($)	Emission (kg)				CPU time (s)
			NO_x	SO_2	CO_2	$NO_x+SO_2+CO_2$	
1	LGM [2]	96421.702	14384.101	44176.312	242456.004	300984.76	-
	EPSO [2]	96421.725	14384.108	44176.324	242456.109	300986.33	-
	γ-PSO [2]	96421.46	14384.03	44176.195	242454.92	300984.762	-
	HLN	96465.712	14328.17	44181.95	241776.424	300286.544	0.1
2	LGM [2]	851.208	572.235	4992.707	2923.986	8488.928	-
	EPSO [2]	851.079	572.264	4992.547	2923.061	8487.872	-
	γ-PSO [2]	852.388	571.97	4994.167	2923.301	8489.438	-
	HLN	850.065	572.723	4991.026	2927.027	8490.776	0.8
3	LGM [2]	54337.014	21745.127	74144.989	373165.02	469025.136	-
	EPSO [2]	54337.027	21745.138	74115.007	373165.186	469025.331	-
	γ-PSO [2]	54336.888	21745.021	74114.821	373163.42	469023.262	-
	HLN	55158.62	20031.652	73731.958	351363.758	445127.368	0.3

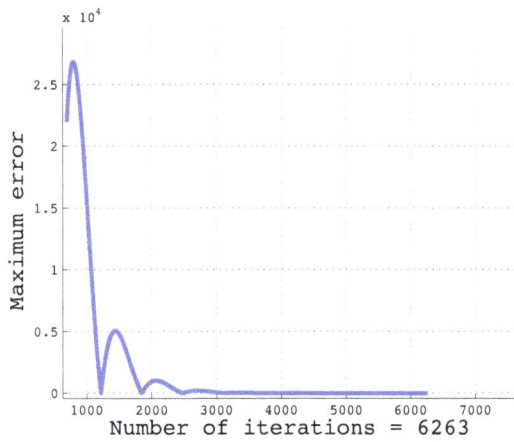

Fig. 1: Convergence characteristic obtained by HLN for economic dispatch of System 1.

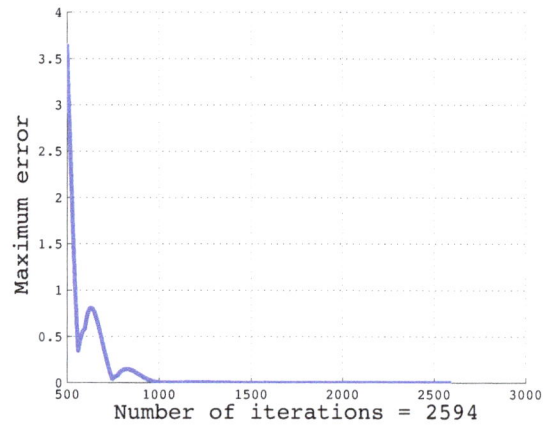

Fig. 3: Convergence characteristic obtained by HLN for economic dispatch of System 3.

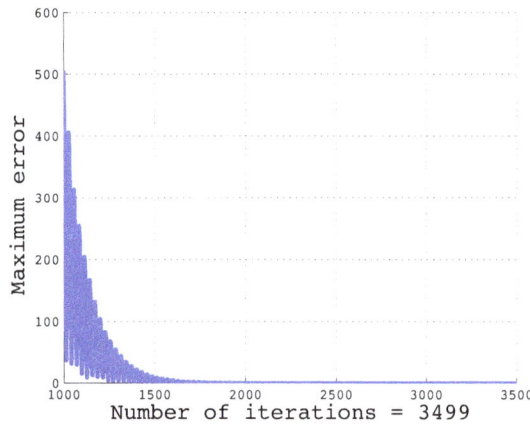

Fig. 2: Convergence characteristic obtained by HLN for economic dispatch of System 2.

Tab. 4: Computational time comparison for the first three systems.

Method	System 1	System 2	System 3
LGM [2]	14.83	11.46	12.26
EPSO [2]	95.36	83.73	105.0
γ-PSO [2]	43.44	39.27	49.01
HLN	1.51	3	0.740

5.2. The Fourth System

The test system in this case includes one total cost function and one emission function scheduled in three subintervals with eight hours for each [12]. The proposed HLN method is applied for obtaining the opti-

mal solutions for the economic, emission and economic emission dispatches.

The values of w_1, w_2 and w_3 in Eq. (13), Eq. (14), Eq. (15) are fixed at 1, 0 and 0, respectively. The value of ψ in Eq. (16) is set to one and zero for the economic and emission dispatches, respectively. For the case of economic emission dispatch, we have determined 11 non-dominated solutions to form Pareto optimal front with the change of weight factor ψ from 0 to 1. The best compromise solution from the obtained 11 non-dominated solutions is determined by the fuzzy based mechanism in Section 4. The obtained results in terms of fuel cost, emission and computational time for the three cases from the proposed method are compared to those from PSO, PSO with penalty method (PSO-PM), predator-prey optimization (PPO), PPO with penalty

Tab. 5: Result comparison for the three cases of dispatch of the fourth system.

Method	Economic dispatch		Emission dispatch		Compromise dispatch		
	Cost ($)	CPU (s)	Emis. (lb)	CPU (s)	Cost ($)	Emis. (lb)	CPU (s)
PSO-PM [13]	65741	18.25	585.67	18	65821	620.78	18.98
PSO [13]	65241	18.32	579.56	18.31	65731	618.78	19.31
PPO-PM [13]	64873	16.14	572.71	15.93	65426	612.34	16.53
PPO [13]	64718	15.99	569.73	15.18	65104	601.16	16.34
PPO-PS-PM[13]	64689	15.98	568.78	15.92	65089	600.24	16.15
PPO-PS [13]	64614	15.89	564.92	15.45	65058	594.18	16.74
HLN	64576	0.3	579.12	0.68	64807	617.64	0.74

method (PPO-PM), PPO-PS with penalty method (PPO-PS-PM), and PPO-PS in [13] as given in Tab. 5. As observed from the table, the proposed method can obtain better cost than other methods for the two cases of economic and combined economic emission dispatch. However, HLN gets lower emission than PSO-PM and PSO only and higher emission than rest of methods for emission dispatch and economic emission dispatch. Furthermore, as seen in Tab. 5 HLN has been run under one second for each dispatch case while it has taken from 15 to 20 seconds for other methods. Obviously, HLN is much faster than these methods although no computer has been reported for the methods in [13] and computer processor comparison has not been performed. Figure 4 shows the convergence characteristic obtained by HLN for economic dispatch of the system. Obviously, the applied HLN method can obtain the optimal solution for the case with fewer number ofiterations than that for three systems above and therefore the execution time for the system is shorter than that for the three systems.

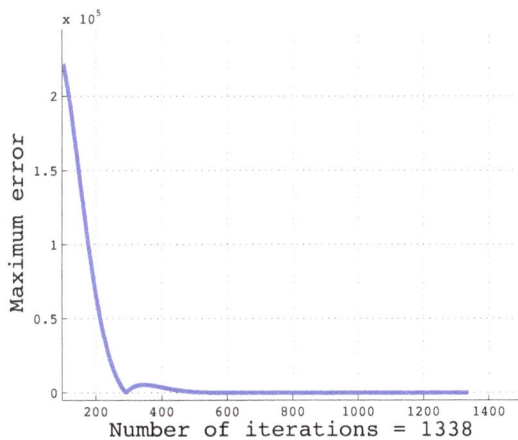

Fig. 4: Convergence characteristic obtained by HLN for economic dispatch of System 4.

6. Conclusions

In this paper, a Hopfield Lagrange network based method has been efficiently implemented for solving the economic emission short-term hydrothermal scheduling problem. The proposed method is a combination of Lagrange function and continuous Hopfield neural network for solving optimal single-objective dispatch problem and a fuzzy based mechanism for obtaining the best compromise solution among several non-dominated solutions. The Hopfield Lagrange network is an improvement of the continuous Hopfield neural network by using the Lagrange function as its energy function. The advantages of the Hopfield Lagrange network are that it is simple, fast, and efficient for solving optimization problems. The proposed method has been tested on four systems with different number of objectives and the obtained results have been compared to those from other methods in the literature. The result comparisons have indicated that the proposed method can obtain better solution than many other methods with shortercomputational time. Therefore, the proposed method can be very favored for solving economic emission dispatch of short-term fixed-head hydrothermal problems.

References

[1] WOOD, A. J. and B. F. WOLLENBERG. *Power Generation, Operation and Control.* 3rd ed. New York: John Wiley & Sons, 1996. ISBN 0-471-58699-4.

[2] SASIKALA, J. and M. RAMASWAMY. PSO Based Economic Emission Dispatch for Fixed Head Hydrothermal Systems. *Electrical Engineering.* 2012, vol. 94, iss. 4, pp. 233–239. ISSN 1432-0487. DOI: 10.1007/s00202-012-0234-x.

[3] RASHID, A. H. A. and K. M. NOR. An Efficient Method for Optimal Scheduling of Fixed Head Hydro and Thermal Plants. *IEEE Transactions on Power Systems.* 1991, vol. 6, iss. 2, pp. 632–636. ISSN 0885-8950. DOI: 10.1007/0.1109/59.76706.

[4] YANG, J. and N. CHEN. Short Term Hydrothermal Coordination Using Multi-Pass Dynamic Programming. *IEEE Transactions on Power Systems.* 1989, vol. 4, iss. 3, pp. 1050–1056. ISSN 0885-8950. DOI: 10.1109/59.32598.

[5] SALAM, M. S., K. M. NOR and A. R., HAMDAN. Hydrothermal Scheduling Based Lagrangian Relaxation Approach to Hydrothermal Coordination. *IEEE Transactions on Power Systems*. 1998, vol. 13, iss. 1, pp. 226–235. ISSN 0885-8950. DOI: 10.1109/59.651640.

[6] LI, C., A. J. SVOBODA, C. L. TSENG, R. B. JOHNSON and E. HSU. Hydro Unit Commitment in Hydro-Thermal Optimization. *IEEE Transactions on Power Systems*. 1997, vol. 12, iss. 2, pp. 764–769. ISSN 0885-8950. DOI: 10.1109/59.589675.

[7] BENHAMIDA, F. and R. BELHACHEM. Dynamic Constrained Economic/Emission Dispatch Scheduling Using Neural Network. *Advances in Electrical and Electronic Engineering*. 2013, vol. 11, no. 1, pp. 1–9. ISSN 1804-3119. DOI: 10.15598/aeee.v11i1.745.

[8] DIEU, V. N. and W. ONGSAKUL. Hopfield Lagrange for Short-Term Hydrothermal Scheduling. In: *IEEE Russia Power Tech*. St. Petersburg: IEEE, 2005, pp. 1–7. ISBN 978-5-93208-034-4. DOI: 10.1109/PTC.2005.4524597.

[9] KINCL, Z. and Z. KOLKA. Test Frequency Selection Using Particle Swarm Optimization. *Advances in Electrical and Electronic Engineering*. 2013, vol. 11, no. 6, pp. 507–513. ISSN 1804-3119. DOI: 10.15598/aeee.v11i6.912.

[10] GEORGE, A., M. C. REDDY and A. Y. SIVARAMAKRISHNAN. Multi-Objective, Short-Term Hydro Thermal Scheduling Based on Two Novel Search Techniques. *International Journal of Engineering Science and Technology*. 2010, vol. 2, no. 11, pp. 7021–7034. ISSN 0975-5462.

[11] FARHAT, I. A. and M. E. EL-HAWARY. Multi-Objective Short-Term Hydro-Thermal Scheduling Using Bacterial Foraging Algorithm. In: *Electrical Power and Energy Conference*. Winnipeg: IEEE, 2011, pp. 176–181. ISBN 978-1-4577-0405-5. DOI: 10.1109/EPEC.2011.6070190.

[12] BASU, M. Economic Environmental Dispatch of Fixed Head Hydrothermal Power Systems Using Nondominated Sorting Genetic Algorithm-II. *Applied Soft Computing*. 2011, vol. 11, no. 3, pp. 3046–3055. ISSN 1568-4946. DOI: 10.1016/j.asoc.2010.12.005.

[13] NARANG, N., J. S. DHILLON and D. P. KOTHARI. Multiobjective fixed head hydrothermal scheduling using integrated predator-prey optimization and Powell search method. *Energy*. 2012, vol. 47, iss. 1, pp. 237–252. ISSN 0360-5442. DOI: 10.1016/j.energy.2012.09.004.

[14] SAKAWA, M., H. YANO and T. YUMINE. An Interactive Fuzzy Satisfying Method for Multi-Objective Linear Programming Problems and Its Applications. *IEEE Transactions on Systems Man and Cybernetics*. 1987, vol. 17, no. 4, pp. 654–661. ISSN 0018-9472. DOI: 10.1109/TSMC.1987.289356.

[15] TAPIA, C. G. and B. A. MURTAGH. Interactive Fuzzy Programming with Preference Criteria in Multi-Objective Decision Making. *Computers & Operations Research*. 1991, vol. 18, no. 3, pp. 307–316. ISSN 0305-0548. DOI: 10.1016/0305-0548(91)90032-M.

[16] DHILLON, J. S., S. C. PARTI and D. P. KOTHARI. Fuzzy Decision-Making in Stochastic Multiobjective Short-Term Hydrothermal Scheduling. *Generation, Transmission and Distribution*. 2002, vol. 149, iss. 2, pp. 191–200. ISSN 1350-2360. DOI: 10.1049/ip-gtd:20020176.

About Authors

Thang NGUYEN TRUNG was born in 6th August 1985. He received his M.Sc. from university of technical education Ho Chi Minh City in 2011. His research interests include optimization of power system, power system operation and control and Renewable Energy.

Dieu VONGOC received his B.Sc. and M.Sc. degrees in Electrical Engineering from Ho Chi Minh City University of Technology, Ho Chi Minh city, Vietnam, in 1995 and 2000, respectively and his Ph.D. degree in Energy from Asian Institute of Technology (AIT), Pathumthani, Thailand in 2007. He is currently a Research Associate at Energy Field of Study, AIT and a lecturer at Department of Power Systems Engineering, Faculty of Electrical and Electronic Engineering, Ho Chi Minh City University of Technology, Ho Chi Minh City, Vietnam. His research interests are applications of AI in power system optimization, power system operation and control, power system analysis, and power systems under deregulation.

Appendix

Tab. A: Control variables for System 1 with four objective function.

Subinterval	Economic dispatch		Emission dispatch		Economic emission dispatch	
	V_{sk} (MW)	V_{hk} (MW)	V_{sk} (MW)	V_{hk} (MW)	V_{sk} (MW)	V_{hk} (MW)
1	231.8904	235.1858	273.9742	191.3175	262.9423	202.7468
2	203.7237	232.3999	255.0545	178.9122	241.6737	192.7512
3	194.3511	231.4743	248.7678	174.7781	234.6012	189.4216
4	186.8589	230.735	243.7456	171.4712	228.9492	186.7587
5	180.3075	230.0889	239.3563	168.578	224.0081	184.4292
6	199.0364	231.9369	251.91	176.8451	238.1364	191.0863
7	262.0162	238.1735	294.2544	204.5551	285.7133	213.4201
8	372.883	249.2464	369.2893	252.9993	369.7242	252.5449
9	431.1394	255.1173	408.9719	278.2835	414.0031	273.0073
10	440.7196	256.0864	415.5146	282.4302	421.2937	276.3661
11	459.9057	258.0303	428.6318	290.7249	435.9021	283.0872
12	469.5116	259.0052	435.2064	294.873	443.2198	286.4496
13	350.0483	246.9553	353.7829	243.0565	352.3935	244.5063
14	373.8355	249.3421	369.9367	253.4137	370.4474	252.88
15	384.3185	250.3959	377.0649	257.9718	378.4086	256.5668
16	419.6542	253.9568	401.1346	273.3081	405.2662	268.9784
17	484.8987	260.569	445.7479	301.5109	454.9472	291.8319
18	503.1996	262.4327	458.3018	309.395	468.9038	298.2275
19	464.7076	258.5175	431.9177	292.7989	439.5598	284.7682
20	443.5954	256.3775	417.4794	283.6742	423.4827	277.3739
21	397.6752	251.7403	386.1555	263.774	388.5566	261.2612
22	354.8016	247.4318	357.0085	245.1277	355.9999	246.1804
23	312.0972	243.1598	328.071	226.4914	323.6217	231.1235
24	277.1106	239.6738	304.4332	211.1759	297.1317	218.7611

Impact of Production from Photovoltaic Power Plants on Increase of Ancillary Services in the Czech Republic

Martin SMOCEK, Zdenek HRADILEK

Department of Electrical Power Engineering, Faculty of Electrical Engineering and Computer Science, VSB–Technical University of Ostrava, 17. listopadu 15, 708 33 Ostrava, Czech Republic

martin.smocek@vsb.cz, zdenek.hradilek@vsb.cz

Abstract. *Renewable energy resources represent a noticeable part of the overall energetic concept development. New integration of renewable energy resources into power grids has a significant impact on the reliability and quality of power supply. The major problem of the photovoltaic and wind power plants is their dependency on weather conditions, since it has a direct effect on their immediate output produced that shows stochastic behaviour. These stochastic outputs result in very adverse impacts on the power grid. Further development of these resources could lead to exceeding of the control and absorption abilities of the power grid. The power grid must be set in balance with respect to the production and consumption of electric power at any time. The operation of photovoltaic power plants impair keeping this balance. That has an adverse impact on the very operation and maintenance of network parameters within the extent required. This survey deals with analysis focused on operation of the photovoltaic power plants with respect to the increase of reserve power in ancillary services in the Czech Republic.*

Keywords

Ancillary services, calculation methodology, computing applications, photovoltaic power plant, real measurement, statistical methods.

1. Introduction

Both photovoltaic power and wind power are sources considerable depended on meteorological conditions [5], [6], [10]. The survey is based on the article [4] with detailed description of the draft methodology for computing of extreme conditions in Photovoltaic Power Plants (PvP). This article elaborates on the survey with draft methodology to assess the increase of reserve power in ancillary services with respect to operation of PvP. These methodologies are then applied to real data collected during operation of the specific PvP and to the actual data of energy regulatory. This analysis shows the period of rapid increase of photovoltaic output across the power grid of the Czech Republic (CZ). The final parts provide definition of output increase to be delivered with respect to the impact of PvP operation on the existing ancillary services MZ15+ (15 minute positive back-up) and MZ15- (15 minute negative back-up) [3]. This is already the case nowadays.

It was elaborated many studies focused on the prediction of photovoltaic and wind power plants such as [7], [8], [9]. This survey differs from others in that it does not predict anything but based on the actual situation.

2. Ancillary Services

The Czech Power Grid Code, Part II [3], approved by the Energy Regulatory Office, defines the method for utilisation of the power grid with respect to ancillary services. One of the prerequisites for determination of summary regulatory backups of ancillary services is the deviation $OD_{RES(t)}$. This is a parameter to establish the increment of balancing deviation caused by production from newly installed renewable resources to electric power. As stated here Eq. (1), this increment is defined by the sum of increase generated by operation of wind $OD_{VtE(t)}$ and photovoltaic $OD_{FVE(t)}$ power plants. The deviation $OD_{RES(t)}$ is then used to determine volumes of ancillary services required for the subsequent period [3]:

$$OD_{OZE(t)} = OD_{VtE(t)} + OD_{OZE(t)}. \qquad (1)$$

Equation (1) implies that the effect of installed capacity of PvP has a certain impact on calculation of ancillary services. This fact initiated elaboration of this survey.

The impact brought by increase of the installed capacity of PvP on remote control of the power grid and the relevant utilisation of Ancillary Services (AS) could be observed since the year 2011, when the installed capacity of PvP grew from approximately 450 MW in the preceding year to approximately 1950 MW in this year [2].

The period from 2010 to 2012 showed an increase of the system output deviation. The system deviation is used for assessing the surplus or lack of electric power in the grid. The standard deviation of this system deviation then grew by up to 50 MW during afternoon hours to reach the extreme value of approximately 200 MW [2]. The reason for this increase may be the effect of PvP, as prediction models applicable to production of electric power by PvP cannot provide an accurate estimate of the weather effects on commissioning of production using these resources. The outcome from prediction models has a direct effect on behaviour of market players, including determination of system deviation.

3. Extreme Conditions of PvP

The very assessment of extreme conditions of PvP is based on the annual monitoring of a specific photovoltaic power plant at the break of 2010/2011. The PvP with peak power output of 1.1 MWp subject to measurement here is connected to the power grid and it represents a sole resource in the network during the measurement period. As far as topology is concerned, this grid is looped, although operated as a radial one [1].

The assessment methodology is based on values obtained from average five-minute outputs. When compared to the above mentioned article published earlier, we have performed slight changes to statistical regressions of extreme conditions of PvP, yet the principles of this assessment have remained the same.

3.1. Maximum Extreme Conditions

The result produced by methodology for assessment of maximum extreme conditions of PvP comprises the Gauss's equation with parameters A, B, C and D for each month of the year for the best approximation of

these maximum values. This formula is generally expressed as follows [7]:

$$f(x) = A \cdot e^{-\frac{(t-B)^2}{2 \cdot C^2}} + D, \qquad (2)$$

where A, B, C and D are parameters of Gauss's equation.

The general function of Gauss's equation can be applied to the output curve showing maximum extreme conditions as follows (e.g. for April):

$$P_{FVE\,\max} = -1274810 \cdot e^{-\frac{(t-0.4928263)^2}{2 \cdot 0.17422^2}} + 383794, \quad (3)$$

where $P_{FVE\,\max}$ is maximum extreme condition of PvP (W), t is time (hh:mm).

When assessing other values of peak output from PvP, one has to multiply the parameter "A" from Eq. (2) with the relevant value determined by the ratio of peak output from another PvP to the peak output of this PvP (1.1 MWp).

The time t and time lines for graphic results are always defined in the so-called Universal Time Coordinated (UTC). Conversions between the UTC and the Central European Time (CET) or the Central European Summer Time (CEST) are defined by Eq. (4) and Eq. (5). However, it is necessary to bear in mind that MS Excel converts the time data into numerical values. That means time is represented by decimal numbers from 0.0 to 1.0 (0.00 to 24.00 o'clock). For example, the numerical value 0.5 is matched by time 12.00 o'clock [3], [6].

$$t_{cet} = \frac{h_{utc} \cdot 60 + m_{utc}}{1440} + \frac{1}{24}, \qquad (4)$$

$$t_{cest} = \frac{h_{utc} \cdot 60 + m_{utc}}{1440} + \frac{2}{24}, \qquad (5)$$

where t_{cet} is Central European Time, t_{cest} is Central European Summer Time, t_{utc} is Universal Time Coordinated, h_{utc}/m_{utc} is value of hour/minute according to UTC.

Tab. 1: Parameters of Gauss's equation of maximum extreme conditions of PvP for individual months of the year.

Months	Parameters of Gauss's Equation			
	A	**B**	**C**	**D**
1	-3600950	0.49921	0.262209	2932140
2	-2089680	0.503123	0.211292	1317490
3	-1956580	0.499046	0.223049	1112310
4	-1274810	0.4928263	0.17422	383794
5	-1357560	0.4936243	0.188437	440029
6	-1180160	0.4983373	0.18539	307280
7	-1180020	0.5057943	0.188995	338914
8	-1384590	0.5036093	0.198893	535948
9	-2212550	0.4949763	0.252074	1389050
10	-5041080	0.4860943	0.367736	4274150
11	-3235550	0.484282	0.268296	2548530
12	-299418	0.484282	0.0412953	-64029.2

For individual parameters of Gauss's Eq. (2) for particular months of the year see the Tab. 1. Similarly to the Eq. (3) for April, these parameters are inserted into the Eq. (2) to determine the curve of maximum extreme conditions of each month.

Maximum extreme conditions of PvP production are described by the Fig. 1 and Fig. 2. Figure 1 includes months in the period from December to June. The courses of individual curves show full match provided the results depend on intensity of solar radiation, the duration of sunshine during the day and the impact of temperature of solar panels. One can observe that the dispersion of production time is extended proportionally to the increasing duration of sunshine within particular months and the amplitude is rising with increasing intensity of solar radiation. The maximum occurs in May rather that during the summer solstice, as might be expected. That is due to the impact of direct proportion between panel efficiency and temperature.

The panel efficiency drops proportionally to the rise of its temperature. Figure 2 shows the maximum extreme curves from July to December, i.e. between the summer and winter solstice. There was also a minor sign of higher panel temperature in July, when compared to August, it means a greater extreme reached in August.

3.2.   Minimum Extreme Conditions

Although the visual assessment of graphic illustration of production from PvP showed evident certain proportion of the diffusion element of radiation, the objective determination of their overall share is very difficult. Accurate assessment would require measurement of the ratio between the direct and diffusion elements of radiation that was not subject to measurement. Approximation of minimum extreme conditions in production by means of a constant function equal to zero is not a convenient approach either, since the graphic output shown by Fig. 3 expresses an evident share of the diffusion radiation element and it needs to be respected. It is therefore necessary to determine this threshold of minimum extreme conditions by other means to respect the diffusion radiation element that is 100 % identifiable upon visual inspection of the Fig. 3. To serve the purpose of this survey, the share of diffusion radiation element has been defined as the so-called guarantee share of diffusion radiation element. This information provides far better informative value about the minimum production at PvP than the sole diffusion radiation element. That is due to the fact it reflects the actual condition of PvP operation with respect to the minimum production, i.e. both the diffusion and direct elements. Figure 3 shows the scatter chart with one-minute outputs per specific month for the year 2010 with clear evidence of the guaranteed diffusion radiation. It is located in the area missing any output values. There is not even one value observed in this field under

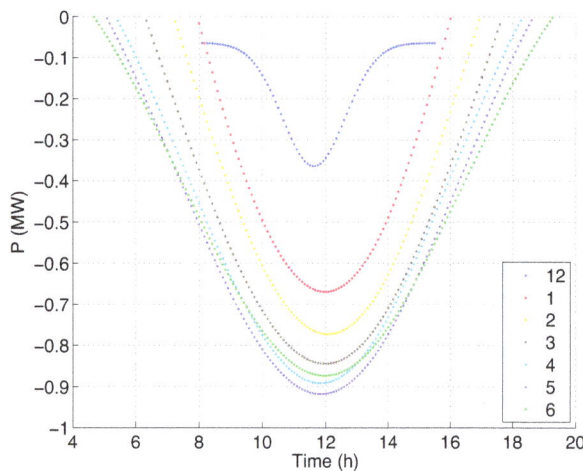

Fig. 1: Regression curves of maximum extreme outputs for individual months (December to June).

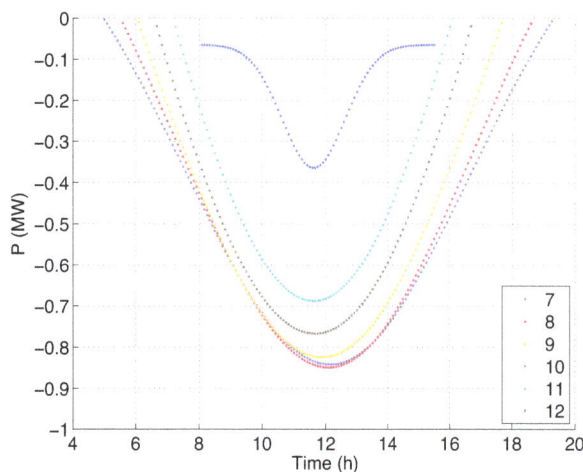

Fig. 2: Regression curves of maximum extreme outputs for individual months (July to December).

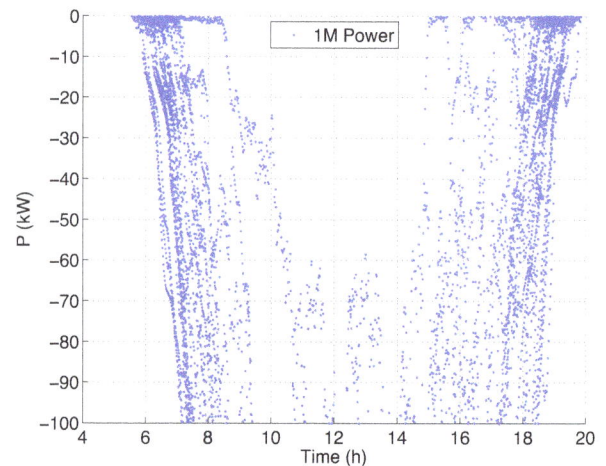

Fig. 3: Guaranteed share of diffusion radiation.

the specific circumstances. That is why this field can be defined as the guaranteed share of diffusion radiation.

Assessment of the minimum extreme conditions follows the same methodology applied in case of the maximum extreme conditions of production delivered by PvP. The outcome from assessment of the minimum extreme conditions of PvP is the Gauss's curve once again; it provides convenient approximation of these minimums. The Gauss's equation of the curve expressing the minimum extreme conditions of production from PvP for April 2011 is defined as:

$$P_{PvP\,min} = -124533 \cdot e^{-\frac{(t-0.494557)^2}{2 \cdot 0.100209^2}} + 1201.97, \quad (6)$$

where $P_{PvP\,min}$ is minimum extreme condition of PvP, t is time (hh:mm).

Same as for regression curves of the maximum extreme conditions, even this equation refers to t defined by the so-called Universal Time Coordinated (UTC). Conversions between the UTC and the Central European Time (CET) and the Central European Summer Time (CEST) are defined by the above mentioned Eq. (4) and Eq. (5).

For individual parameters of Gauss's Eq. (2) for specific months of the year see the Tab. 2. Same as in the Eq. (6) for April, these parameters are inserted into the Eq. (2) to plot the curves of minimum extreme conditions of PvP.

Tab. 2: Parameters of Gauss's equation of minimum extreme conditions of PvP for individual months of the year.

Months	Parameters of Gauss's Equation			
	A	B	C	D
1	-65199.2	0.498540444	0.0941881	14166
2	-123893	0.501989222	0.127915	38423.7
3	-115800	0.499487444	0.12374	10454.8
4	-124533	0.494557	0.100209	1201.97
5	-144600	0.491698444	0.120717	3399.95
6	-151374	0.500168	0.146449	12303.4
7	-118188	0.508342667	0.142813	8777.12
8	-104815	0.503144778	0.11419	1134.17
9	-96112.4	0.496590778	0.101559	594.665
10	-89190.1	0.485495667	0.127737	18992.3
11	-60925.1	0.507583	0.123552	20331.2
12	-18368	0.513792	0.0529752	2237.18

For graphic illustration of the minimum extreme conditions of production from PvP, see the Fig. 4 and Fig. 5. Figure 4 includes months from December to July. The course of individual curves matched as expected, depending on the intensity of solar radiation, yet they would not match with respect to duration of sunshine. That was due to the insufficient size of database with the data measured. To suppress the negative impact, curves for particular months were centered against one another at the same ratio as for

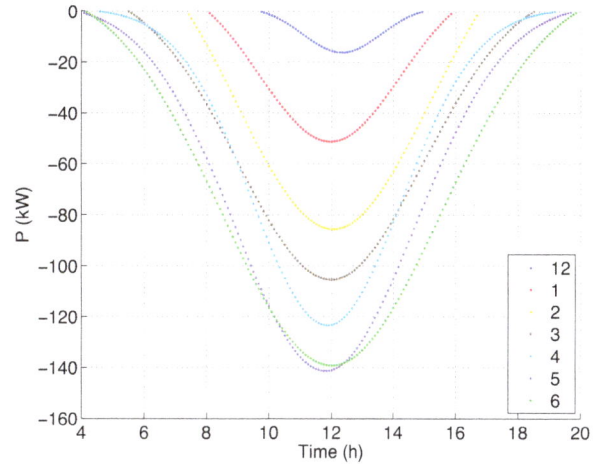

Fig. 4: Regression curves of minimum extreme outputs for individual months (December to June).

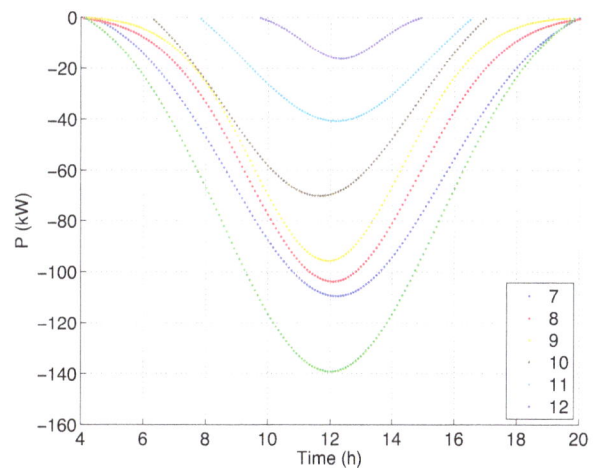

Fig. 5: Regression curves of minimum extreme outputs for individual months (July to December).

curves showing the maximum extreme conditions. Figure 5 illustrates the maximum extreme curves from July to December. Even these have been changed accordingly with respect to the maximum extreme conditions.

The defined maximum and minimum extreme conditions of PvP approximated by the relevant regression curves can be used to establish the difference between these two curves. This is called the difference of active power from PvP and it differs for each month or time interval. This survey concerns one-hour intervals. These results are vital for optimisation of control outputs for ancillary services, for more details see next Section 4.

4. Analysis of Output Deviation Changes in the Power Grid of Czech Republic

The period from 2009 to 2013 brought a significant increase of the installed capacity of photovoltaic output within the power grid in CZ. The rising number of these resources with stochastic output also leads to interference with the capacity balance between the immediate production and consumption of electric power. That results in an increase of the system output deviation as well as deviations of regulatory energy to cover this capacity imbalance. The system deviation defines the immediate difference between the electric power produced and consumed across the power grid, whereas deviations of regulatory energy represent the amount of reserve power used, mainly by ancillary services, to cover the deflection of the system deviation. Keeping the system output deviation in balance is necessary to ensure safe and reliable operation of the power grid and maintenance of specified standards.

4.1. Dependency of System Deviation on the Installed Capacity of PvP

The first step is to prove the dependency between the increase of system deviation and the rise of installed capacity of PvP. This assessment has been carried out using data from hourly averages of the system deviation as well as the data referring to installed capacity of PvP within the territory of Czech Republic to 31st December each year during the period from 2008 to 2012 [3]. This assessment is based on correlation analysis of the increase of the system deviation and the increase of installed capacity of PvP. Those are the following parameters:

- SyS_{diff} - relative maximum daily difference of system deviation,

- $p_{inst.}$ - relative installed capacity of PvP for each year to 31st December 2013.

The relative maximum daily difference of the system deviation has been defined as the ratio between the maximum daily differences of system deviations with hourly averages during the individual years from 2009 to 2014 and the top value of maximum daily difference throughout the observed period. The relative installed capacity of PvP represents a simple ration of installed capacity with respect to the top value of installed capacity during the observed period. For correlation between these curves refers to the Fig. 6.

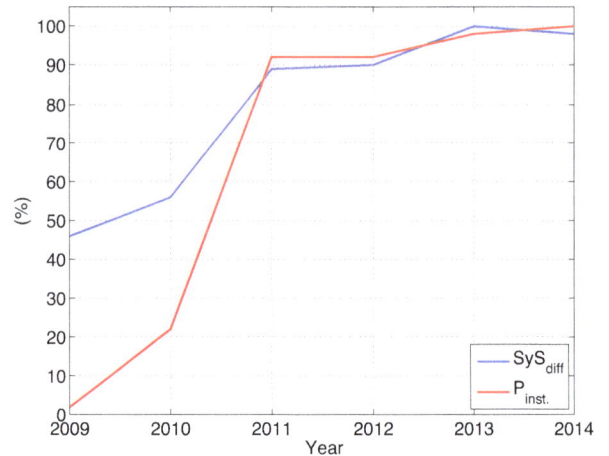

Fig. 6: Correlation of the relative increase of system deviation difference and the installed capacity of PvP.

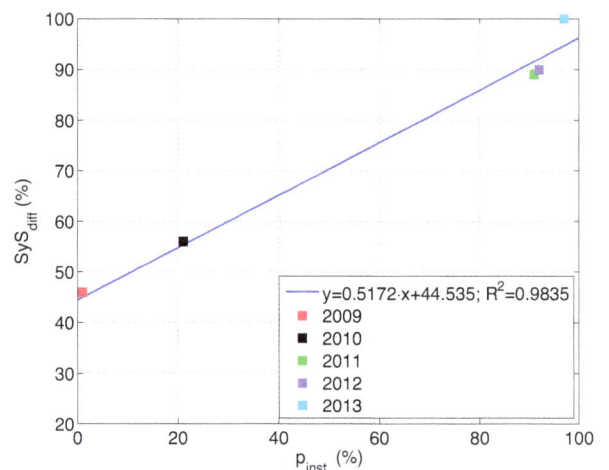

Fig. 7: Correlation analysis of the difference in system deviation and installed capacity of PvP.

There is an evident correlation between these curves based on the visual evaluation. The annual increase of new installed capacity in the power grid results in rise of the relative difference of system deviation. To rule out any speculation regarding the visual evaluation, these parameters have been subject to correlation analysis, see Fig. 7. The most convenient regression is the linear regression with determination index R^2 over 98 %, which proves that the selected regression model is correct. There is a very strong dependency.

This Subsection 4.1. provides clear evidence of changes to system deviation depending on operation of PvP.

4.2. Analysis of Regulatory Deviations

This chapter deals purely with the change to control energy that has a direct effect on provision of ancillary

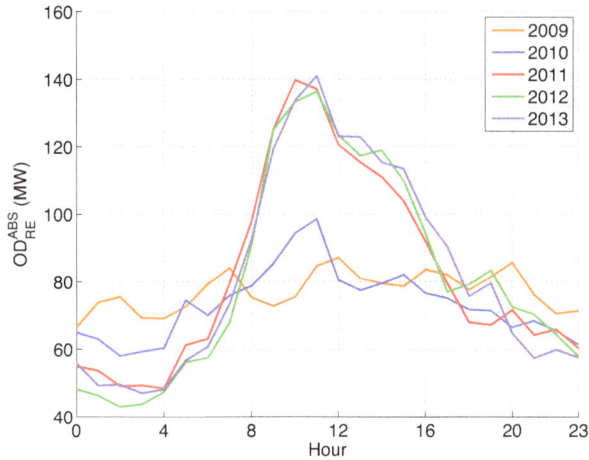

Fig. 8: Hourly course of absolute values of regulatory energy in particular years.

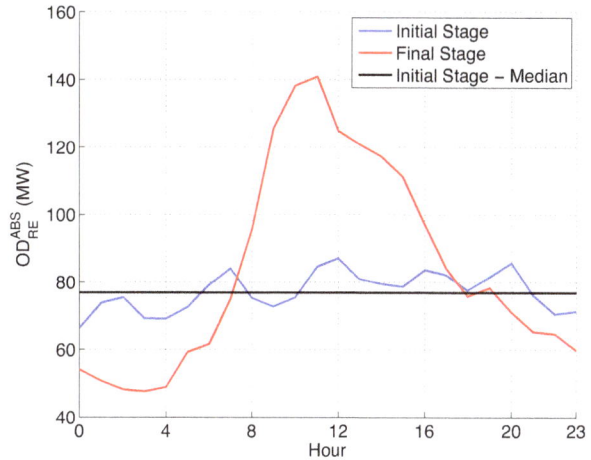

Fig. 9: Course of regulatory energy, initial and final stages.

services. The aim is to enumerate the increase of regulatory energy in relevance with the rise of photovoltaic output.

The average hourly course of absolute values of regulatory deviations in particular years from 2009 to 2013 is presented by Fig. 8. Those are average values of output really used to cover the imbalance between production and consumption of electric power. That reflects the actual intervention of power grid operators with this fact; it has a direct impact on procurement of ancillary services.

The optimal type of service purchased for balancing of capacity in this situation is the Positive Minute Regulatory Back-Up achievable within 15 minutes (MZ15+) together with the Negative Minute Regulatory Back-Up achievable within 15 minutes (MZ15-) [3]. It has been concluded pursuant to professional consultation with personnel from the Czech Energy Transmission System and observation of the nature of data considered with respect to regulatory energies.

The objective is to define a specific value for capacity increase in the ancillary service MZ15+ and MZ15-. The assessment methodology is based on comparison of imbalance of regulatory energy depending on gradual integration of photovoltaic power plants within the power grid in CZ.

The sole assessment is highly restricted to the data available. That is defined by the method for connecting PvP to the power grid. There has actually been a jump increase of these resources within one to two years only; followed by further decay that brings a serious impairment to detailed analysis of this issue. Another fact to be considered is the change to both production from PvP as well as the daily load diagram with direct impact on regulatory energy values.

Same as for PvP, even this issue has been evaluated with respect to the optimal period of one month. The limited amount of data restricts the option to determine the evaluation methodology pursuant to data for individual months. That is why the draft of this methodology is based on annual average data to be applied on the data for particular months later on.

Figure 8 shows the course of absolute values of regulatory energy over the observed period. The course details for years 2011, 2012 and 2013 do not show any statistical difference. That is because the magnitude of installed capacity of PvP in the power grid is almost identical. Further assessment will be carried out with an average of these three years, while the resultant values will explain the increase of system deviation caused by operation of PvP. It is the so-called final stage, as far as this survey is concerned. The year 2009 represents the period, when the impact of PvP operation on the installed capacity was very low and it stands

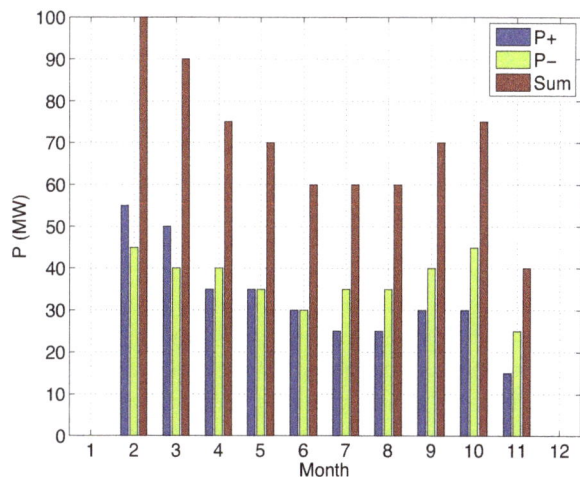

Fig. 10: Monthly values of positive and negative values of the regulatory output.

Tab. 3: Values of increase of the ancillary service MZ15+ and MZ15- with respect to PvP.

Months	2	3	4	5	6	7	8	9	10	11
4:00	0/0	0/0	0/0	0/0	5/5	0/0	0/0	0/0	0/0	0/0
5:00	0/0	0/0	0/0	5/5	5/5	5/5	5/5	0/0	0/0	0/0
6:00	0/0	10/5	10/10	10/10	10/10	10/10	5/5	10/10	5/10	0/0
7:00	10/10	20/20	20/20	15/15	15/15	10/15	10/15	15/20	15/20	5/10
8:00	20/20	35/30	35/30	25/25	20/20	15/25	15/20	25/30	20/30	10/15
9:00	40/35	45/35	40/35	30/30	25/25	20/30	20/30	25/35	25/40	10/20
10:00	50/40	50/40	45/40	30/35	30/30	20/30	25/30	30/40	30/40	15/25
11:00	55/45	50/40	50/40	35/35	30/30	25/35	25/35	30/40	30/45	15/25
12:00	55/45	50/40	50/40	35/35	30/30	25/35	25/35	30/40	30/40	15/25
13:00	50/40	50/40	50/40	30/35	30/30	25/35	25/35	30/40	25/40	15/25
14:00	40/35	45/35	45/40	30/30	25/25	20/30	20/30	25/35	20/35	10/15
15:00	20/20	35/30	40/35	20/25	20/20	20/25	15/20	20/30	15/25	5/10
16:00	5/5	25/20	30/25	15/15	15/15	15/20	10/15	15/20	10/10	5/5
17:00	0/0	10/10	20/15	10/10	10/10	10/15	5/10	5/10	0/0	0/0
18:00	0/0	0/0	5/5	5/5	5/5	5/10	5/5	0/0	0/0	0/0
19:00	0/0	0/0	0/0	0/0	5/5	5/5	0/0	0/0	0/0	0/0

for the so-called initial stage. The distribution of data shows a clear tendency towards convergence to a constant value. That is why the initial stage values will be replaced with its median.

The difference between extreme of the final stage curve and the initial stage curve (median) represents the change in increase of regulatory energy over the observed period. This value represents the maximum absolute value of regulatory energy, for more details refer to Fig. 9.

This methodology is applied to individual months of the year to determine the maximum absolute values of regulatory energy. These absolute values are further divided to positive and negative regulatory energies. The division is based on database of regulatory energy data for the year 2013, with definition of the ratio between the positive and negative regulatory energies. The absolute values of regulatory energy for individual months were further divided proportionately.

The outcome has been plotted in the column chart shown by Fig. 10 that defines the increase of ancillary services for individual months due to integration of PvP. The worst impact of PvP was experienced in February. That is due to the fact that the weather change in that period is dynamic, resulting in significant demands to keep the balance between production and consumption of electric power. Similar weather can be also expected for September and October showing greater increase of regulatory energy. The paradoxically lowest demand for support to ancillary services with respect to operation of PvP comes during summer months. That is due to the more stable nature of weather, when the production from PvP is easier to predict. The winter months did not prove any dependency between production from PvP and ancillary services, since the production from PvP is generally lower

during winter, while the load curve rises significantly. It means the share of PvP on increase of ancillary services is very small.

4.3. Optimisation of Regulatory Energy

The current data shows the increase of regulatory energy for individual months as constant throughout each day. However, this is not wise when considering the distribution of production delivered by PvP, since the change of production from PvP during the day must be paralleled by changes to increase of control energy induced by operation of PvP.

The subsequent step will therefore comprise the optimisation process. The principle is that the data already evaluated by difference of PvP capacities (the difference between the maximum and the minimum extremes of PvP, see Section 3.) has been plotted in hourly histogram. The bar with the highest value for particular month represents the greatest difference and the highest stochastic nature of PvP possible. The highest bar at specific hour therefore represents the worst impact on control energy during the day and it corresponds with the value of control energy determines by the difference between the initial and final stage of regulatory energy for the particular month. The demand for regulatory energy during other hours is distributed proportionally, as shown in the bar chart with PvP capacity.

The outcome from this optimisation process is the specific known value of regulatory energy, both positive and negative for each hour throughout the day of the particular month. These values have been summarised in Tab. 3, where the value before slash corresponds with MZ15+ and the value behind slash refers to MZ15-.

5.   Conclusion

This survey determines the impact produced by integration of PvP into the power grid of the Czech Republic. It is specifically focused on stochastic evaluation of production delivered by these resources leading to interference with the balance between production and consumption of electric power. There has been a clear proof of relevance to the increase of system deviation amplitude. During their peak production periods, photovoltaic power plants represent approximately 20 % share of the total capacity delivered by all power plants. This situation generates two impacts - technical and economical. The power grid operator must comply with engineering criteria defined by international requirements regarding network parameters in order to ensure stability, reliability and safety of operation. All these aspects then affect the economic issues, since there is a greater need for procurement of reserve power in terms of ancillary services. The extent of its direct impact on increase of costs incurred by purchase of ancillary services is rather speculative and this survey does not deal with such issues at all. The survey emphasizes mainly the said technical issues.

Acknowledgment

This work was supported by the Ministry of Education, Youth and Sports of the Czech Republic (No. SP2015/192).

References

[1] Department of Electrical Power Engineering, Faculty of Electrical Engineering and Computer Science, VSB–Technical University of Ostrava. *Data source*. Ostrava: VSB–Technical University of Ostrava, 2013, unpublished.

[2] Czech Transmission System Operator (CEPS). *Data source*. Prague: CEPS, 2013, unpublished.

[3] Grid code. *CEPS, a.s.* [online]. 2014. Available at: https://www.ceps.cz/ENG/Data/Legislativa/Pages/Kodex.aspx.

[4] SMOCEK, M. and Z. HRADILEK. Extreme Production Conditions of Photovoltaic Power Plant Operated in Distribution Grid. *Advances in Electrical and Electronic Engineering*. 2014, vol. 12, no. 3, pp. 185–191. ISSN 1804-3119. DOI: 10.15598/aeee.v12i3.1075.

[5] HRADILEK, Z. and T. SUMBERA. Simulator of Power Forecasting Gained from Wind Power Plants. *Przeglad Elektrotechniczny*. 2010, vol. 86, iss. 8, pp. 196–199. ISSN 0033-2097.

[6] HRADILEK, Z. and T. SUMBERA. Reliability and Predictions of Power Supplied by Wind Power Plants. In: *International Conference on Renewable Energies and Power Quality*. Las Palmas de Gran Canaria: University of Vigo, 2011, pp. 254–259. ISBN 978-84-614-7527-8.

[7] PIOTROWSKI, P. Analiza statystyczna danych do prognozowania ultrakrotkoterminowego produkcji energii elektrycznej w systemach fotowoltaicznych. *Przeglad Elektrotechniczny*. 2014, vol. 90, iss. 4, pp. 1–4. ISSN 0033-2097.

[8] SHI, J., W.-J. LEE, Y. LIU, Y. YANG and W. PENG. Forecasting power output of photovoltaic system based on weather classification and support vector machine. In: *IEEE Industry Applications Society Annual Meeting (IAS)*. Orlando: IEEE, 2011, pp. 1–6. ISBN 978-1-4244-9498-9. DOI: 10.1109/IAS.2011.6074294.

[9] YONA, A., T. SENJYU, A. Y. SABER, T. FUNABASHI, H. SEKINE and C.-H. KIM. Application of neural network to 24-hour ahead generating power forecasting for PV system. In: *IEEE Power and Energy Society General Meeting - Conversion and Delivery of Electrical Energy in the 21st Century*. Pittsburgh: IEEE, 2008, pp. 1–6. ISBN 978-1-4244-1905-0. DOI: 10.1109/PES.2008.4596295.

[10] GOETZBERGER, A. and V. HOFFMANN. *Photovoltaic solar energy generation*. New York: Springer, 2005. ISBN 978-3-540-23676-4.

About Authors

Martin SMOCEK was born in Hranice. In 2011 he graduated VSB–Technical University of Ostrava, Faculty of Electrical Engineering and Computer Science, Department of Electrical Power Engineering. Today he is Ph.D. student in the Department of Electrical Power Engineering, VSB–Technical University of Ostrava and he applies himself to the issue of photovoltaic power plant.

Zdenek HRADILEK was born in Brno. After graduation of college education at Faculty of Electrical Engineering and Computer Science at Brno University of Technology in 1962 he worked as a technician in company Southern Moravian power plants in Brno, than he worked as a major power-supply director in Heat-supply Ostrava and from 1966 until now he is at the VSB–Technical University Ostrava. His scientific preparation graduated by his candidate dissertation defending at the Brno University of Technology in 1972. In 1988, he defended his doctoral thesis at the Czech Technical University in Prague and was appointed as professor.

Permissions

All chapters in this book were first published in AEEE, by VSB-Technical University of Ostrava; hereby published with permission under the Creative Commons Attribution License or equivalent. Every chapter published in this book has been scrutinized by our experts. Their significance has been extensively debated. The topics covered herein carry significant findings which will fuel the growth of the discipline. They may even be implemented as practical applications or may be referred to as a beginning point for another development.

The contributors of this book come from diverse backgrounds, making this book a truly international effort. This book will bring forth new frontiers with its revolutionizing research information and detailed analysis of the nascent developments around the world.

We would like to thank all the contributing authors for lending their expertise to make the book truly unique. They have played a crucial role in the development of this book. Without their invaluable contributions this book wouldn't have been possible. They have made vital efforts to compile up to date information on the varied aspects of this subject to make this book a valuable addition to the collection of many professionals and students.

This book was conceptualized with the vision of imparting up-to-date information and advanced data in this field. To ensure the same, a matchless editorial board was set up. Every individual on the board went through rigorous rounds of assessment to prove their worth. After which they invested a large part of their time researching and compiling the most relevant data for our readers.

The editorial board has been involved in producing this book since its inception. They have spent rigorous hours researching and exploring the diverse topics which have resulted in the successful publishing of this book. They have passed on their knowledge of decades through this book. To expedite this challenging task, the publisher supported the team at every step. A small team of assistant editors was also appointed to further simplify the editing procedure and attain best results for the readers.

Apart from the editorial board, the designing team has also invested a significant amount of their time in understanding the subject and creating the most relevant covers. They scrutinized every image to scout for the most suitable representation of the subject and create an appropriate cover for the book.

The publishing team has been an ardent support to the editorial, designing and production team. Their endless efforts to recruit the best for this project, has resulted in the accomplishment of this book. They are a veteran in the field of academics and their pool of knowledge is as vast as their experience in printing. Their expertise and guidance has proved useful at every step. Their uncompromising quality standards have made this book an exceptional effort. Their encouragement from time to time has been an inspiration for everyone.

The publisher and the editorial board hope that this book will prove to be a valuable piece of knowledge for researchers, students, practitioners and scholars across the globe.

List of Contributors

Petr Chlebis, Martin Tvrdon, Katerina Baresova and Ales Havel
Department of Electronics, Faculty of Electrical Engineering and Computer Science,
VSB–Technical University of Ostrava, 17. listopadu 15, 708 33 Ostrava–Poruba, Czech Republic

Ahmed Thabet
Nanotechnology Research Center, Faculty of Energy Engineering, Aswan University, 81528 Aswan, Egypt

Youssef Mobarak
Nanotechnology Research Center, Faculty of Energy Engineering, Aswan University, 81528 Aswan, Egypt Department of Electrical Engineering, Faculty of Engineering, Rabigh, King Abdulaziz University, 21589 Jeddah, Kingdom of Saudi Arabia

Martin Kaspirek
Management of Grid, E.ON, F. A. Gerstnera 2151/6, 37001 Ceske Budejovice, Czech Republic

Petr Krejci
Department of Electrical Power Engineering, Faculty of Electrical Engineering and Computer Science, VSB–Technical University of Ostrava, 17. listopadu 15/2172, 708 33 Ostrava, Czech Republic

Pavel Santarius
Department of Cybernetics and Biomedical Engineering, Faculty of Electrical Engineering and Computer Science, VSB–Technical University of Ostrava, 17. listopadu 15/2172, 708 33 Ostrava, Czech Republic

Karel Prochazka
EGC EnerGoConsult CB, Cechova 727, 37001 Ceske Budejovice, Czech Republic

Radim Kuncicky, Lacezar Licev, Michal Krumnikl, Karolina Feberova and Jakub Hendrych
Department of Computer Science, Faculty of Electrical Engineering and Computer Science, VSB–Technical University of Ostrava, 17. listopadu 15, 708 33 Ostrava, Czech Republic

Marek Kukucka
Institute of Automotive Mechatronics, Faculty of Electrical Engineering and Information Technology, Slovak University of Technology, Ilkovicova 3, 81219 Bratislava, Slovak Republic

Andreas Weisze
Dr. Ing. h.c. Porsche AG, Porscheplatz 1, 70435 Stuttgart, Germany

Daniela Durackova, Zuzana Krajcuskova and Viera Stopjakova
Institute of Electronics and Photonics, Faculty of Electrical Engineering and Information Technology, Slovak University of Technology, Ilkovicova 3, 81219 Bratislava, Slovak Republic

Stanislav Misak and Lukas Prokop
Centre ENET, VSB–Technical University of Ostrava, 17. listopadu 15, 708 33 Ostrava, Czech Republic

Michal Kratky
Department of Computer Science, Faculty of Electrical Engineering and Computer Science, VSB–Technical University of Ostrava, 17. listopadu 15, 708 33 Ostrava, Czech Republic

Masoud Jokar Kouhanjani and Ali Reza Sei i
Department of Power and Control Engineering, School of Electrical and Computer Engineering, Shiraz University, Zand Street, Shiraz, Iran

Hemza Medoukali, Mossadek Guibadj and Boubaker Zegnini
Laboratoire d'Etude et Developpement des Materiaux Semi Conducteurs et Dielectriques, LEDMaScD, Amar Telidji University-Laghouat, BP 37G road of Ghardaia, Laghouat 03000, Algeria

Fateh Ounis and Noureddine Golea
Electrical Engineering Department, Sciences and Applied Sciences Faculty, Larbi Ben M'hidi University, 04000 Oum El Bouaghi, Algeria

El Hassane Margoum, Nissrine Krami and Hassan Mharzi
Department of Electrical Engineering, National School of Applied Science, Ibn Tofail University, Kenitra, Morocco

Luis Seca and Carlos Moreira
Institute for Systems and Computer Engineering, Technology and Science (INESC TEC), Faculty of Engineering, University of Porto, R. Dr. Roberto Frias, 4200 Porto, Portugal

Chourouk Bouchareb and Mohamed Said Nait Said
Electrical Engineering Department, Laboratory LSPIE
Batna 2000, Batna University, Route de Biskra, 05078,
Algeria

Lubomir Ivanek
Department of Electrical Engineering, Faculty of
Electrical Engineering and Computer Science, VSB–
Technical University of Ostrava, 17. listopadu 15, 708
00 Ostrava-Poruba, Czech Republic

Vladimir Mostyn
Department of Robotics, Faculty of Mechanical
Engineering, VSB–Technical University of Ostrava, 17.
listopadu 15, 708 00 Ostrava-Poruba, Czech Republic

Karel Schee and Jan Grun
Prvni signalni a.s., Bohuminska 368/172, 712 00
Ostrava-Muglinov, Czech Republic

**Viera Stopjakova, Martin Donoval, Martin Daricek,
Jozef Mihalov, Michal Hanic and Vladimir Tvarozek**
Institute of Electronics and Photonics, Faculty of
Electrical Engineering and Information Technology,
Slovak University of Technology, Ilkovicova 3, 812 19
Bratislava, Slovakia

Helena Svobodova
Institute of Medical Physics, Biophysics, Informatics
and Telemedicine, Faculty of Medicine, Comenius
University, Sasinkova 2, 813 72 Bratislava, Slovakia

Erik Vavrinsky
Institute of Electronics and Photonics, Faculty of
Electrical Engineering and Information Technology,
Slovak University of Technology, Ilkovicova 3, 812 19
Bratislava, Slovakia
Institute of Medical Physics, Biophysics, Informatics
and Telemedicine, Faculty of Medicine, Comenius
University, Sasinkova 2, 813 72 Bratislava, Slovakia

Khalid Dahi, Soumia El Hani and Ilias Ouachtouk
Departement of Electrical Engineering, Ecole Normale
Superieure de l'Enseignement Technique, Mohammed V
University in Rabat, Avenue des Nations Unies, Agdal,
Rabat, Morocco

Jitendra Mohan and Bhartendu Chaturvedi
Department of Electronics and Communication
Engineering, Jaypee Institute of Information
Technology, Sector-62, 201304 Noida, Uttar Pradesh,
India

Triwiyanto
Department of Electrical Engineering & Information
Technology, Faculty of Engineering, Universitas
Gadjah Mada, Grafika No. 2, Yogyakarta, Indonesia

Department of Electromedical Engineering, Politeknik
Kesehatan Surabaya, Pucang Jajar Timur No. 10,
Surabaya, Indonesia

Oyas Wahyunggoro and Hanung Adi Nugroho
Department of Electrical Engineering & Information
Technology, Faculty of Engineering, Universitas
Gadjah Mada, Grafika No. 2, Yogyakarta, Indonesia

Herianto
Department of Mechanical & Industrial Engineering,
Faculty of Engineering, Universitas Gadjah Mada,
Grafika No. 2, Yogyakarta, Indonesia

Farideh Ebrahimi
Department of Computer, Faculty of Electrical and
Computer Engineering, Babol Noshirvani University
of Technology, Shariati Avenue, Babol, Mazandaran,
Iran
Control and Intelligent Processing Center of Excellence,
School of Electrical and Computer Engineering,
College of Engineering, University of Tehran, North
Kargar Street, Tehran, Iran

Seyed Kamaledin Setarehdan
Control and Intelligent Processing Center of Excellence,
School of Electrical and Computer Engineering,
College of Engineering, University of Tehran, North
Kargar Street, Tehran, Iran

Radek Martinek
Department of Cybernetics and Biomedical
Engineering, Faculty of Electrical Engineering and
Computer Science, VSB–Technical University of
Ostrava, 17. listopadu 15, 708 33 Ostrava, Czech
Republic

Homer Nazeran
Department of Electrical and Computer Engineering,
College of Engineering, University of Texas El Paso,
500 W University Ave, El Paso, TX 79968, United
States of America

Mayank Srivastava
Department of Electrical, Electronics and
Communication Engineering, The Northcap University,
HUDA Sector-23 A, 122017 Gurgaon, India

Dinesh Prasad
Department of Electronics and Communication
Engineering, Faculty of Engineering and Technology,
Jamia Millia Islamia, 110025 New Delhi, India

Thang Nguyen Trung
Department of Electrical Engineering, Faculty of
Electrical and Electronics Engineering, Ton Duc Thang
University, 19 Nguyen Huu Tho Street, Ho Chi Minh
City, Vietnam

Dieu Vo Ngoc
Department of Power Systems, Faculty of Electrical and Electronics Engineering, Ho Chi Minh City University of Technology, 268 Ly Thuong Kiet Street, Ho Chi Minh City, Vietnam

Martin Smocek and Zdenek Hradilek
Department of Electrical Power Engineering, Faculty of Electrical Engineering and Computer Science, VSB–Technical University of Ostrava, 17. listopadu 15, 708 33 Ostrava, Czech Republic

Index